# Fault Tolerant
# System Design

## Other McGraw-Hill Books of Interest

| ISBN | AUTHOR | TITLE |
|---|---|---|
| 0-07-911158-0 | L. Baker | *C Mathematical Function Handbook* |
| 0-07-003358-7 | L. Baker | *More C Tools for Scientists and Engineers* |
| 0-07-004259-4 | W. Beam | *Systems Engineering* |
| 0-07-010924-9 | C. H. Chen | *Computer Engineering Handbook* |
| 0-07-037491-0 | S.-T. Levi, A. Agrawala | *Real-Time System Design* |
| 0-07-049433-9 | D. Perry | *VHDL* |
| 0-07-051216-7 | J. Ranade, S. Zamir | *C++ Primer for C Programmers* |
| 0-07-059180-6 | J. Smith | *C++ for Scientists and Engineers* |

# Fault Tolerant System Design

**Shem-Tov Levi**
*Independent Consultant*

**Ashok K. Agrawala**
*University of Maryland at College Park*

McGraw-Hill, Inc.
New York  San Francisco  Washington, D.C.  Auckland  Bogotá
Caracas  Lisbon  London  Madrid  Mexico City  Milan
Montreal  New Delhi  San Juan  Singapore
Sydney  Tokyo  Toronto

**Library of Congress Cataloging-in-Publication Data**

Levi, Shem-Tov.
    Fault tolerant system design / Shem-Tov Levi, Ashok K. Agrawala.
       p.   cm.
    Includes index.
    ISBN 0-07-037515-1
    1. Fault tolerant computing.  2. System design.  I. Agrawala, Ashok K. II. Title.
QA76.9.F38L48     1994
004'.36—dc20                                                93-23181
                                                                                           CIP

Copyright ©1994 by McGraw-Hill, Inc. All rights reserved. Printed in the United States of America. Except as permitted under the United States Copyright Act of 1976, no part of this publication may be reproduced or distributed in any form or by any means, or stored in a data base or retrieval system, without the prior written permission of the publisher.

1 2 3 4 5 6 7 8 9 0   DOC/DOC   9 9 8 7 6 5 4 3

**ISBN 0-07-037515-1**

*The sponsoring editor was Daniel A. Gonneau and the production supervisor was Donald Schmidt. This book was set in Century Schoolbook by Electronic Technical Publishing Services.*

*Printed and bound by R. R. Donnelley & Sons Company.*

---

Information contained in this work has been obtained by McGraw-Hill, Inc., from sources believed to be reliable. However, neither McGraw-Hill nor its authors guarantee the accuracy or completeness of any information published herein, and neither McGraw-Hill nor its authors shall be responsible for any errors, omissions, or damages arising out of use of this information. This work is published with the understanding that McGraw-Hill and its authors are supplying information but are not attempting to render engineering or other professional services. If such services are required, the assistance of an appropriate professional should be sought.

# Contents

List of Figures   xiii
List of Programs   xvii
Preface   xix

**Chapter 1.  Introduction**   1

   1.1  Computer and Computation Distribution   2
       1.1.1  Distributed computer structure   2
       1.1.2  The hierarchical model   2
       1.1.3  Principles of distribution   4
   1.2  System Models   5
       1.2.1  System definition   6
       1.2.2  Layered system model   6
       1.2.3  Object based model   7
       1.2.4  System composition life cycle   11
   1.3  Fault Models   12
   1.4  Objectives   13
   1.5  Book Organization   13

**Part 1   Architecture**   15

**Chapter 2.  Hardware Pieces**   17

   2.1  Current Systems   17
       2.1.1  Examples of general purpose processors   17
       2.1.2  High availability systems   24
       2.1.3  Examples of reduced instruction set architectures   32
   2.2  A Quad-Processor Architecture   39
       2.2.1  Shadow systems architecture   39
       2.2.2  Thread oriented architecture   41
   2.3  Concluding Remarks   46

## Chapter 3. Language and Tools — 49

- 3.1 Semaphores — 50
- 3.2 Monitors — 51
- 3.3 Resources — 52
- 3.4 Examples — 53
  - 3.4.1 The bounded buffer problem — 53
  - 3.4.2 The readers writers problem — 55
  - 3.4.3 Monitors and semaphores — 56
- 3.5 Communicating Sequential Processes — 58
  - 3.5.1 CSP concepts and primitives — 58
  - 3.5.2 Examples in CSP design — 64
- 3.6 Concluding remarks — 67

## Chapter 4. Networks — 69

- 4.1 The Layered Architecture — 70
  - 4.1.1 Layers, services and functions — 71
- 4.2 Service Definition and Protocol Specification — 76
- 4.3 Layer Functions — 77
- 4.4 Protocol Specifications — 78
- 4.5 OSI Reference Model — 80
  - 4.5.1 The Application layer — 81
  - 4.5.2 The Presentation layer — 83
  - 4.5.3 The Session layer — 84
  - 4.5.4 The Transport layer — 84
  - 4.5.5 The Network layer — 86
  - 4.5.6 The Link layer — 88
  - 4.5.7 The Physical layer — 89
- 4.6 The IEEE 802 Standard — 90
  - 4.6.1 The Physical layer — 91
  - 4.6.2 The Medium Access Control layer — 92
  - 4.6.3 The Logical Link Control layer — 93
- 4.7 Network Concepts and Characteristics for Embedded Systems — 93
  - 4.7.1 The network interface — 93
  - 4.7.2 Internetwork communication — 95
  - 4.7.3 The internetwork interface — 96
- 4.8 Concurrency, Commitment and Recovery — 97
  - 4.8.1 Commitment — 97
  - 4.8.2 Concurrency — 99
  - 4.8.3 CCR services — 99
- 4.9 Concluding Remarks — 102

# Part 2  Distribution — 103

## Chapter 5. Concepts and Measures — 105

- 5.1 Terminology, Definitions, and Conventions — 105
  - 5.1.1 Programs and their images — 105
  - 5.1.2 Computers and systems — 106
- 5.2 Distributed System Model — 109

| 5.3 | Distribution Measures | 112 |
|---|---|---|
| | 5.3.1 System control sensitivity | 112 |
| | 5.3.2 Sensitivity to environment | 115 |
| | 5.3.3 Comparison criteria | 117 |
| | 5.3.4 Network topology dependence | 119 |

## Chapter 6. Mutual Exclusion  121

| 6.1 | Problem Definition | 121 |
|---|---|---|
| | 6.1.1 The axiomatic approach model | 122 |
| | 6.1.2 The dining philosophers problem | 122 |
| 6.2 | Token Ring Algorithms | 125 |
| | 6.2.1 The generic token ring solution | 125 |
| | 6.2.2 Two tokens solution | 126 |
| | 6.2.3 Logical tokens | 127 |
| 6.3 | Time Ordering Algorithm | 130 |
| | 6.3.1 Lamport's timestamp ordering | 131 |
| | 6.3.2 Ricart and Agrawala | 132 |
| | 6.3.3 Carvalho and Roucairol | 133 |
| | 6.3.4 Suzuki and Kasami | 136 |
| | 6.3.5 Mamoru Maekawa | 137 |
| 6.4 | Path Reversal Based | 141 |
| 6.5 | Concluding Remarks | 143 |

## Chapter 7. Election Algorithms  145

| 7.1 | Problem Definition | 145 |
|---|---|---|
| 7.2 | Election in Ring Architecture | 146 |
| | 7.2.1 Unidirectional ring | 146 |
| | 7.2.2 Bidirectional ring | 148 |
| | 7.2.3 Bidirectional ring simulated by unidirectional ring | 151 |
| 7.3 | Broadcast Elections | 155 |
| 7.4 | Concluding Remarks | 157 |

## Chapter 8. Deadlock and Termination  159

| 8.1 | The Deadlock Problem | 159 |
|---|---|---|
| | 8.1.1 Deadlock in resource allocation | 160 |
| | 8.1.2 Deadlock in communication | 162 |
| | 8.1.3 Axiomatic definition of deadlock | 163 |
| | 8.1.4 Optimistic and pessimistic solutions | 164 |
| 8.2 | Deadlock Prevention in Multiple Resource Allocation | 164 |
| | 8.2.1 Basic principles | 164 |
| | 8.2.2 The Global Graph solution | 165 |
| | 8.2.3 The Local Graph solution | 166 |
| | 8.2.4 "On Each Resource" solution | 167 |
| 8.3 | Deadlock and Termination Detection | 169 |
| | 8.3.1 Computation deadlock | 169 |
| | 8.3.2 Communication deadlock | 170 |
| | 8.3.3 Termination detection | 171 |
| | 8.3.4 Detection algorithms | 173 |
| | 8.3.5 Quiescence | 187 |
| | 8.3.6 Algorithm for termination and communication deadlock detection | 192 |

viii    Contents

    8.4  Concluding Remarks    200

## Chapter 9.  Agreement Protocols    201

  9.1  Commit    202
      9.1.1  Crash recovery orientation    202
      9.1.2  Two-phase commit    204
      9.1.3  Nonblocking three-phase commit    207
  9.2  Weighted Voting    209
  9.3  Consensus    212
      9.3.1  Decentralized commit and consensus    212
      9.3.2  History-based quorum consensus    214
      9.3.3  Quorum consensus with nested transactions    216
  9.4  Approximate Agreement    221
      9.4.1  Synchronous approximation algorithm    223
      9.4.2  Asynchronous approximation algorithm    223
  9.5  Byzantine Agreement    226
      9.5.1  Unsigned messages solution    227
      9.5.2  Signed messages solution    227
  9.6  Concluding Remarks    228

# Part 3  Fault Tolerance    231

## Chapter 10.  Tolerating Faults    233

  10.1  Fault Tolerance Concepts    233
      10.1.1  Definitions    233
      10.1.2  Fault tolerance scenario    235
      10.1.3  Hardware and software faults    238
  10.2  Recovery in Time and Space    240
  10.3  Fault Detection Techniques    241
      10.3.1  Detection of errors in data management    242
      10.3.2  Component fault detection    245
      10.3.3  Detection tests at system level    246
  10.4  Performability Measures    247
  10.5  Modeling Fault Tolerant Systems    248
      10.5.1  Elemental units and computation graphs    249
      10.5.2  Fault handling modeling with elemental units    249
      10.5.3  User view of elemental unit model    250

## Chapter 11.  Roll-back Mechanisms    253

  11.1  Roll-Back Mechanisms    253
      11.1.1  Acceptance tests    253
      11.1.2  Recovery blocks    255
      11.1.3  Nested recovery blocks    255
      11.1.4  The domino effect in roll-back recovery    257
  11.2  Checkpointing    261
      11.2.1  Analytical models for checkpoint performance    261
  11.3  Concluding Remarks    269

## Chapter 12.  Modular Redundancy    271

|  |  |  |
|---|---|---|
| 12.1 | N-Version and Modular Redundancy | 271 |
| 12.2 | SIFT | 274 |
| 12.3 | Replicas | 277 |
| 12.4 | Alternatives | 278 |
| 12.5 | Dynamics of Replicas and Alternatives | 280 |
|  | 12.5.1 Forker | 282 |
|  | 12.5.2 Joiner | 282 |
|  | 12.5.3 Resilient Computation Graph Using Elemental Units | 282 |
| 12.6 | Concluding Remarks | 284 |

## Chapter 13.  Handling Exceptions 285

|  |  |  |
|---|---|---|
| 13.1 | Interrupts and Traps | 286 |
|  | 13.1.1 A general model | 287 |
|  | 13.1.2 Real-time performance | 290 |
| 13.2 | Reaction to Exceptions | 295 |
|  | 13.2.1 A single level approach | 296 |
|  | 13.2.2 A multilevel approach | 298 |
|  | 13.2.3 Comparison | 300 |
| 13.3 | Exception Handling Model | 301 |
|  | 13.3.1 Standard and exception domains | 301 |
|  | 13.3.2 Modeling exception handling with elemental units | 302 |
|  | 13.3.3 Dynamic adjustment of exception handling EUs | 302 |
|  | 13.3.4 Error reporting hierarchies | 303 |
| 13.4 | Concluding Remarks | 304 |

## Chapter 14.  Consistency 305

|  |  |  |
|---|---|---|
| 14.1 | Concurrency Control | 305 |
|  | 14.1.1 Serializability criteria | 306 |
|  | 14.1.2 Locking | 307 |
| 14.2 | Atomicity and Transactions | 309 |
|  | 14.2.1 Transaction model description | 310 |
|  | 14.2.2 Approaches to recoverability | 311 |
|  | 14.2.3 Approaches to indivisibility | 314 |
| 14.3 | Partitioning | 315 |
|  | 14.3.1 Pessimistic strategies | 316 |
|  | 14.3.2 Optimistic strategies | 318 |
| 14.4 | Broadcasting Solutions | 322 |
|  | 14.4.1 Model | 322 |
|  | 14.4.2 Tolerating exclusion faults | 323 |
|  | 14.4.3 Tolerating timing faults | 325 |
|  | 14.4.4 Tolerating Byzantine faults | 326 |
| 14.5 | Concluding Remarks | 328 |

## Chapter 15.  Safe Systems 329

|  |  |  |
|---|---|---|
| 15.1 | Safety Measures | 330 |
|  | 15.1.1 The system model | 331 |
|  | 15.1.2 Safety state monitors | 333 |
|  | 15.1.3 Safe states | 335 |
| 15.2 | Safety Aspects in Resiliency | 336 |
|  | 15.2.1 System's distinct configurations | 336 |

|  |  |  |
|---|---|---|
|  | 15.2.2 Faulty configurations | 338 |
|  | 15.2.3 Resiliency compensation for safety | 339 |
| 15.3 | Design Examples | 341 |
|  | 15.3.1 Monitoring the safety | 341 |
|  | 15.3.2 Decreasing failure probability | 344 |
|  | 15.3.3 Coping with common mode faults | 347 |
| 15.4 | Engineering Practices | 349 |
|  | 15.4.1 Qualification | 349 |
|  | 15.4.2 Software engineering | 350 |
| 15.5 | Concluding Remarks | 351 |

## Chapter 16.  Fault Tolerant Allocation    353

|  |  |  |
|---|---|---|
| 16.1 | Problem Definition | 354 |
| 16.2 | Definitions and Formulation | 354 |
|  | 16.2.1 Model description | 354 |
|  | 16.2.2 Conditions and formulation | 356 |
| 16.3 | Allocation Algorithm | 358 |
|  | 16.3.1 Message types used | 359 |
|  | 16.3.2 Principles of allocation initiation | 359 |
|  | 16.3.3 Principles of an algorithm for allocator | 360 |
| 16.4 | Concluding Remarks | 363 |

# Part 4  Fault Tolerance in Real-Time Systems    365

## Chapter 17.  Allocation in Real-Time Systems    367

|  |  |  |
|---|---|---|
| 17.1 | Additional Conditions and Formulation | 369 |
| 17.2 | Allocation Algorithm under Real-Time Constraints | 369 |
|  | 17.2.1 Message types used | 371 |
|  | 17.2.2 Local and external variables | 372 |
|  | 17.2.3 The allocation algorithm | 372 |
| 17.3 | Concluding Remarks | 375 |

## Chapter 18.  Protocols for Real-Time Communication    377

|  |  |  |
|---|---|---|
| 18.1 | Protocols with Contention | 377 |
| 18.2 | Synchronous Protocols | 380 |
| 18.3 | Bounded Semantic Links | 383 |
|  | 18.3.1 Passive links | 384 |
|  | 18.3.2 Agents | 385 |
| 18.4 | Fault-Tolerant Real-Time Communication | 386 |
| 18.5 | Concluding Remarks | 387 |

## Chapter 19.  Fault-Tolerant Time Services    389

|  |  |  |
|---|---|---|
| 19.1 | Algorithm I: Local Resynchronization | 389 |
| 19.2 | Example | 390 |
| 19.3 | Algorithm II: Byzantine Clock Broadcast | 392 |
| 19.4 | Algorithm III: Complete Time-Service | 394 |
| 19.5 | Achievement of Fault Tolerance | 395 |

## Part 5  Epilog     397

**Chapter 20.  Conclusion**     399

  Bibliography   401
  Index   409

## ABOUT THE AUTHORS

SHEM-TOV LEVI is an independent consultant with 18 years of experience in design, development, implementation, testing and deployment of large, complex, dependable systems for mission critical applications. In the past he worked for Israel Aircraft Industries where he developed inertial navigation systems, space systems and the Arrow system.

ASHOK K. AGRAWALA is a professor in the Department of Computer Science, University of Maryland, researching in the design of hard real-time, distributed fault tolerant systems. He has made many research contributions to the field of computer science over the past twenty years. He has served as a consultant to many industrial concerns, as well as to the government of India as part of the United Nations Development Program, and as a member of the Aerospace Research Subcommittee under the NASA Advanced Space Technology Advisory Committee.

# List of Figures

| | | |
|---|---|---|
| 1.1 | Hierarchical control structure | 3 |
| 1.2 | Information hiding concept | 4 |
| 1.3 | Layered structure: peers relations | 7 |
| 1.4 | Layers transparency principle | 8 |
| 1.5 | A single access server object | 9 |
| 1.6 | Relations between different types of objects | 11 |
| | | |
| 2.1 | Motorola 68020 block diagram | 18 |
| 2.2 | Motorola 68020 coprocessor configuration | 19 |
| 2.3 | Intel 80186 with 82586 network coprocessor | 21 |
| 2.4 | MPC: message passing coprocessor | 22 |
| 2.5 | Intel 82389 message passing coprocessor (MPC) | 23 |
| 2.6 | Architecture of Tandem-16 computer system | 25 |
| 2.7 | Tandem-16 shadow processor concept | 26 |
| 2.8 | Architecture of the Stratus-32 system | 27 |
| 2.9 | Architecture of a typical electronic switch system (ESS) | 28 |
| 2.10 | Duplicated modules in ESS | 29 |
| 2.11 | Memory organization in ESS | 31 |
| 2.12 | Architecture of a typical Synapse N+1 system | 32 |
| 2.13 | IDT R3000 RISC system block diagram | 34 |
| 2.14 | The R3000 in a multiprocessor configuration | 35 |
| 2.15 | Intel i960 system block diagram | 37 |
| 2.16 | Intel i960 master/checker configuration | 38 |
| 2.17 | Intel i960 master-checker main-shadow configuration | 40 |
| 2.18 | Required threads of actions | 42 |
| 2.19 | CPU processors | 44 |
| 2.20 | Multiprocessor Network Configuration | 47 |
| | | |
| 3.1 | The bounded buffer problem | 53 |
| 3.2 | The readers-writers problem | 56 |
| 3.3 | Matrix multiplication in CSP | 64 |
| 3.4 | Ordered set partitioning | 67 |

## List of Figures

| | | |
|---|---|---|
| 4.1 | Logical messages model of communication primitives | 70 |
| 4.2 | OSI layer transparency principle | 72 |
| 4.3 | Quality of Service parameters related to performance over a connection | 77 |
| 4.4 | OSI interlayer relations | 80 |
| 4.5 | Layers append head-tail data | 82 |
| 4.6 | RS422 connector and pin assignment | 90 |
| 4.7 | IEEE layers architecture | 91 |
| 4.8 | OSI versus IEEE layers model | 91 |
| 4.9 | IEEE medium access control classification | 92 |
| 4.10 | Network interface unit (NIU) | 94 |
| 4.11 | The principle of bridging networks | 96 |
| 4.12 | CCR service primitives | 100 |
| 4.13 | CCR APDUs and their parameters | 101 |
| 5.1 | Network topology examples | 107 |
| 5.2 | Distributed computer | 108 |
| 5.3 | Distributed computer-embedded system | 109 |
| 5.4 | Distributed system model | 110 |
| 5.5 | Resource deadlock | 113 |
| 5.6 | Object replication | 114 |
| 5.7 | Event ordering principle | 116 |
| 6.1 | Dining philosophers problem | 123 |
| 6.2 | Token-ring mutual-exclusion algorithm structure | 126 |
| 6.3 | Logical-token mutual-exclusion algorithm | 129 |
| 6.4 | Another logical-token mutual-exclusion algorithm | 130 |
| 6.5 | Intersecting subsets of processes | 132 |
| 7.1 | Extinction of messages in unidirectional ring | 146 |
| 7.2 | Hierarchical elections: "oil-slick" behavior | 148 |
| 7.3 | Unidirectional-ring in election algorithm | 152 |
| 7.4 | Unidirectional-ring election example | 154 |
| 8.1 | Deadlock in resource allocation | 161 |
| 8.2 | Release blocking in an OR graph | 162 |
| 8.3 | Deadlock prevention in local graph with timestamps | 166 |
| 8.4 | "Wait & die" prevention with timestamps | 168 |
| 8.5 | Token race with underlying computation message | 178 |
| 8.6 | Successor dependence on predecessor in a circuit | 185 |
| 9.1 | A basic step for commit communication | 202 |
| 9.2 | Sequence of crash-recoverable committing | 203 |
| 9.3 | Two-phase commit protocol | 205 |
| 9.4 | Reachable states of two-phase commit protocol | 206 |
| 9.5 | Symmetric two-phase commit protocol | 207 |
| 9.6 | Symmetrical three-phase commit protocol | 207 |
| 9.7 | Symmetrical consensus commitment protocol | 213 |

| | | |
|---|---|---|
| 10.1 | System decomposition: subsystems and components | 234 |
| 10.2 | Fault tolerance scenario | 235 |
| 10.3 | Repetitions versus alternatives effects | 241 |
| 10.4 | Feedback shift register | 244 |
| 10.5 | Elemental unit of a fault-tolerant system model | 249 |
| 10.6 | Scope zooming from subgraph to elemental unit | 250 |
| 10.7 | Synchronization and communication elemental units | 250 |
| 10.8 | Fault handling structure based on elemental units | 251 |
| 10.9 | User view of elemental unit model | 251 |
| | | |
| 11.1 | Recovery block | 255 |
| 11.2 | Process interactions in recovery blocks | 257 |
| 11.3 | Occurrence ordering | 259 |
| 11.4 | Occurrence graph with recoverable subgraphs | 260 |
| 11.5 | Audit trail and roll-back recovery | 261 |
| 11.6 | Overhead versus intercheckpoint time | 264 |
| 11.7 | Queueing model for checkpointing and recovery | 264 |
| | | |
| 12.1 | Recovery block versus modular redundancy | 273 |
| 12.2 | SIFT hardware architecture | 275 |
| 12.3 | SIFT clock ticks | 276 |
| 12.4 | SIFT logical hierarchy | 277 |
| 12.5 | Replicated procedure calls | 278 |
| 12.6 | Multi-connected alternatives | 279 |
| 12.7 | Forkers and joiners as links and objects | 281 |
| 12.8 | Modeling redundancy with elemental units | 283 |
| | | |
| 13.1 | Single interrupt request, daisy-chain prioritized acknowledge | 287 |
| 13.2 | Multiple interrupt requests, daisy-chain acknowledge | 289 |
| 13.3 | Interrupt service without preemption | 291 |
| 13.4 | "First-deadline" scheduling with preemption | 292 |
| 13.5 | Single level exception handling | 296 |
| 13.6 | Multilevel exception handling | 299 |
| 13.7 | Standard and exception domains | 301 |
| 13.8 | Exception handling structure based on elemental units | 303 |
| | | |
| 14.1 | Serializability in single phase locking | 308 |
| 14.2 | Two phase locking | 308 |
| 14.3 | Graph representation of locking dependencies | 309 |
| 14.4 | Acyclic precedence graph: no partitions' conflicts | 319 |
| | | |
| 15.1 | Damage containment in a system | 332 |
| 15.2 | Monitoring fault-isolated blocks | 334 |
| 15.3 | Temporal and physical distinct configurations | 337 |
| 15.4 | Common-mode faults | 340 |
| 15.5 | Sensing and actuating system | 341 |
| 15.6 | The example system with monitoring blocks | 343 |

| | | |
|---|---|---|
| 15.7 | The example system with modified actuation | 346 |
| 15.8 | The example solution to common mode fault | 348 |
| 16.1 | Temporal $(a,b)$ and physical $(c/d)$ redundancy | 355 |
| 16.2 | Wrong use of resorces: 0-resiliency | 356 |
| 16.3 | Forward wave of ALLOC_REQ messages | 361 |
| 16.4 | Backward wave of ALLOC_REP messages | 361 |
| 18.1 | Duration of semantic link | 384 |
| 19.1 | Example results of resynchronization algorithm | 392 |

# List of Programs

| | | |
|---|---|---:|
| 3.1 | The bounded buffer problem: a semaphore solution | 54 |
| 3.2 | The bounded buffer problem: a monitor solution | 55 |
| 3.3 | The readers writers problem: a semaphore solution | 57 |
| 3.4 | The readers writers problem: a monitor solution | 58 |
| 3.5 | The readers writers problem: resource solution | 59 |
| 3.6 | A semaphore built with monitors | 60 |
| 3.7 | A monitor based procedure built with semaphores | 61 |
| 3.8 | Matrix multiplication in CSP | 65 |
| 3.9 | The bounded buffer problem: CSP solution | 65 |
| 3.10 | The cookie shop problem: CSP solution | 66 |
| 3.11 | The ordered set partitioning problem: CSP solution | 68 |
| 6.1 | Solution to dining philosophers problem | 123 |
| 6.2 | Two-token mutual-exclusion algorithm | 128 |
| 6.3 | Logical-token mutual-exclusion algorithm | 129 |
| 6.4 | Another logical-token mutual-exclusion algorithm | 130 |
| 6.5 | Lamport's time-stamp mutual-exclusion algorithm | 132 |
| 6.6 | Ricart and Agrawala's mutual-exclusion algorithm | 134 |
| 6.7 | Carvalho and Roucairol's mutual-exclusion algorithm | 135 |
| 6.8 | Suzuki and Kasami's mutual-exclusion algorithm | 137 |
| 6.9 | Mamoru Maekawa's mutual-exclusion algorithm | 139–140 |
| 7.1 | Unidirectional ring election algorithm | 147 |
| 7.2 | Bidirectional ring election algorithm | 149–150 |
| 7.3 | Simulated bidirectional ring election algorithm | 153 |
| 7.4 | Broadcast election algorithm | 155 |
| 9.1 | Crash-recoverable commiting algorithm | 204 |
| 9.2 | Weighted voting algorithm: read quorum | 209–210 |
| 9.3 | Weighted voting algorithm: write quorum | 211 |
| 9.4 | History-based quorum consensus algorithm | 215–216 |
| 9.5 | Serial scheduler for nested transactions | 218 |

| | | |
|---|---|---|
| 9.6 | Basic object with read/write accesses | 219 |
| 9.7 | Transaction manager with read access | 220 |
| 9.8 | Transaction manager with write access | 221 |
| 9.9 | Synchronous approximation algorithm | 224 |
| 9.10 | Asynchronous approximation algorithm | 225 |
| 9.11 | Byzantine Generals solution: unsigned messages | 227 |
| 9.12 | Byzantine Generals solution: signed messages | 228 |
| | | |
| 11.1 | Recovery block | 256 |
| 11.2 | Nested recovery block | 256 |
| 11.3 | Optimal availability algorithm | 259 |
| | | |
| 13.1 | Interrupt handler example | 293 |
| 13.2 | Exceptions in division procedure | 297 |
| 13.3 | Roll-back by exception | 298 |
| | | |
| 14.1 | Forward conflict validation algorithm | 320 |
| 14.2 | Backward conflict validation algorithm | 321 |
| 14.3 | Broadcast algorithm for tolerating exclusion faults | 324 |
| 14.4 | Broadcast algorithm for tolerating timing faults | 325 |
| 14.5 | Broadcast algorithm for tolerating Byzantine faults | 327 |
| | | |
| 17.1 | Typical joint variables in a calendar | 368 |

# Preface

This book concentrates on *fault tolerance* issues that we find in the design of reliable systems. Namely, this book refers to the design of systems that have the *ability to function in the presence of faults*. Detection, recovery, prevention, consistency, and other issues are discussed to give the reader a comprehensive overview of what is involved in designing fault tolerant systems. It is needless to state that we could not cover every issue to the depth some experts may feel is necessary, and thus, we recommend that these readers consult the references.

However, we did include a thorough discussion on *distributed systems*, whose role in fault tolerance design is fundamental. Therefore, we begin the book by establishing a basis of knowledge in distributed systems. In this background we equip the reader with tools used in discussions that relate to architectural and algorithmic issues that follow.

In this book we present a detailed discussion of the current state of the art of methodologies and algorithms required for designing reliable and fault-tolerant distributed systems. We also include a number of major recent advances as well as a complete presentation of the various aspects of fault tolerant system design from basic design principles to detailed implementation examples. We present various perspectives of recovery from failure, managing hot and cold redundancy, and handling exceptions and maintaining consistency. In particular, performance in the presence of faults under severe safety restrictions is discussed and methods are proposed for the implementation of such safety constraints.

We deal with components from the system point of view rather than on the basis of technology dependent decomposition. Although the book's focus is primarily fault-tolerant system design, it provides many methods that should be applicable for any system design. We include many examples from computer embedded systems. We elaborate, when necessary, on real-time systems whose complexity and dynamics add a new dimension to the issue of fault tolerance.

The book is a reflection of our experience in the industrial and the academic world.

## Acknowledgments

The writing of this book required a great deal of support from many friends and colleagues. Results of research work performed by the System Design and Analysis Group of the Department of Computer Science at the University of Maryland have contributed to the material presented in this book. In particular we recognize the contributions of Daniel Mosse, and his Maruti team at Maryland.

Years of design and development of very large systems in the Israeli industries have given us the practical aspects of systems that must continue to function correctly even when faults occur.

*Shem-Tov Levi*

*Ashok K. Agrawala*

# Chapter 1

# Introduction

Computers have been around now for about five decades. We have seen very rapid growth in this industry during this period, with the most remarkable growth being in the last decade. In addition to the usual general purpose computing, computer technology is finding its way in a large variety of systems and products ranging from the microwave oven to space stations. With this proliferation comes the challenge of making all such systems both highly reliable and safe.

Reliability and safety properties of these systems have often been treated as statistical properties. Such an approach reflects a lack of knowledge about the details of the system behavior and its environment and in most cases, there is a direct relationship between a potential fault and a resulting failure. When we find the consideration of this relationship too complex or intractable, we may be able to capture some aspects of it through a probabilistic model. Clearly, once a failure occurs, it occurs with probability one. For a highly reliable system, it is essential that all known fault causes, and fault-to-failure relationships be considered explicitly throughout the entire life cycle of the system, beginning with the earliest design phases.

The current state of computer technology allows the contruction of rather complex systems at relatively low costs, of compact size and relatively modest power requirements. The networking technology permits the interconnection of not only a very large number of computers in a world wide network, but also permits the implementation of a system consisting of several computers able to communicate at very high speeds. This permits the implementation of parallel and distributed systems.

Clearly, the occurrence of a failure results in some component of the system becoming inoperable. The functions assigned to that component have to be carried out by some other component of the system in order to tolerate this fault and not effect the output. This requires the availability of redundant components in the system. Distributed system organization offers a natural framework within which such redundancy can be implemented.

## 1.1 Computer and Computation Distribution

### 1.1.1 Distributed computer structure

Concurrent programming provides the foundation for potential parallelism. The main characteristics that differentiate sequential and concurrent programs originate in the ability to order events.

- Sequential program: each event has a unique predecessor event which is completed before its successor starts its execution.
- Concurrent programs: events do not always assure the certainty of the completion of their predecessors' events.

It is, therefore, evident that the main difference between concurrent and sequential systems is in their traces of events. We identify the thread of program control as the sequential trace of events caused by the program execution, and the address space of a program as the space into which the program is given access during execution. Based on this approach we identify a distributed program as a program with multiple threads and multiple address spaces which do not intersect. A parallel program has multiple threads and uses intersecting address spaces.

In this book we focus on computerized systems. In our approach a computer has at least processors, address spaces, and communications capabilities.

A distributed computer consists of a set of multiple processors. Each of these processors is allocated with a distinct address space and the processors are interconnected by a communication subsystem. A distributed system contains a distributed program running on a distributed computer.

### 1.1.2 The hierarchical model

Generally, a system can be viewed as a collection of components which cooperate to accomplish specified intentions. The accomplishment takes place in a particular environment and within a given time constraint. The components of the system are its logical elements which carry out its designated functions. The cooperation is controlled by a central controller, which can be either a physical or a conceptual entity.

The components themselves demonstrate hierarchical relations in their control structure. Figure 1.1 depicts this nature of hierarchy. A higher level process controls and coordinates the activities performed by its lower level neighbors. The top-level process is viewed as the supervisor of the entire system activities, and naturally of its observed behavior.

The hierarchical decomposition of modules obeys several characteristic properties.

1. *Data-flow oriented design.* Let us consider a reactive system in which we want to respond to events by producing output. Let each triggering event be defined through data items that arrive at a process receiver. This type of systems demonstrates why a data-flow characterization of the system provides a powerful way to determine the logical and temporal behavior of the system.

   In data-flow oriented design, the major streams of data are identified as they flow, transformed from external input to external output. This approach is also very useful where the data flow consists mainly of control information. In other words, it is applicable where data passed to a component initiates an action (or a sequence of actions) based on the incoming data.

2. *Component communication and synchronization.* In order to support relationships between system components, one must require their components to communicate and synchronize with each other. There are two synchronization primitives that are the most common in current systems:

   - Semaphores: used to control concurrent access to a data item shared between processes, while imposing mutual exclusion in critical sections.

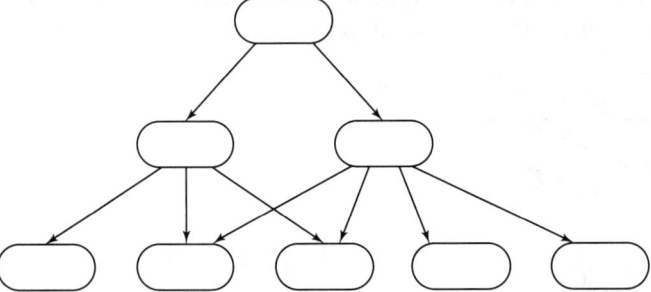

**Figure 1.1** Hierarchical control structure.

- Signals: used in a rendezvous synchronization, where one process awaits a signal from another in order to proceed.

These primitives and others are further discussed in the following parts of this book.

3. *State dependency*. Incorporating state dependency in addition to data dependency allows validation of legality of actions according to the system state. It also allows us to express conditions on executions of such actions.

An important concept in any structured design method is the *information hiding concept* [110, 111], which enforces a consistent use of the data. One way to enforce such consistency is to let only one module know each key, thereby keeping the shared data to a minimum. Modules are therefore more self contained, and fault tolerance performance and autonomy are enhanced. In addition, the system becomes more modifiable and more easily maintainable. The cost involved in this approach is the overhead in accessing data via a function rather than directly.

Figure 1.2 demonstrates the information hiding principle. Processes $P_1$, $P_2$, and $P_3$ must access modules $B$ and $C$ via a function $a$ that hides them. In this example, $a$ is responsible for the consistency of $B$ and $C$. Due to this single-key approach, access may involve wait-time in a queue as long as $a$ is busy. For example, consider the case of a simultaneous access by $P_1$, $P_2$, and $P_3$ where only one of them can be served at a time, and therefore the other two must wait.

The above principles of hierarchical composition and information hiding have given rise to the object orientation and layered architecture discussed hereafter. If we consider the scope of each module in the intermediate layer in Figure 1.1, we find that each layer is essentially an abstraction level of the system.

### 1.1.3 Principles of distribution

Our focus is on a distributed system which consists of a distributed program running on a distributed computer. A distributed computer

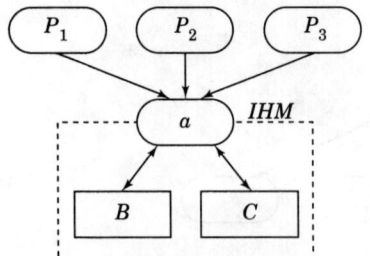

**Figure 1.2** Information hiding concept.

consists of a set of multiple processors, each allocated with a distinct address space. The processors are interconnected by a communication subsystem. Let us start with the basic principles of distributed system architecture, as they are reflected on hardware design and components.

1. Computation nodes are located to support processing according to data locality [34]. In other words, data processing before data migration is advisable, and generally reduces communication traffic significantly. As there may be various data entry points to a system, some processing can be expected at these points.

2. Communication arcs provide connectivity between the computation nodes. These arcs and their topology outline the system performance limitations with respect to data transfer as well as robustness with respect to arc failures. For example, the routing selected for a particular data transfer can be blocked at some slow communication arc.

3. Layered protocols support distributed abstraction levels and provide locality independence. Thus, a design of services supported by one node can be replaced by another with no effect on the client which receives the service.

4. Redundancy of nodes and arcs supports fault tolerance on a system wide basis, relying on the assumption that faults are local and not global. This dependence requires comprehensive support from the hardware for fault confinement and contamination control.

Therefore, a major parameter of concern in hardware support for a distributed computer focuses on connectivity of computation nodes. Connectivity characteristics should provide support for arc performance on one hand and for fault confinement on the other. "Hot" (on line) or "cold" (stand by) redundancy support is also required at some level from the hardware. In the following sections we introduce several approaches of hardware architecture which support the above principles.

## 1.2 System Models

This book deals with systems, and in order to be able to focus the discussion we use system models which capture system properties. The models we use maintain the system properties and the ordering of events in the system along with the system structural decomposition. But first, we have to define what a system is.

### 1.2.1 System definition

The basic element of activity in any mechanism is called an *action*. In [82], a class of such actions is defined as an *operation*, a specific instance of executing such a set of actions is defined as an *operation execution*, and a set of operation executions with precedence relations which obey certain axioms is denoted *system execution*.

Systems are implemented with components which carry out these system executions. Since we want to discuss properties of systems based on the way they are implemented, and on the impact of their behavior on the environment affected by them, let us define a system as follows:

> DEFINITION 1. A *system* is an identifiable mechanism that maintains an observable behavior at its interface with its environment, as a result of the set of all its possible *system executions*. □

We divide the components with which a system is implemented into hardware resources connected by physical links and software components connected via logical links. The software components which implement the operation executions are processes, which are the execution instances of programs. The thread of program control is the sequential trace of events caused by the program execution, maintaining the events in a total order.

> DEFINITION 2. A *process* is a communicating execution-instance of a program, whose implementation is achievable by a single-thread processor with a bounded address space and a set of resources. □

In other words, a process reflects no concurrency in its execution as observed by the environment. A number of processes (more than one) which interact with each other are required for concurrent execution.

### 1.2.2 Layered system model

The hierarchical composition of the system components allow dealing with the system requirements at different levels of detail. In the example described in Figure 1.1, we show a central controller (the root node) which relies on the services of two components. These components form a layer of abstraction with enough detail to satisfy the controller's needs. However, the needs of the components of this layer, involve another layer of other five components.

Figure 1.3 describes a relation between two entities at the $n$th layer, based on agreed upon data exchange and synchronization. Each $n$-entity implements functions of the $n$th layer, and has peer protocols to communicate with other $n$-entities. The peer protocol in each layer

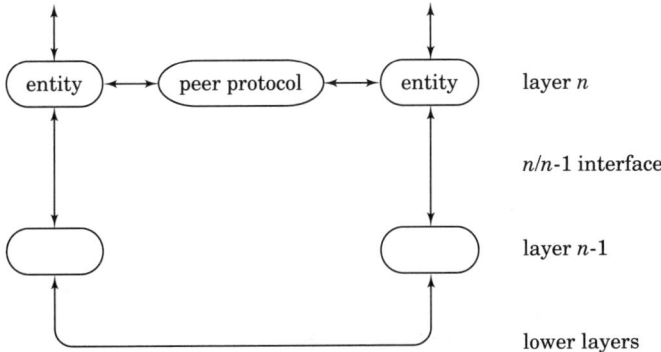

**Figure 1.3** Layered structure: peers relations.

is a set of rules which enforce syntax, semantics, and timing. Each entity communicates with entities of layers above and below with interface protocols, as depicted by Figure 1.3. The set of rules which governs the relation is dictated by the peer protocol. However, the actual data exchange and synchronization are implemented by the lower layers. Various protocols based on this approach are described in Section 4.5.

Another fundamental idea in layered architectures is the notion of transparency [34]. The implementation of a relation between two application entities is independent of the detailed specifics of the mechanism which carries it out. Based on this idea of mechanism and locality independence, the goals of concurrency and parallelism are significantly enriched.

The example in Figure 1.4 illustrates this principle. We first consider two application entities with a defined relation. If the entities can access each other, then the relation is implemented as a direct relation. If however the entities do not allow direct access, then the relation is translated to a peer protocol and a communication mechanism which carries out the synchronization and the data exchange. This translation, described in Figure 1.4(2), maintains the same conventions and restrictions imposed by the relation. The communication mechanism itself has a layered architecture. Its upper layer, detailed in Figure 1.4(3) as layer $n+1$, maintains predefined rules which activate a lower layer (e.g., the physical communication). The transparency principle, therefore, imposes strict rules in in-layer peer relations, along with interface rules for layer interconnections.

### 1.2.3 Object based model

The layered design approach subjects components to strict rules in peer relations and layer interconnections. It further emphasizes the

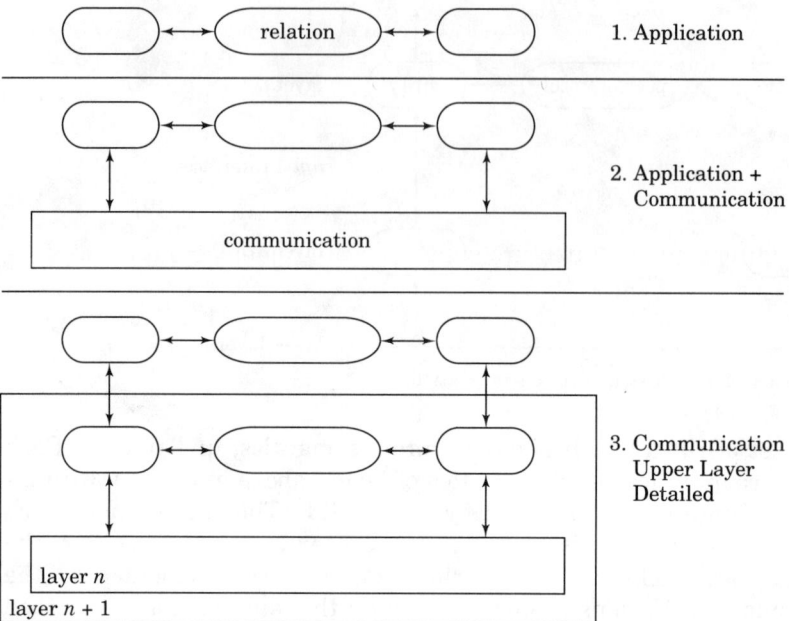

**Figure 1.4** Layers transparency principle.

information hiding principle as a major model characteristic. It is therefore just natural to establish a system model on object based architecture [24] in which each of the system components is an object.

The result of an object-oriented design approach [35] yields the following properties for the system objects:

- An object is identified by its name.
- An object has a state.
- There is a set of operations to which an object may be subjected and a set of operations it requires from other objects.
- An object has restricted visibility of (as well as by) other objects.

An object is a distinct and selectively accessible element whose logic (software) resides on one of the storage resources of the system. Logic elements may hide hardware elements which they control or serve. The object model defines the objects as the elements that construct the system, their classification, and the relations between them. It also defines the set of operations they are subjected to and the execution parameters that permit scheduling them for execution and access. An object based architecture is an implementation of a system based on object entities.

Triggering an object operation is called an *invocation*. The invocation is syntax and semantics restricted as defined by the peer protocol and the interfaces. It satisfies information hiding as the invokee allows connection only through it service access points (SAPs). Thereby, containment is significantly enhanced and contamination (in case of a failure) is considerably confined.

We have proposed [94] an architecture in which we handle object manipulation (creation, deletion, access, and protection) such that access-time behavior supports the objectives of real-time constraints. In addition, the architecture supports achievement of fault tolerance goals through architectural damage containment. Furthermore, this proposed architecture demonstrates a high level of determinism in object execution time, a property that enables predictability of its behavior. This architecture defines a classification of object types, the set of operations each of the object types is associated with, and their relations.

An object consists of an executable body and a joint, as described by Figure 1.5. The *joint* of each object contains the following parts:

- A context independent pointer to the object's body, to give the naming network ability to permit the selective sharing of the object.

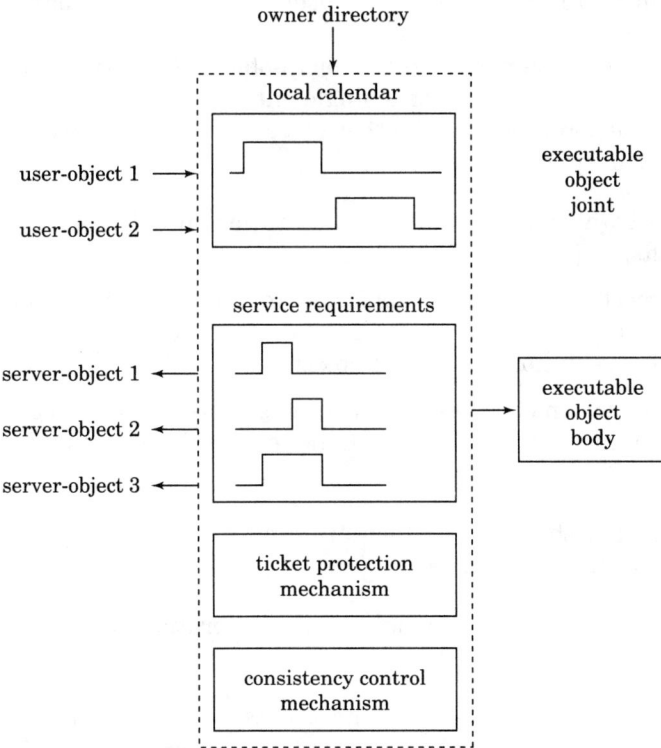

**Figure 1.5** A single access server object.

- An owner/user justification structure.
- Resource (and/or server) requirements.
- A ticket based mechanism for the protection scheme.
- A time constraint for an executable object.
- A replica/alternative control mechanism for the fault tolerance scheme.

Using the mechanisms that this architecture supports, let us consider a possible computation model. In our model, objects that relate to each other are connected via the owner/user justification structure in the joint. These relations are compatible with the visibility restrictions and the set of operations to which the justificand is subjected and the justifier is operating on. Operations in this set can change an object state, evaluate the current state of an object, and allow the visiting of parts of an object. One can carry out these operations on object bodies as well as on object joints. We can model a system as a graph whose nodes are objects and whose directed arcs go from justifier to justificand representing the owner/user relationship. In order to support one-to-many and many-to-one relations between objects, we can group objects of the same type into a meta-object. The rest of the objects may refer to such a group as a single entity.

Depending on their relations with other objects, we can divide executable objects into three classes [35]:

- *Actor*: An object which is subjected to no operation. It only operates on other objects.
- *Server*: An object which is only subjected to operations by other executable objects. It does not operate on other executable objects, but it may operate on non-executable objects.
- *Agent*: An object which operates on (one or more) executable objects on behalf of another executable object. Other executable objects may operate on it.

In addition there are the objects which are only subjected to operations, and thus are non-executable.

- *Passive*: An object which is only subjected to operations and does not operate on others.

Remote invocations are performed through agents, which take care of heterogeneity, communication link access, and reservations. A general scheme of object relations is depicted by Figure 1.6.

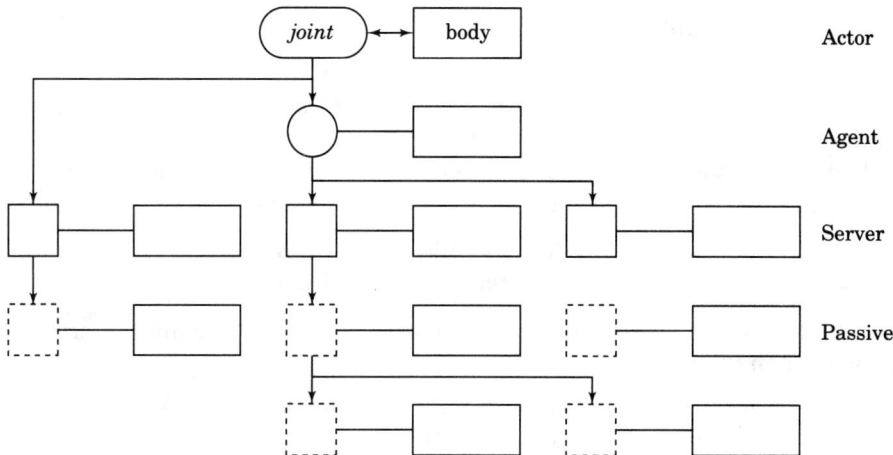

**Figure 1.6** Relations between different types of objects.

### 1.2.4 System composition life cycle

We can identify five phases in the cycle of composing a computerized system.

1. *Design.* The first phase in developing an application in a computerized system is its design. It includes such activities as requirement specification, conceptual design, decomposition, and detailed design. This phase ends with detailed definition of each object in the system along with inter-object interfaces and relations.

2. *Compilation.* In this phase, each object is created, supporting its detailed design specification. Resource requirements are identified and static requirements are marked. Dynamic requirements are not known at this point, and placeholders are kept in the object's joint for them.

3. *Integration.* A computation graph of the system is created by interconnecting the objects and their service requirements with directed arcs. The information exchange channels (links) are defined and verified syntactically and operationaly. The remaining resource requirements are identified and recorded with the application.

4. *Allocation.* In this phase resource allocation is carried out, and reservations are made in the proper calendars to support future schedulability [94].

5. *Execution.* During this phase the operating system loads the allocated objects and performs dispatching, message passing, and reservation enforcement.

## 1.3 Fault Models

Erroneous behavior of a system is undesirable and this book tries to provide some solutions. It is important to define the sources and mechanisms of this undesirable behavior and to understand the problem before solutions are suggested. The first thing we need is the definition of an entity to use as a reference for correctness.

> DEFINITION 3. An *ASR* (authoritative system reference) is a fictitious entity which produces a correct behavior of the system. □

Once a reference has been established, we need to define the erroneous behavior.

> DEFINITION 4. An *error* is a difference between the actual system behavior and that produced by an ASR. □

> DEFINITION 5. A *failure* is an event which corresponds to the first occurrence of the generated error. □

> DEFINITION 6. A *fault* is a source which has the potential of generating errors. □

Potential sources of generated errors may be divided into classes according to their existence duration.

> DEFINITION 7. A *permanent* fault is a potential source of generated errors whose life span coincides with the system's life span. □

> DEFINITION 8. A *transient* fault is a potential source of generated errors which has a life span significantly shorter than the system's recovery requirements. □

In other words, a transient fault may be treated as a noise in the system, while the permanent fault calls for the execution of maintenance procedures. Another classification of faults can rely on the faults origin or technology. One usually distinguishes design faults from erosion, mainly because design faults are permanent by nature and erosion may have some statistical properties which are time dependent. Another distinction commonly made is between hardware and software faults, and again for the same reasons. A measure often used for system failures depends on the probability of errors: the mean time between failures (MTBF). This measure is used when the status of faults in the system is not deterministic, but rather stochastic.

Another aspect which corresponds to failure consequences is risk involved. Ideally, we want failure to result in a neutral state. However, predictability of behavior after failure is not always feasible.

> DEFINITION 9. A *fail-stop* process is a process which reacts to any failure by an execution halt. □

Therefore, fail-stop processes have a predictable behavior after failure. Other types of processes can result in erratic state transitions, and therefore are difficult to handle.

## 1.4 Objectives

Equipped with the above definitions, conventions, approaches and terminology, the reader can continue to the next parts of the book. These parts introduce the principles of fault tolerance in distributed systems along with major approaches to accomplish fault tolerance properties. The book is intended for the use of hardware, software and system engineers for acquiring skills during their academic career. However, it is also organized as a reference book to help practicing engineers design fault tolerant systems.

The approach taken in the following parts starts with the establishment of a common ground for hardware and software engineers. Hardware insight is given for software engineers, and software tools are then presented for hardware and system engineers. Then, the distribution principles are provided, covering a variety of approaches and algorithms, and establishing criteria to compare them. Fault tolerance principles are then introduced, and various issues involved with the presence of faults are discussed.

In order to support the introductory objectives, major topics are covered with generalizing approaches. The organization of the detailed discussions in each chapter is aimed at solving specific problems, and not necessarily at the method of solution.

## 1.5 Book Organization

A designer of a fault tolerant system must have a comprehensive view of various aspects of the normal functioning of the system as well as its failure modes and the corrective actions that can be taken by the system. In this book we introduce the aspects which we found essential for a designer to know. In addition, we have included the background material to lay the foundation for the material discussed. Therefore, the book should help those who are getting started in the design of fault tolerant systems, as well as serving as a ready reference for the experts. The material is well suited for a graduate course on this topic. However, the aim of this book has been to collect the most useful information for the practitioners in the field. While much literature exists in the area of fault tolerance, our experience indicates that such a collection of useful techniques is missing.

This book is organized in four parts. The first part addresses the architectural issues relating to both, hardware and software. The primary concepts and techniques for distributed systems are presented in the second part. The following two parts make use of the information developed in these two parts in discussing fault tolerance for systems. Part III presents the fault tolerance issues and concepts in general while Part IV concentrates on real-time systems. The first chapters of Part II and Part III include the basic concepts and definitions. These chapters provide the foundation on which the rest of the part and the book is built. New students may pay particular attention to these chapters before continuing into the rest of the material.

We have included a number of algorithms in this book. For most of them a pseudo-code version of a program is included. These algorithms may be used by the designers in direct implementations. An easy way to locate them is through the list provided in the beginning of the book or by going to the chapter discussing the subject material of interest.

# Part 1

# Architecture

Following the models and definitions of the introduction, the first part of our book describes the architecture and environmental support we find in computer and computerized systems which are used in real life systems. It begins with a presentation of hardware elements which allow building computation nodes that communicate with each other. The hardware is shown to support various high availability architectures, both for reduced and complex instruction set processors. Then, we describe several problems in software design concerning distributed programming, namely, classical synchronization problems with tools to solve them as well as language support for communicating sequential processes. The architecture topics are concluded with the presentation of communication networks, their concepts and characteristics. Layered models and open system standards provide the reader with insight to various communication procedures. Thus, this part covers discussions of building blocks in hardware, software, and communication networks, their inter-relationships and the likely results of selecting one approach over the other.

# Chapter 2

# Hardware Pieces

This chapter relates basic hardware architectures and approaches to the system and fault models previously introduced. The chapter's main objective is to provide the reader with insight to the hardware on which our systems and solutions are implemented.

## 2.1 Current Systems

We divide the discussion on current computing systems into three parts according to their architecture: the general purpose processors, the high availability systems, and the reduced instruction set (RISC) computers.

### 2.1.1 Examples of general purpose processors

Currently, the major portion of general purpose processor architectures are based on a complex instruction set computer (CISC). These architectures are based on 16 and 32 bit microprocessors and their family of support components: arithmetic coprocessors, memory management units, interface controllers, and communication controllers. The following examples have been chosen due to the high frequency of their use and their outstanding properties. We do not describe the processors in detail, but rather focus on ways to employ distributed architectures based on them.

**The Motorola 680x0 family.** The Motorola M68000 family of microprocessors is a good example of CISC architecture. Motorola's 68020 processor [105] is a 32 bit implementation which followed its 16 bit predecessor, the 68000 processor. Figure 2.1 describes the MC68020

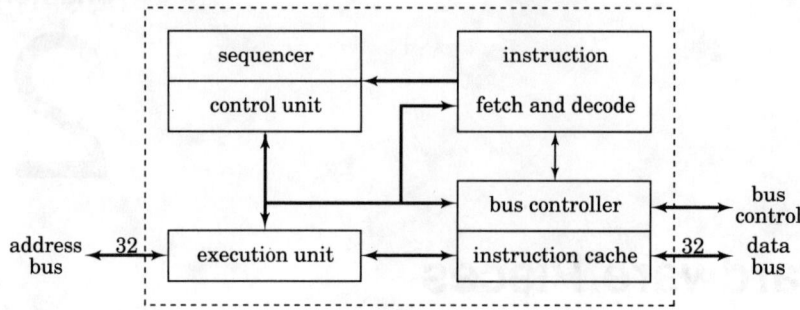

**Figure 2.1** Motorola 68020 block diagram.

block diagram. The upper left block, which coordinates the internal activities of the processor, includes a sequencer and a control unit. The instructions are brought from the data bus to the decoding segment and to the local cache. An observable property of the 68020 is its nonmultiplexed bus architecture. It uses a 32 bit data bus and a different 32 bit address bus. The processor includes 16 general purpose 32 bit registers to manipulate data and addresses, 5 control registers, a program counter, and stack pointers.

The instruction fetch & decode segment uses a three stage pipelined architecture. Being a CISC processor, its instruction set includes the following instruction classes:

- Data movement operations for memory, stack and register data transfers
- Integer arithmetic operations
- Binary coded decimal (BCD) arithmetic
- Logical operations
- Shift and rotate operations for registers and memory
- Bit and bit-field manipulations for testing, clearing, setting, and changing a bit or a bit field of up to 32 bits
- Program flow controls
- System control
- Multiprocessor operations

The multiprocessor operations include compare-and-swap (CAS) and test-and-set (TAS) operations, which are indivisible read-modify-write operations. In addition, it includes a set of coprocessor instructions, that activate the coprocessor as well as monitor its results and condition code selector.

Figure 2.2 describes the MC68020 with its MC68881 coprocessor in a multiprocessor configuration. The function codes ($FC_0$ to $FC_2$) distinguish data and program segments as well as user, supervisor, and

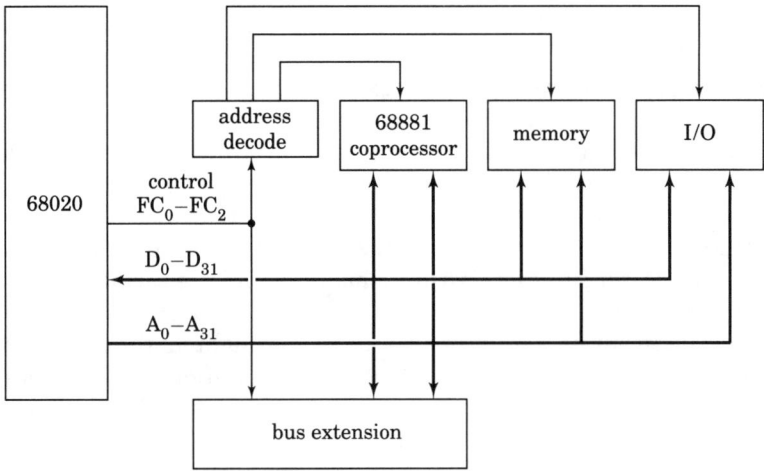

**Figure 2.2**  Motorola 68020 coprocessor configuration.

CPU spaces. A set of control signals support bus arbitration control, asynchronous bus control, and bus exception control.

The coprocessor communicates with the main processor through the asynchronous bus, and such communication is initiated once a coprocessor instruction has been fetched. Such instruction contains the coprocessor identifier and the instruction type. Coprocessor identifiers allow user defined coprocessors, in addition to Motorola's units such as MC68851 memory management unit (MMU) for paged memory and the MC68881 floating point coprocessor (FPC).

Interrupts are handled in a seven priority level scheme. The requesting interrupt level appears on these lines and if the level is higher than the current mask, the interrupt is acknowledged and the interrupt vector number is acquired. Exceptions are processed in four steps: current status register is saved and modified to exception, then exception vector is determined, the processor state information is stacked, and finally the new processor state is obtained and processing continues. After the exceptions stack is emptied, a return from exception procedure is carried out.

**The 80x86 family.**  Intel began its 8 bit microprocessors with the 8080 processor, which was followed by the 16 bit 8086, and then by its 32 bit 80386. The sixteen bit processors have had intermediate family members such as the 80186 and 80286 processors. As this processor family expanded, the network support requirements broadened significantly. Distributed systems needed the network support from its hardware components to allow performance figures to be comparable to those of a single processor. Network and communication coprocessors emerged as a good solution. They relaxed the communication burden on the main processor, functioning concurrently with it, and thus increased the overall system performance.

Figure 2.3 describes the 80186 processor with 82586 network coprocessor in a configuration that supports both serial and multiplexed system bus activities. This approach allows the network coprocessor to manage and control the serial communication independently of the main processor's state. The main processor initiates the IO activity, and the coprocessor interacts with the memory and the serial communication controller.

Further distribution requirements have highlighted the need for multiple buses in the system architecture. In addition, performance requirements necessitated parallel buses operating side by side. Intel's Multibus-II (MB-II), for example, justifies two spaces: the local space and the system space. These spaces are partitioned such that the local space is totally transparent with respect to the system space. Communication in the local space is kept private from the system technology-independent space.

The system space in MB-II is divided into two sections:

- *Message space* in which communication is standardized and thus supports high level protocols.
- *Interconnect space* in which initialization, configuration, and diagnostics requirements of the global system are accomplished.

The local space in MB-II is divided to two sections as well:

- *CPU/memory expansion space* is entirely technology dependent and processor related. Expansion properties must support consistency and coherence, especially in processor clusters that maintain the local caches.
- *Incremental I/O space* through which additional input and output devices are linked to the system.

The MB-II approach isolates the system level standardized communication from the technology dependent local implementation. The system level communication uses a 32 bit wide address/data parallel bus, and each word is accompanied by four parity bits. The bus is centrally controlled in a synchronous design through control signals generated by a central service module (CSM). Ten system control lines are used for control (request phase) and status (reply phase). Two bus arbitration algorithms are reflected in the MB-II scheme: fair and prioritized. Arbitration result is transferred to the boards on six identification lines.

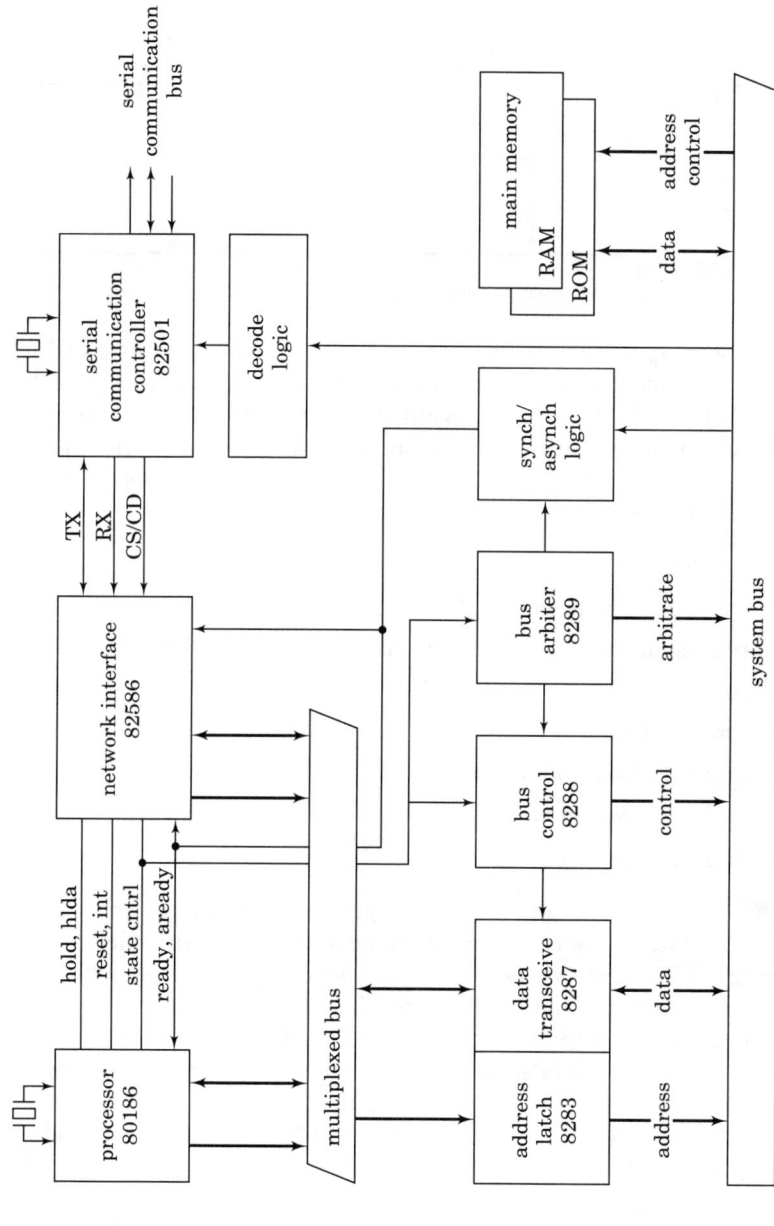

**Figure 2.3** Intel 80186 with 82586 network coprocessor.

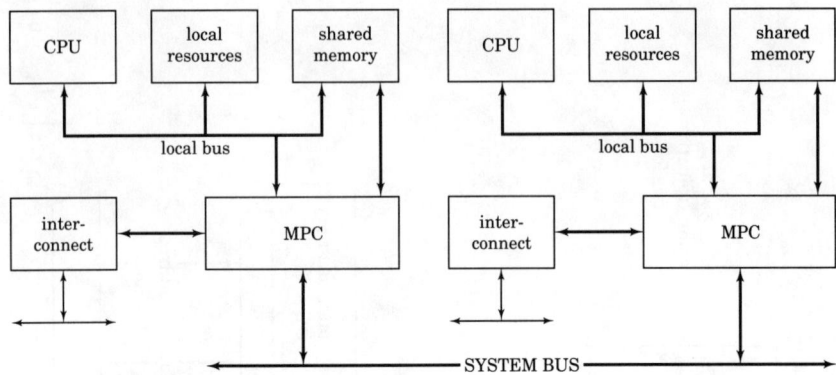

**Figure 2.4** MPC: message passing coprocessor.

The above approach yields to specially designed coprocessors which uniformly support message exchanges across the system. Figure 2.4 describes the MPC (message passing coprocessor) concept as reflected for the MB-II architecture. The system bus interconnects the system components through a set of MPCs, and these coprocessors perform the following tasks:

- Bus arbitration
- Bus speed matching and buffering
- Bus representation (word width) matching and buffering
- Bus control
- Bus interrupt buffering
- Parity generation and checking
- Shared memory control
- BIT and bus error reporting

System message communication is generally based on hardware-dependent data types, commonly called *packets*, to obtain efficient data exchange. There are two categories of packets: unsolicited, which are always unexpected events at the receiver side, and solicited, which are events whose arrival is expected. Unsolicited packets are used for interrupts and signals, as well as for negotiating a subsequent solicited message. Solicited messages are used for larger data exchange. In the solicited mode the packing, unpacking, error detection, and error resolution are all done by the MPC, without involving the local processor.

Figure 2.5 describes the Intel 82389 MPC that is compatible with the MB-II architecture. It supports message passing buffering with nine integrated high speed FIFOs:

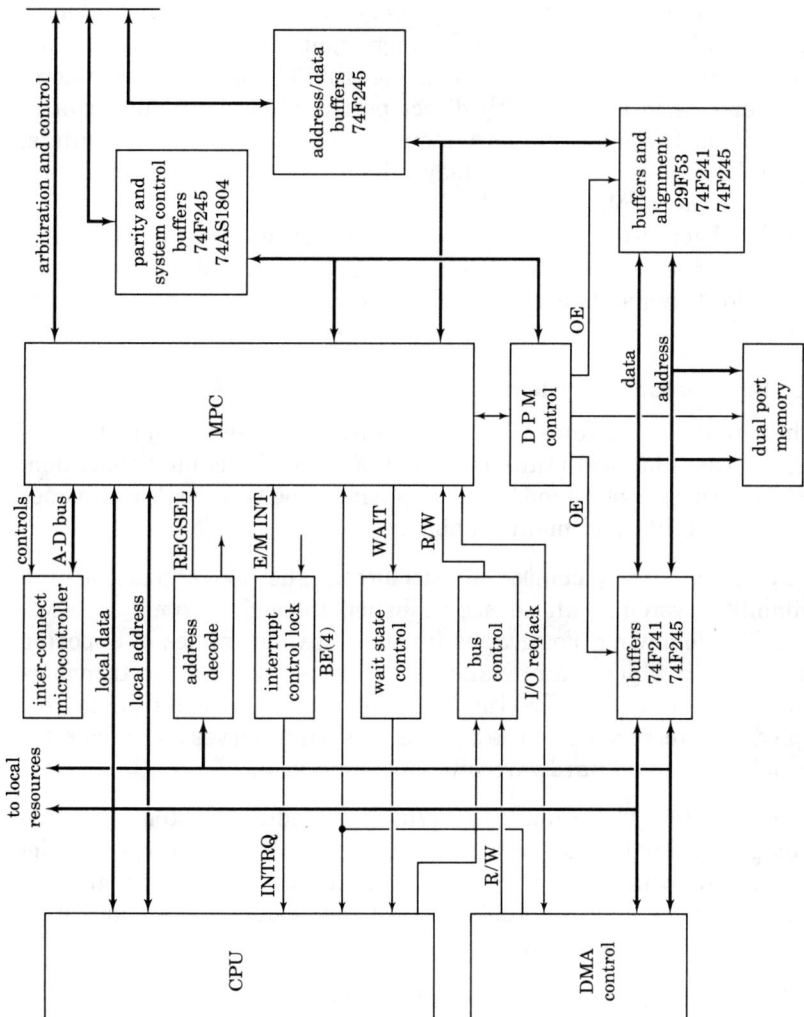

**Figure 2.5** Intel 82389 message passing coprocessor (MPC).

- Four interrupt inputs
- One interrupt output
- Two block transmissions
- Two block receivings

The four interrupt input and one interrupt output buffers support incoming unsolicited packets, and the four block exchange buffers support solicited data exchange.

The 82389 MPC interfaces the system bus through external address and data buffers, and parity and system control buffers, and it deals directly with arbitration and central control. The interconnect microcontroller is also treated with direct ports. Shared memory control is interfaced through dual-port memory (DPM) control and buffers, and the IO request and acknowledge signal exchanges with the direct memory access (DMA) controller.

The local activities are interfaced mainly on the processor's local bus control and bus enable (BE) signals, wait state control, and interrupt control. MPC registers are selected (REGSEL) through address decode logic.

### 2.1.2 High availability systems

High availability systems [71, 117] ensure system-wide robust structures for cases of local faults in execution of a particular application. This robustness is obtained by accomplishing the twin goals of dependability: reliability and maintainability.

**Tightly versus loosely coupled architectures.** The above goals of high availability systems can be accomplished through various architectures. The following paragraphs illustrate two examples of opposing approaches in which maintenance is performed on line and transparent to the application. The Tandem 16 system represents the loosely coupled architecture while the Stratus 32 system serves as an example of a tightly-coupled hardware-intensive approach.

**The Tandem 16.** The Tandem 16 [75] is a modular system of transaction processing (introduced in mid 1970s) with no hardware single-point failure which can impose any threat to the data integrity in the system. The goals and objectives of the design of the Tandem 16 computer have been as follows.

1. High maintainability, achieved through
   - autonomous fault detection,
   - system reconfiguration,
   - system repair and component replacement,
   
   while the rest of the system continues performing correctly.

## Hardware Pieces 25

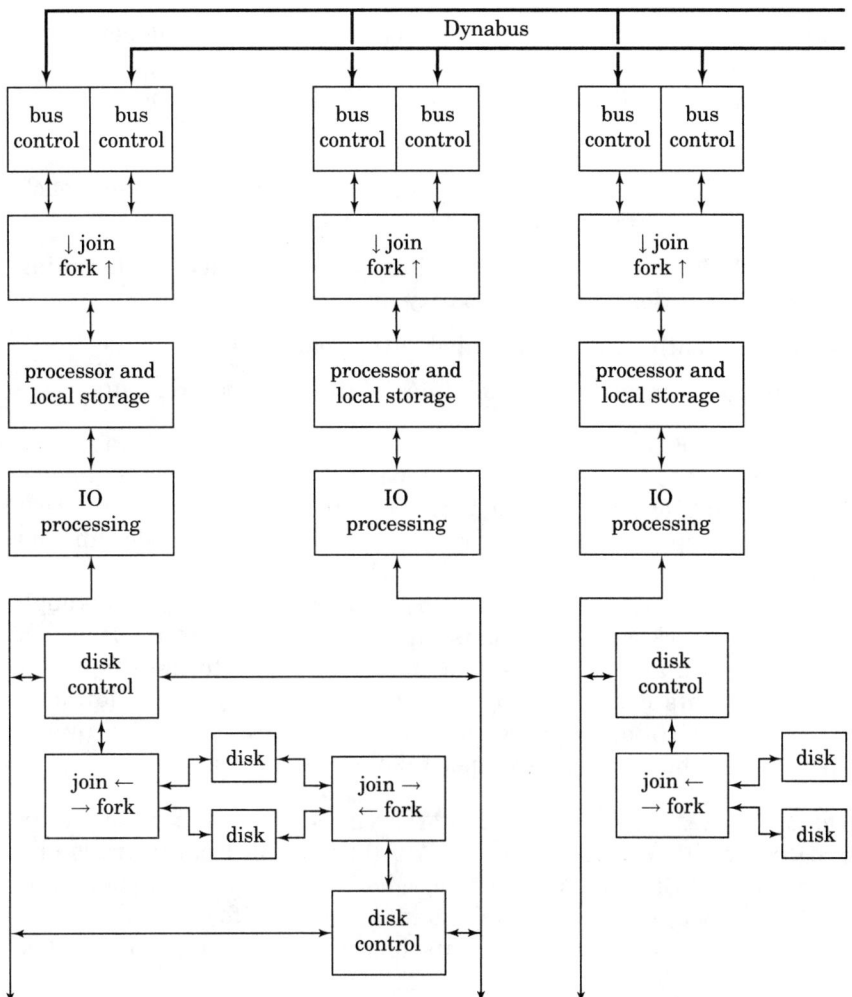

**Figure 2.6** Architecture of Tandem-16 computer system.

2. Elimination of all hardware single point failures, providing a robust environment for transaction processing of large data volumes.

3. High degree of modularity to support maintainability as well as easy expandability, supporting on-line device replacement and resource addition.

The Tandem architecture supports up to sixteen processing nodes, interconnected via a dual redundant bus, called Dynabus. Figure 2.6 depicts the Tandem 16 architecture and its building blocks. The right node is independent, and the two left nodes are interconnected as an

IO process pair. Each element in this architecture is independently powered, thereby supporting the on-line replacement objective.

Autonomous fault detection is accomplished through the following mechanisms:

1. Interprocessor transmissions are composed of data packets, each of which includes its own calculated checksum.
2. Each 16 bit word in the memory is appended with a 5 bit Hamming error correction code and a parity bit.
3. All data paths are accompanied with a parity bit.
4. Watchdog timers trigger violations of liveness requirements.

The major recovery mechanism in Tandem 16 is based on checkpointing, and "cold" redundancy supported by a master and shadow IO processing organized as a pair. Figure 2.7 depicts the relation between an application which runs in one processing node, and its shadow which runs in another. The shadow processes are not executed concurrently, and are in a rather inactive standby state, through which only checkpoint tracking is supported. Checkpoint information, including data and status, is sent from the master to the shadow at recovery points designated by the application designer. Additional IO operations (from last checkpoint to failure) are designated by the operating system to assure proper take over by the shadow.

**The Stratus 32.** The Stratus 32 [124] architecture uses a different approach than that of the Tandem. A tightly coupled hardware is employed in the Stratus architecture, where each of the components is continuously compared to a redundant component for error detection. Each of the compared components has another spare, which takes

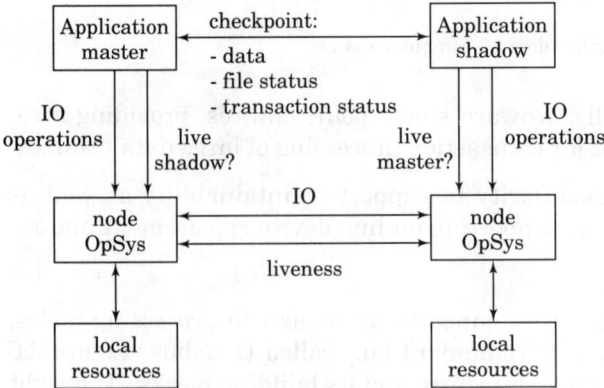

**Figure 2.7** Tandem-16 shadow processor concept.

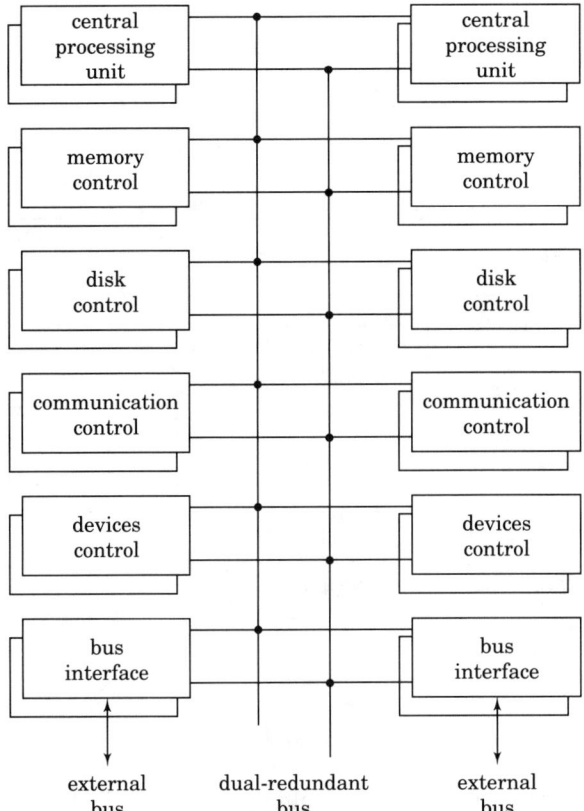

**Figure 2.8** Architecture of the Stratus-32 system.

over when an error is detected. Figure 2.8 depicts this approach, emphasizing the intensive use of hardware showing that four CPUs are employed to carry out functions of a single CPU.

Once a pair is replaced by its duplicates, diagnostic procedures verify the fault nature as transient or permanent. If a transient fault is verified, the replaced pair becomes a spare pair itself. We note that this approach is used across the board within Stratus 32, for various controllers as well as for CPUs. The disk controller's pair must reach mutual agreement before every modification of the protected data.

Such an architecture requires a tight synchronization of its elements in order to support the on-line replacement requirement. Since bus failures can also occur, a dual redundant bus is used for interconnection.

**The Electronic Switch System (ESS).** The Electronic Switch Systems (ESS) [117, 133] have been introduced by Bell Telephone Laboratories for the telephone switching applications. These systems try to achieve

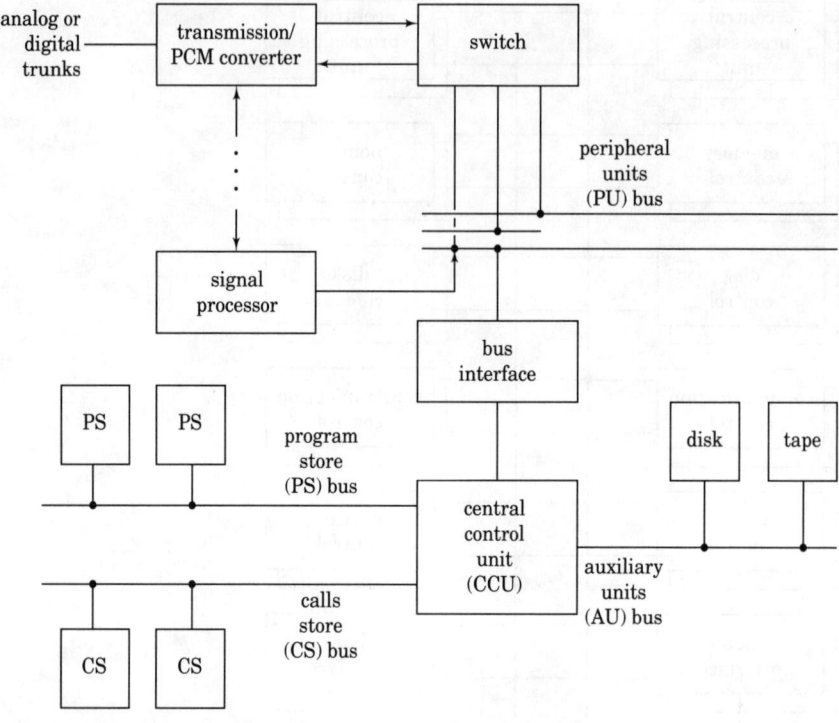

**Figure 2.9** Architecture of a typical electronic switch system (ESS).

a minimal hardware originated down time, along with significant reduction in incorrectly handled telephone calls[1].

Figure 2.9 depicts a typical ESS block diagram. The transmission/PCM converter interfaces the incoming and leaving telephone trunks, which can be analog or digital. It converts them to an eight bit PCM (pulse coded modulation) stream, which is switched to the peripheral unit (PU) buses. A signal processor detects the various activity changes and informs the central control unit (CCU), thereby reducing its load significantly. The CCU is in charge of four major tasks in the ESS:

- ESS control
- Call processing (the 3A processor is designed for 500-5000 calls)
- ESS maintenance (fault detection, reconfiguration support, diagnosis support, etc.)
- System administration services

---

[1] In other words, high availability and fault tolerance requirements.

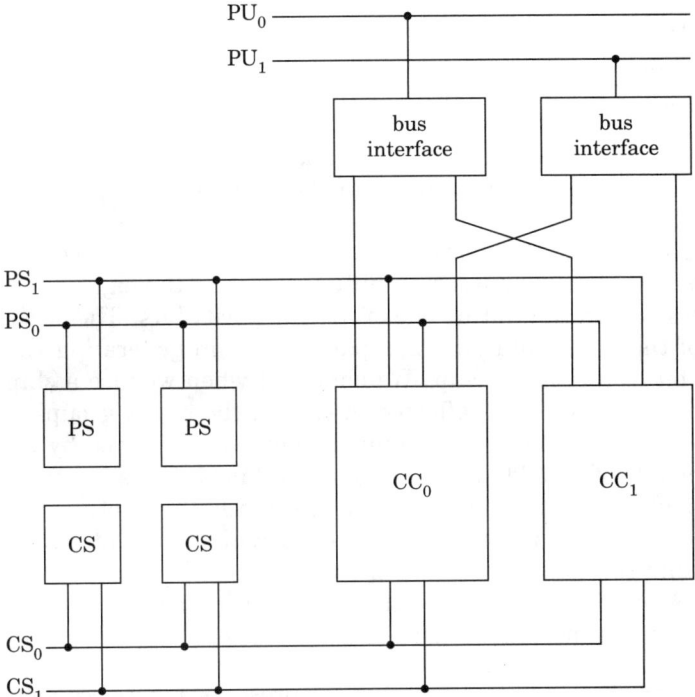

**Figure 2.10** Duplicated modules in ESS.

Different segments of the memory are allocated to program storage (PS) which is permanent and call storage (telephone calls data, routing data, current configuration data) whose nature is transitory.

A basic technique in ESS is duplication of elements, such that the whole processor is switched (replaced) upon detection of failure in any of its elements. Figure 2.10 depicts the duplicated processors architecture of a typical ESS. In the ESS-2 processor comparisons and self-tests are used to detect failures. However, the ESS-3A does not compare two processors to detect failures, instead, each processor performs its own fault checking. Thus, the tightly coupled synchronization requirement is omitted.

In the ESS-3A cold redundancy architecture, the on-line processor writes to both stores. Once a fault is detected by the self checking logic, the standby processor (kept at a halt state) assumes the responsibility for processing the calls. Error detection and correction are accomplished by the following means:

- 4-of-8 encoded control word in the microprogram
- Parity on address word in the microprogram
- Self checking decoders

- Two bit parity on memory and registers
- Watch-dog timers
- CCU duplication

Figure 2.11 presents the error detection and correction mechanisms employed for the CS memory along with those of the microprogram store.

The ESS-3A memory is organized in two bit slices: nine two bit chips compose each memory word. A bit from each of the eight data-bit chips takes part in generating one of the two parity bits. The other bit of each of the eight data-bit chips participates in generating the other parity bit in the parity chip. We note that when we face a chip failure, say $U_4$, only one bit is affected in each of the parity groups.

The microprogram memory (read-only memory) is organized by interleaving the bits of its address field and control field. Thus, an error affecting two adjacent bits affects the different mechanisms of the the different fields. The address field is encoded with a one bit parity mechanism, and the control uses a 4-of-8 code.

The ESS-3A architecture invests 25 percent of its logic circuits in self checking, and 14 percent in maintenance access.

**The Synapse.** The Synapse N+1 architecture presents a more economical solution than that provided by duplication. Instead of heaving each executing element duplicated (2N), it uses only one additional copy (N+1) of each resource (processor, memory, bus, disk, etc.). Figure 2.12 depicts a typical Synapse N+1 block diagram. This system has also been introduced (1984) for transaction processing applications.

The Synapse N+1 system is composed of

- General purpose processors
- IO processors
- Shared memory modules
- Redundant bus

The system may contain up to 28 processors (general purpose and IO), which are interconnected via the bus for memory sharing procedures of data exchange. The memory modules (1 megabyte segments) can be extended to contain up to 16 segments in 4 modules. As both application programs and operating system reside in the shared memory and can be executed by any one of the general purpose processors, a single redundant processor makes sense.

Task allocation is perform by all processors. Each idle processor selects its next task from a queue in the shared memory. Each processor

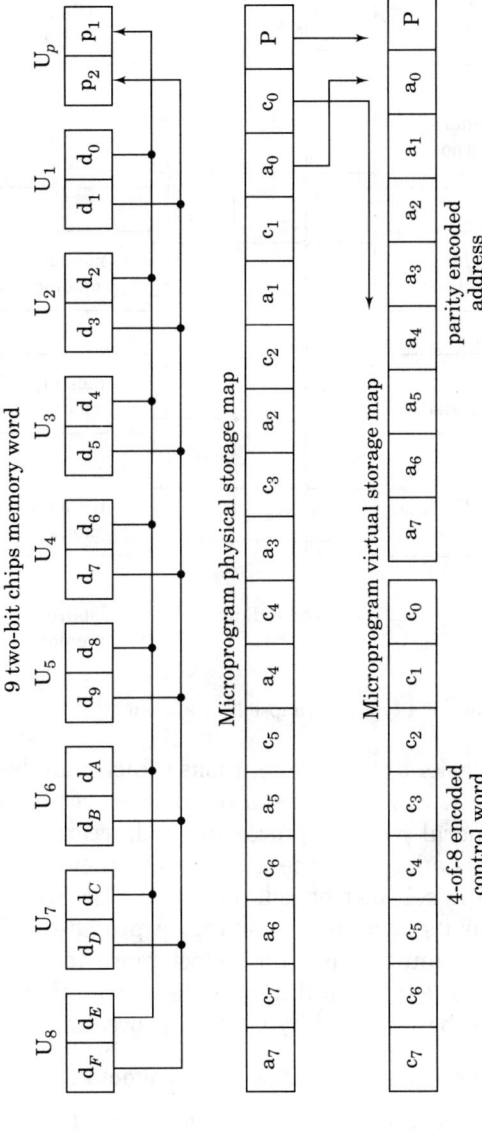

**Figure 2.11** Memory organization in ESS.

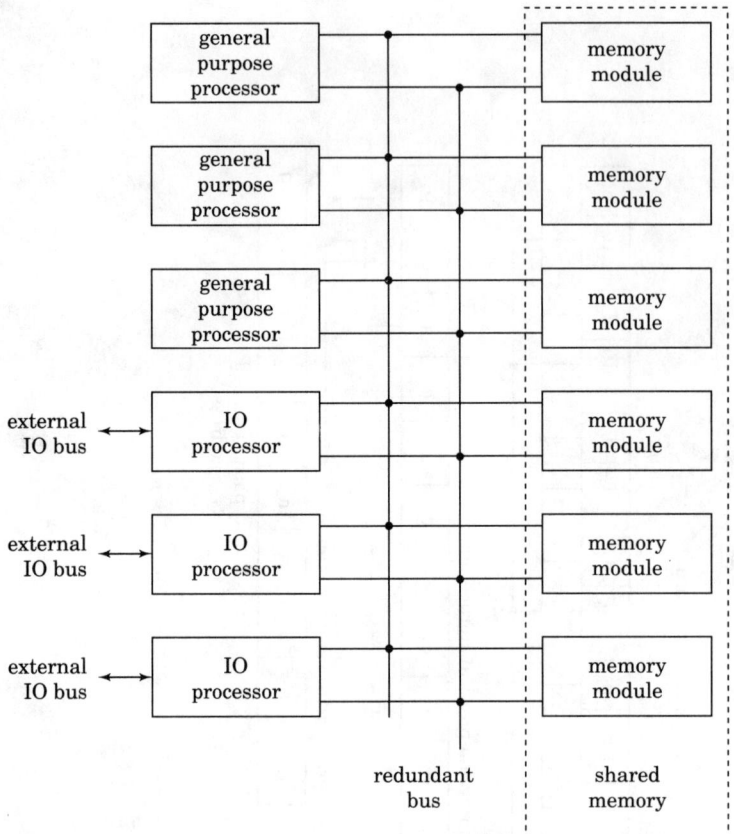

**Figure 2.12** Architecture of a typical Synapse N+1 system.

can also append tasks to the queue, tasks which may be executed by different processors. Since the tasks themselves reside in the shared memory, each general purpose processor (and accordingly, IO processor) can access this address space and execute them.

Fault detection is self-test based, as in the ESS-3A case, and each processor performs its own fault checking. A processor whose self test indicates that it is faulty, does not select more tasks. Thus, tasks previously assigned to it will fail to provide information to others, and new instances will be spawned by waiting processors.

### 2.1.3 Examples of reduced instruction set architectures

Developments of the late 1980s introduced significant performance growth (over CISC machines) for processors designed for reduced instruction set computers (RISC). The most important properties of RISC architecture are simplicity and efficiency. In order to support these properties, all RISC processors demonstrate some key characteristics

- Most of their operations are based on register-resident operands. Thus, they have a large set of registers or register windows.
- Most of their instructions have a single execution cycle.
- They have simple addressing modes and simple memory interfaces for read and write operations.
- The implementation of their branch instructions is performance motivated. For example, delayed branch instructions allow execution of the instructions which sequentially follow the branch before branch condition is determined. Such an attribute enhances pipelined machine performance with respect to pipeline flushing.

Nevertheless, some of the RISC processors support additional instruction sets which are more suited for the complex instruction set machines. These include such examples as graphics instructions, bit manipulations, and vector and matrix operations.

**The R3000 and R4000.** Figure 2.13 describes the architecture of the R3000 reduced instruction set processor by Integrated Device Technology (IDT). The R3000 processor contains a 32 bit ALU, a set of 32 general purpose 32 bit registers, and a 32-bit integer multiplier/divider. The memory management segment of the processor uses a translation lookaside buffer (TLB) of 32 entries. The TLB maps the memory cache of the most recently used page table entries from the user's virtual terms to physical terms. Each entry translates an even/odd address pair of physical entries. The R3010 floating point processor operates concurrently with the main processor, and the latter coordinates its relationship with the cache.

The extensive use of caches raises the issue of cache coherence with the memory. The methodology employed in the use of these caches, stores data written by the processor to the cache, and memory is updated only when one cache line is replaced by another.

R4000MP by MIPS Computer Systems Inc. contains the CPU, the floating point coprocessor, and the primary instruction and data caches. It also supports additional external secondary instruction and data caches in an architecture which separates instructions and data as in its predecessor's primary caches. It must, therefore, maintain the secondary cache coherent with the memory as well.

Figure 2.14 describes a multiprocessor configuration based on the R3000 RISC by IDT. This configuration requires a secondary cache as an interface with the system bus. The R4000MP by MIPS Computer Systems Inc. uses a complementary integrated circuit, the R4000CP or R4050, to interface the R4000MP with the system bus, and to carry out a cache coherence protocol. A cache line which does not agree with

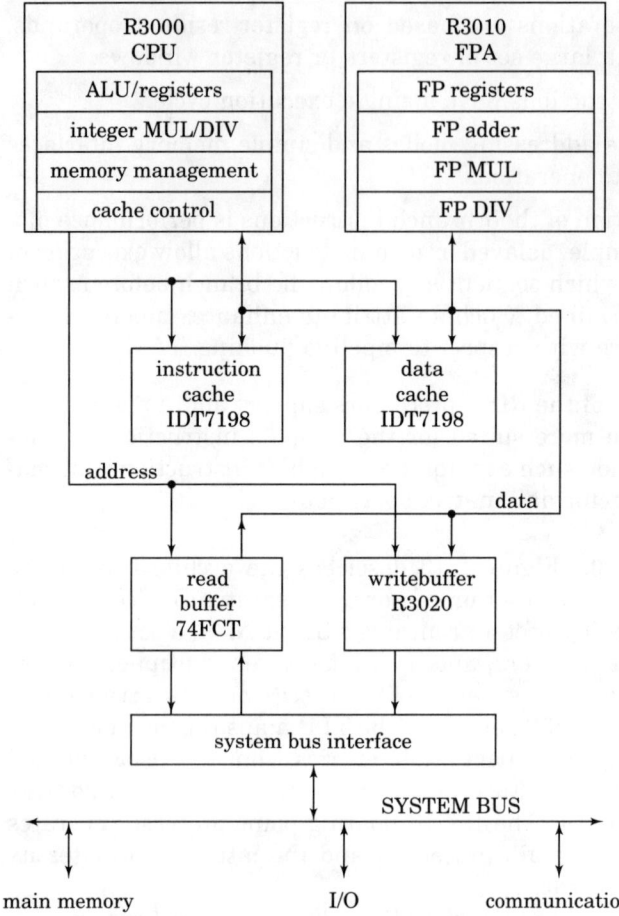

**Figure 2.13** IDT R3000 RISC system block diagram.

the corresponding memory is marked "dirty," and is marked "clean" for the opposite case. The cache coherence protocol monitors the cache line state on every read and write operation, and carries out the proper state transitions to obtain a coherent data transfer. The R4000CP can also be used as an interface between the memory or an IO and the system bus.

In a multiprocessor environment, caches are accessed by external processors as well. Thus, in addition to monitoring the "clean" or "dirty" state of a cache line, the coherence protocol must check to see if the cache line is "shared" or "exclusive," and react accordingly. For example, in addition to the memory updates already discussed, additional updates are needed when external access is detected by the secondary cache which interfaces the bus accesses.

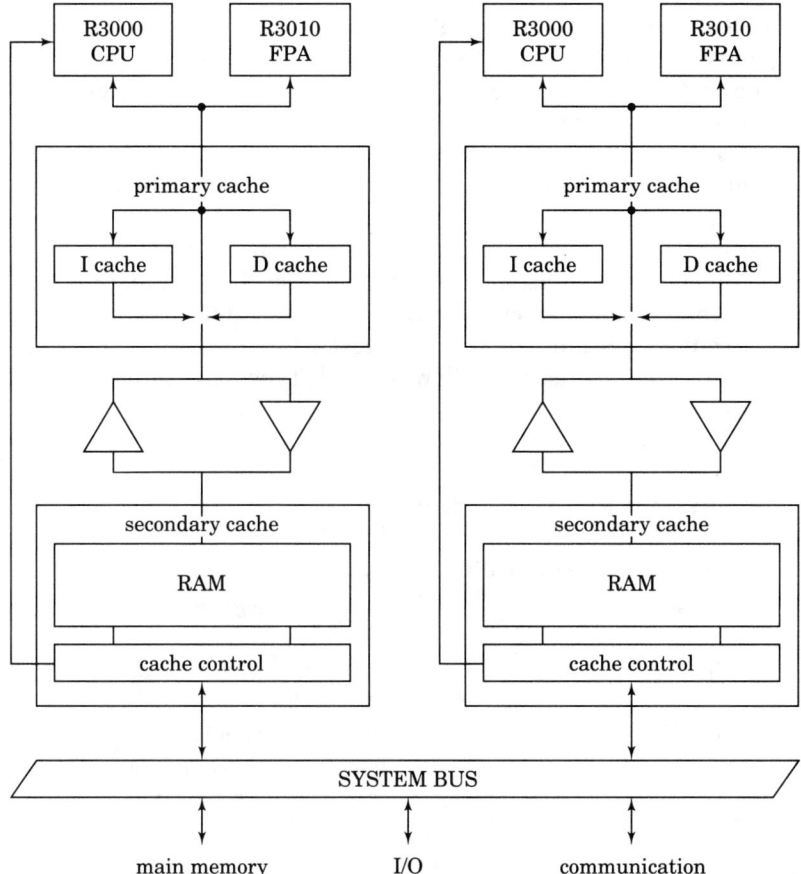

**Figure 2.14** The R3000 in a multiprocessor configuration.

Interprocessor accesses are events at a system level and the hardware must support the means to conduct them properly. Therefore, in addition to the 32 bit data bus and 32 bit address bus (and their check bits), command and data identifiers are transmitted between the processor and the external agent. The external request, ready state, and release indicator are needed to arbitrate the external flow. These requirements are satisfied by the R4000 system interface.

**The i80960.** The 80960 by Intel [68] presents an architecture which tries to achieve the execution of multiple instructions per clock cycle. It does this by allowing parallel and out-of-order execution of instructions. The internal architecture of the processor can therefore be viewed as a large set of registers, accessed in parallel by processing elements, each of which executing an instruction scheduled from the instruction cache.

The 80960 processor has a fault mechanism for an environment in which out-of-order execution takes place. As several instructions are executed at the same clock cycle, it may be impossible to identify the fault source on detection. A special fault mechanism, the synchronize-fault instruction (SYNCF), halts the execution to assure the completion of all pending operations and the signaling of the corresponding faults. Inherent fault detection is provided for division by zero, integer overflow, inappropriate operand, and inappropriate opcode. Fault table management defines a fault handler for each fault, and the handler is invoked after detection with a record of information on that fault. A conditional fault instruction (FAULT_IF) allows fault signaling on conditions. Thus, the following fault types are defined for the 80960 architecture:

| Fault Type | |
|---|---|
| Arithmetic (overflow, underflow) | For integer and zero division. Distinct floating point faults |
| Conditional (on FAULT_IF) | $=, <, \leq, \neq$, etc. |
| Access | Call out of range, memory access failure |
| Operation type | Mismatch of supervisor vs user mode |
| Invalidity | Opcode or operand |
| Trap (breakpoint) | Programmable detection of occurrence of event to trace |

Figure 2.15 describes the architecture of the 80960 32 bit register-rich processor by Intel, and its interfaces with memory and IO. The processors can be aided by Intel's 82965 Bus Extension Units (BXUs). Each BXU can replace the processor in interfacing at the dotted line shown in Figure 2.15. This approach provides an interesting building block for distributed, and in particular, fault tolerant architectures. For example, if we let one component contain a CPU and a BXU and the other contain a BXU and memory, we can access the remote memory through the system bus. In order to support these modes of operations, a BXU consists of seven major subunits:

1. Interface for system bus.
2. Interface for local bus.
3. Support logic for inter-agent communication (IAC) between the 80960 and the BXU.
4. Support logic for IO prefetch operations.

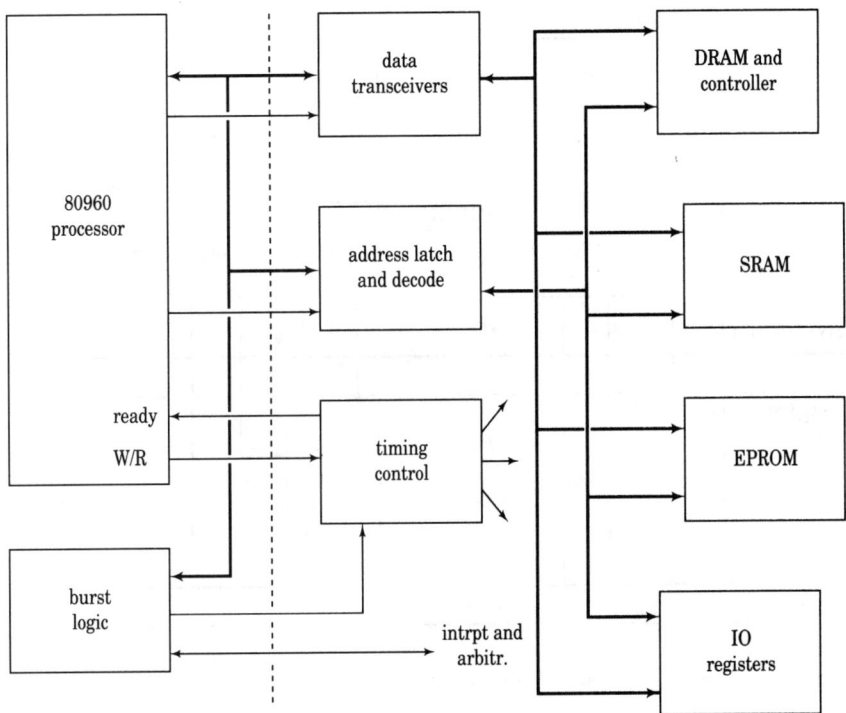

**Figure 2.15** Intel i960 system block diagram.

5. Support logic for memory control.
6. Support logic and management for cache.
7. Support for fault tolerance and backup BXUs.

The BXU can operate in processor mode and memory mode. In the processor mode it allows operating as a local bus slave, while in the latter it operates only as a local bus master.

Figure 2.16 describes the fault tolerant architecture of the i960 in a master and checker configuration, a configuration based on Intel's older 432 systems (1981). The operating system here joins two processors as a pair out of which it appoints one as master and the other as checker. The two processors run the same software, but only the master issues system-bus requests. The checker monitors the bus activity of the master and the master monitors the checker. When a conflict occurs, the detector generates an error report. This dual redundant configuration is an example of an error detection mechanism at the processor-bus level. However, it does not support recovery because detection points to the pair as erroneous with no isolation of the faulty processor. We note here that initialization of such a configu-

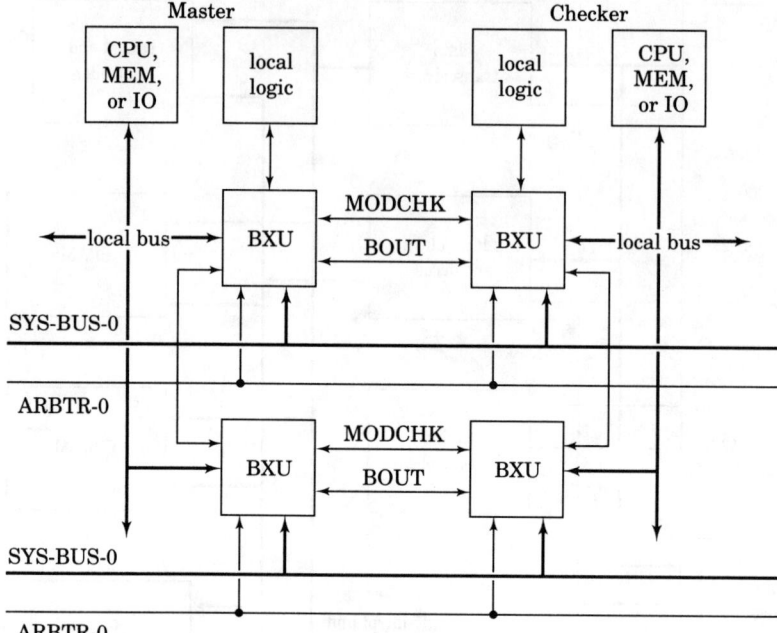

**Figure 2.16** Intel i960 master/checker configuration.

ration must point each processor to its mate processor, and appoint them to properly carry out the master and checker tasks.

The appropriate BXUs are closely coupled through two signals to allow continuous comparison of the corresponding system-bus activities. The checker BXU's internal image of its system-bus information is compared to the information actually observed on the bus when the master sends its information. These events are synchronized by the buffer-out (BOUT) signal shown in Figure 2.16. When there is a contradictory BOUT request, the detector issues a "Bus Arbitration" error report for the master checker pair. Another signal, the Module Check (MODCHK), informs a BXU that its mate has been locally requested to access the system bus. When there is a contradictory MODCHK request, the detector issues an "Unsafe Confinement Area" error report for the master checker pair.

Dual system-bus redundancy is also supported by the BXU. In order to properly communicate between the local BXUs, various synchronization and control signals are used, including queue control of pending requests (POPQUE) coordination (SSBUSY) and error lines (LERL). When a BXU detects a permanent error on its system bus, it can switch off this bus and let its partner continue working with the redundant bus. The system-bus requests are then served by the backup

BXU. Bus switching allows recovery of system-bus failures, detected through parity bits on data and arbitration lines and timeouts.

## 2.2 A Quad-Processor Architecture

We have discussed the Stratus 32 (see Figure 2.8) approach for fault tolerance, a tightly-coupled, hardware-intensive approach. We show here how the cost of this approach can be significantly reduced, with various quad-processor architectures.

### 2.2.1 Shadow systems architecture

Intel's 80960 architecture supports a quad modular redundancy (QMR) approach, forming a primary-shadow partnership, where each partner is a master-checker pair. The shadow partner acts as a hot backup, and upon an error detection by the primary's master or checker it takes over. Figure 2.17 describes this fault tolerant architecture as an extension of a master and checker configuration to include a shadow for every primary system.

Setting up a partnership requires the following conditions:

- BXUs must be at the same state
- Partner pointing register of each partner must point the other
- Set the QMR register of each, appropriately marking the primary and shadow
- Synchronize the two with a processor IAC message

After these conditions are satisfied and actions executed, the primary and shadow still carry distinct physical identifiers, and report differently on module errors and bus errors.

**Error types.** Consistent with the definitions in Section 1.3, the 80960 distinguishes between permanent and transient errors. Permanent errors are considered as the primary's or shadow's catastrophe, or an unsafe confinement of the error propagation. The permanence of these three error types is verified upon detection by a comparison to previously detected errors in the error log and error record which the BXU maintains.

**Detection-report-recovery cycle.** The sequence initiated upon error detection in the primary-shadow configuration starts with error reporting at the master-checker level. The detecting BXU reports to its mate as well as to its QMR partner. Then, the partners must resolve a dual system-bus integrity problem: the pending bus requests on the failing

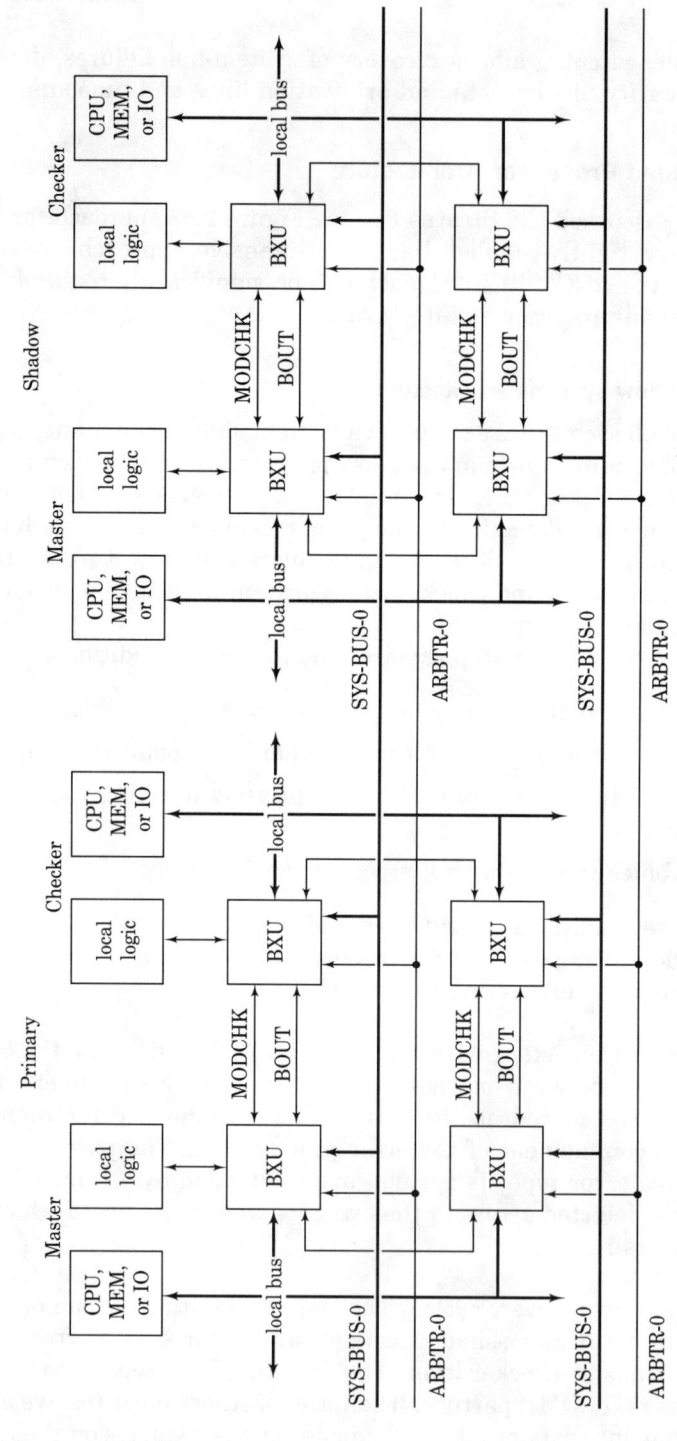

**Figure 2.17** Intel i960 master-checker main-shadow configuration.

BXU. Information on the state of these requests is transferred to the healthy partner to retry servicing them.

If the error is found to be transient, a settling time is allowed to permit the stability to return. Then, all pending system-bus requests are retried, in the order received by the BXU. If an error is detected during the retry period, another error cycle is initiated. A transient error may become a permanent one if it occurs again before the BXU terminates the permanent-error window it initiated after the retry period. On the other hand, if the detected error is permanent, a resource reconfiguration is required in order to recover.

**Resource reconfiguration.** Three basic recovery mechanisms are required in the 80960 QMR architecture:

- Partners take over (primary/shadow)
- Mates splitting (master/checker)
- Bus switching

On a primary module failure, the shadow module takes over after receiving an error report from the primary. It alternates its BXU status in the QMR register, and thus responds to system bus pending requests. The faulty partner blocks itself from the system bus and stops serving bus requests. It declares itself out of partnership, inactive and faulty, by setting the appropriate registers accordingly.

Mates splitting, unlike module shadowing, does not provide recovery which is transparent to the software. It allows disablement of a master-checker relationship, continuing with a single unit while the other is deactivated. We note here that splitting is unable to resolve which unit is faulty (master or checker) unless the detected error is self evident.

Upon detection of a system-bus permanent failure, the corresponding BXUs automaticly switch buses. The BXU on the faulty system-bus blocks its connection to the bus, and stops receiving requests from the local bus. The redundant BXU then starts servicing these requests, after the pending requests are completed. We note here that this switch is transparent to the 80960 CPU.

### 2.2.2 Thread oriented architecture

A different approach of quad-processor architecture emphasizes specializing processors instead of the duplication approach just described. The following discussion concentrates on a real-time, fault-tolerant, and distributed system, and the specialized tasks derived from these attributes.

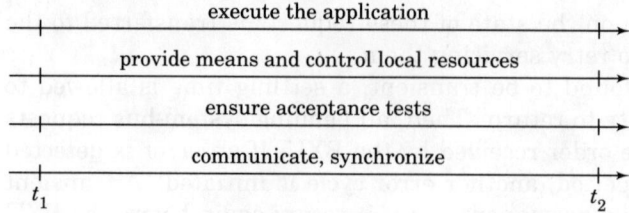

**Figure 2.18** Required threads of actions.

**The site view of the system.** In our system model we describe a distributed computation as a set of processing nodes $\{n_1, \ldots, n_k\}$ and a set of processes $\{P_1, \ldots, P_m\}$ which execute on these nodes. At a particular instance of time, a specific process $P_i$ is executing on a specific processing node $n_i$. A processing node, or in other words a *site*, is a group of processing, storage, and input-output resources, which have a high degree of coupling, mainly due to locality properties. A site's $(n_j)$ view contains parts of distributed computations, which are to be executed on this site at some time. Each of these parts of computation is a nonempty set of processes.

Since our system of interest is a real-time, fault-tolerant, and distributed, each site must perform the following threads of actions for each process in its view:

1. The process has to be executed such that a deadline constraint is met.

2. Means (resources, communication, etc.) are to be provided for the distributed computation. Local resources should be controlled to serve either local or external computations, each represented locally by a process of some type.

3. Each computation is required to pass an acceptance test, to detect an error and allow fault-tolerant actions. We have discussed comparison tests and we discuss later in this book, other types of reasonability and legitimacy tests.

4. Data transfer and synchronization services must be provided, both for local processes and for communicating between two external processes (as dictated by topology constraints).

These threads of actions are summarized in Figure 2.18 emphasizing their continuous nature on the time line between two time constraints.

**Local processing architecture derivation.** The system real-time, distributed, and fault tolerant characteristics imply specific requirements for properties of the system. These properties interact with each other in various ways.

The distributed architecture of the system is used to provide redundancy which enhances fault tolerant parameters of the system. This distributed nature is also used to achieve performance enhancement in the response time of the system by optimizing parallel activities, while minimizing the necessary synchronization overhead. Such a performance enhancement supports real-time objectives through efficient switching and fast response time.

In a real-time system a deadline constraint is to be satisfied for any real-time process. However, a process may fail and the fault tolerant objective must be achieved within the deadline constraint. Therefore, backup processes or alternatives must be activated such that upon failure, the same deadline is met, and the real-time constraint is satisfied. In order to achieve this goal, it is clear that errors should be detected as soon as possible. Therefore, the more extensive checkup we implement, the sooner an error is likely to be detected.

On the other hand, implementation may bring up contradictory effects. Extensive testing may result in a high overhead in switching, control, and synchronization activities, as well as in the testing execution time. We can, therefore, derive our first objective of a processing architecture if a real-time, distributed, and fault tolerant system: *achieving parallel and loosely coupled testing during execution time of an application process.* Another property which is achieved along with this objective is the independence of execution time of an application process and testing.

The requirement of an a priori knowledge of the execution time of a process is of crucial importance for a proper real-time scheduling discipline. Yet, a topology in which the sites are not fully connected may require communication-relay services from other sites. These relay services are also real-time processes, since they contribute a communication delay to the execution time of the message sender and the message receiver. Furthermore, it is impossible to predict the processing load of a site due to such processes, since they are spawned by processes that are executing on other sites.

The communication services decrease the degree of predictability of the execution time of other processes that are allocated to a specific site. A second objective can, therefore, be derived from the relay requirement: *to achieve parallel and loosely coupled communication control and service during the execution of an application process.* This objective is further justified by requiring feasibility of voting procedures, election algorithms, detection of communication wait-for cycles, etc.

Local resources are locally controlled, but they are required to provide services to the entire distributed system. An example of a service requirement in a fault-tolerant distributed system is the maintenance of object replicas, which increase object resiliency. To maintain con-

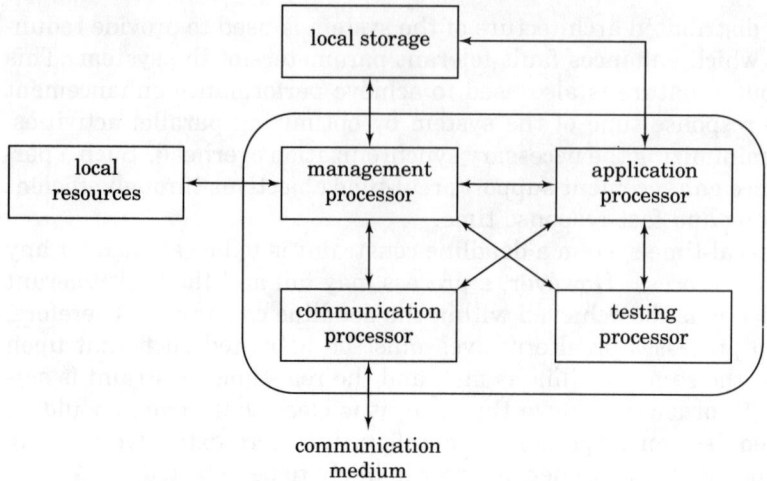

**Figure 2.19** CPU processors.

sistent replicas, updates with check-point mechanisms are generally used, thereby reducing the predictability of the execution time of application processes. This type of requirement yields a third objective of a processing architecture of a real-time, distributed, and fault tolerant system: *to achieve parallel and loosely coupled management of local storage and resources during the execution of an application process.*

This objective is strongly emphasized by multi-processing and multi-user security system requirements, remote input-output services of slow devices in an heterogeneous system (e.g., devices that might require translation), etc. One of the most important jobs of this parallel management in a real-time system is to execute binding procedures of processes that are prepared to execute later, while other real-time constraints are executing. This job is to be carried out without affecting the execution time of the currently executing real-time constraint.

The objectives of the local processing architecture are described in Figure 2.19 and can be summarized as follows:

1. Achieve parallel and loosely coupled testing
2. Achieve parallel and loosely coupled communication control
3. Achieve parallel and loosely coupled local resource and local storage management

A straightforward implementation of such a processor is based on four parallel processing elements: the management processor (MP), the application processor (AP), the testing processor (TP) and the communication processor (CP). These processors use the same data storage, but each accesses different segments of the storage.

**Interprocessor relations.** The four processors are assigned with different tasks. The AP runs the application processes, using application-dedicated memory segments. The MP runs the local operating system kernel (interrupt handler, dispatcher, local resource drivers, remote access control, etc.), operating system services (e.g., allocator, binder, name service, file service, authentication, accounting), and run-time support. The CP runs the interprocess-communication (IPC) layered protocol, network support protocols between data circuit-terminating equipment (DCE) ports (e.g., relaying two X.25 protocols), and time services and interprocessor synchronization mechanisms. The TP runs monitored data testing (range, reasonability, type), agreement protocols, roll-back switch control (e.g., checkpoint management), and logging processes. It may also run comparison processes (as in the master-checker case) which are not necessarily replicas of the application processes. For example, one may consider running a simpler algorithm with imprecise results just for generating independent data for validity tests.

**Application-management-communication** relations are realized through local resources allocated by the management processor. The AP and CP are allocated with a set of buffers and synchronization mechanisms. Each sends and receives data to and from the buffers, using synchronization tools (signal and wait) to support the proper mutual exclusion. The external layered communication is managed by the CP, at the layer accessed internally by the application. The session administration is carried out between the MP and the CP, and is transparent to the application. The relaying protocols (DCE to DCE) involve the CP only, after resources are allocated by the MP.

Interrupt handling is provided by the MP to the CP and the AP. The time service which runs on the CP synchronizes the network clocks, and time-stamps the communications. It provides this timestamp data to the MP, which distributes it locally and uses it for its dispatching process.

**Application-management-testing** relations are also realized through local resources allocated by the management processor. Synchronization tools are used to properly monitor the state of the AP by tests run on the TP. Agreement protocols which run on the TP provide the AP with conclusions, and receive from the CP and the AP inputs for these algorithms. At an error detection, the TP provides the MP with the recovery point data, to appropriately initiate the retry cycle of the AP (probably with another resource combination).

**Hardware support** of this architecture is undoubtedly on-chip implementable. The processors concurrent storage-access is achieved through fetch and execution overlaps as well as by segment allocation which prevents contention. Signal and wait synchronization tools are also hardware supported. The register rich architectures presented above for the RISC processors fits these implementations adequately.

**The overall system view.** Figure 2.20 summarizes the overall system view in this architecture. The system wide activities are interconnected via the communication network, through the communication processors of the various nodes. The hardware support of this architecture significantly releases the unpredictable load of the application processor. The communication coprocessor is in agreement with the BXU and MPC presented above, but provides a stronger support allowing the execution of a more complicated communication protocols (not just the lower layer) along with time services.

System-wide fault tolerance is achieved through protocols which run on testing processors and data exchange which occurs between communication processors. The application itself can therefore be implemented almost as if there is a single alternative, a feature which mitigates design risks significantly.

## 2.3 Concluding Remarks

Current technology provides a number of hardware components which can be used for building a distributed processing environment which supports the execution of applications in a fault-tolerant mode. It is interesting to note that several distinct approaches have been taken in the design of the systems, some examples of which have been presented in this chapter. Different designs are a consequence of the target application domain, application requirements and fault model, and the design tradeoffs.

In this chapter we have introduced various architectures and the hardware elements which allow building computation nodes that communicate with each other. As presented, both CISC and RISC architectures provide a means for building high availability systems, which require redundancy and communication support. High availability system architectures have been discussed and examples based on current systems are brought to demonstrate the principles of such designs. The hardware pieces which serve as building blocks of our systems have advanced to allow message passing between processing nodes, and further, to support communication recovery in the presence of failures by arbitration and redundancy.

The description of current systems is just the first step in providing the reader with the knowledge on hardware necessary for implementing reliable and safe systems. Special attention is therefore given to further architecture analysis, emphasizing approaches of shadow-

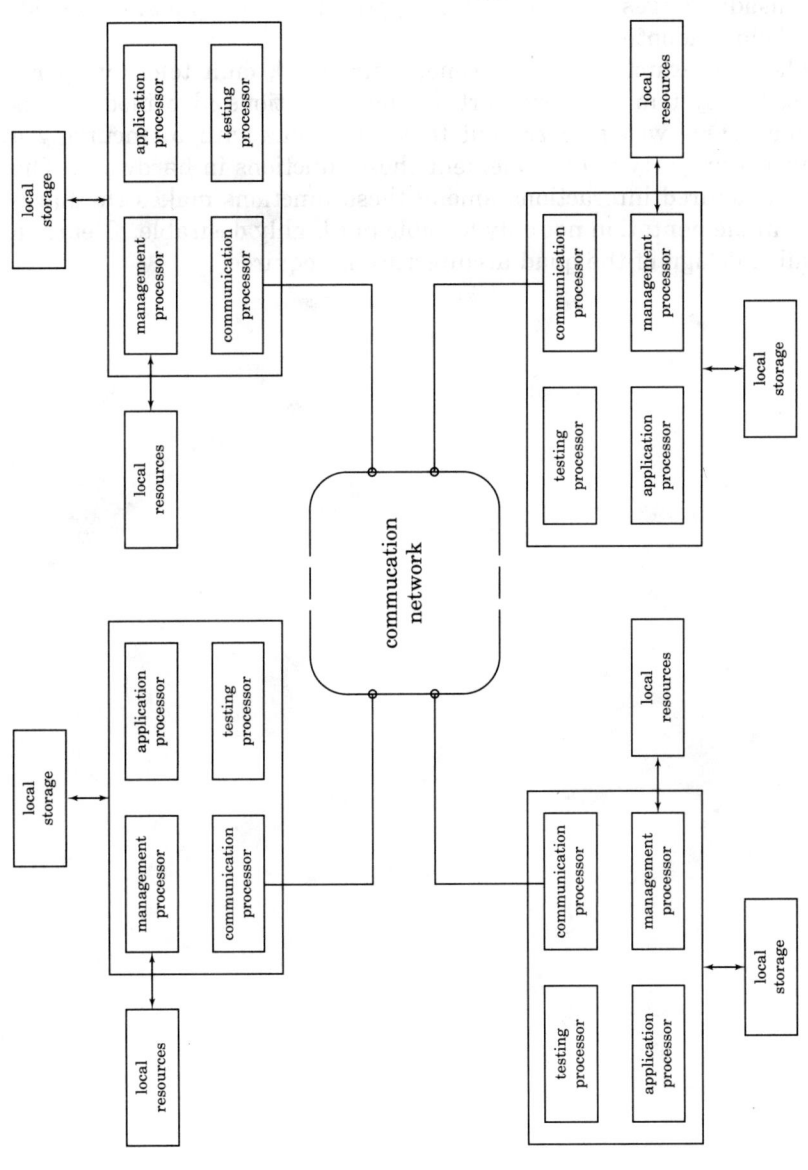

**Figure 2.20** Multiprocessor Network Configuration.

ing, slaving, and thread orientation. Clearly, there are approaches which lead to the specialization of elements as contrasted with the approaches which use multiple standard units. There are advantages and disadvantages of each of these approaches which must be considered before adopting either.

When we examine the real-time, distributed, fault tolerant operations, the system must support the four functions identified in this chapter. One way to carry out these functions without incurring a temporal property is to implement these functions in hardware. The well structured interactions among these functions makes the hardware implementation not only feasible but highly desirable. Clearly a detailed design of the quad architecture is required.

# Chapter 3

# Language and Tools

Sequential programming has been named and characterized by its sequential trace of events resulting from the program execution. In this type of programming processes constitute the execution images of programs, each having a single thread and a single address space[1]. On the other hand a distributed program is a program with multiple threads and multiple address spaces which do not intersect.

Distributed programs are a natural building block for fault tolerant software. However, distributed programming requires several features whose importance in sequential programs is negligible. This requirement arises mainly from the growth in complexity due to concurrency and the difficulties in event ordering and tracing. This chapter ties basic software architectures and approaches to the system and fault models introduced above. The chapter's main objective is to provide insight into distributed programming, with which our systems and solutions are implemented.

The definition of a distributed program states it has multiple threads and multiple address spaces which do not intersect. However, as these threads cooperate in a given program, we need to establish a model of communication with sequential processes with no resource sharing. This model is presented below. Another important issue is the variety of possible solutions to the requirement for synchronization (semaphore, lock, etc.) to implement mutual exclusion in resource sharing. Although the address spaces do not intersect, resource

---

[1] See Chapter 5 also.

sharing is a common requirement. Furthermore, the distribution of resources and programs introduces the notions of duplication and redundancy. If address spaces do not intersect and consequently faults do not migrate, duplication can adequately provide for recovery from faults.

Following this approach, we present various synchronization devices without which concurrent programming is unachievable. We then show how these tools are used to solve classical problems, such as the bounded buffer problem and the readers writers problem. We then present a formal programming language that describes communicating sequential processes (CSP), introduced by Hoare [66]. We show a variety of examples for solving concurrency problems, as reflected in this language concepts.

## 3.1 Semaphores

Semaphore is a synchronization tool, introduced by Dijkstra [47] in 1968. The semaphore is a global variable which can be accessed only by two atomic operations $P(s)$ and $V(s)$, defined as follows:

**var**
        $s$: semaphore
$P(s)$::
    **begin**
        **while** $s \leq 0$ **do** skip ;
        $s \leftarrow s - 1$ ;
    **end**
$V(s)$::
    **begin**
        $s \leftarrow s + 1$ ;
    **end**

In some places $V(s)$ and $P(s)$ are marked $signal(s)$ and $wait(s)$ respectively, names which better explain the operations. Informally, one can describe these atomic operations as follows. When a process enters the execution of $P(s)$, it examines the value of $s$ to be positive, and if so, it decreases its value by one. This operation is atomicly executed by a test-and-set instruction of the hardware. If $s$ is not positive, the process continues monitoring its value until it becomes positive. Since the hardware supports the atomicity of the test-and-set instruction, no two processes can decide concurrently on entrance to the critical section.

Some interesting properties of the semaphore have been proven to hold, independent of the semaphore use by the program. Let $n_P$ count the number of $P(s)$ executions and let $n_V$ count the number of $V(s)$

executions. Say the semaphore $s$ guards a particular critical section, and thus the number of times $P_s$ was passed into the critical section $n_S$ can be expressed with respect to $c$, the initial value given to $s$. Furthermore, the number of processes inside a critical section are always $n_S - n_V$. It turns out [60] that the following relation always holds:

$$c + n_V \geq n_S \leq n_P. \tag{3.1}$$

Hence, an important result comes out:

$$n_S = \min(c + n_V, n_P). \tag{3.2}$$

## 3.2 Monitors

Monitors with an appropriate set of proof rules, have been suggested as a set of building blocks for operating systems and synchronization tools by Hoare [65]. A monitor is a collection of procedures and associated data. Inside the procedure bodies of a monitor, there is a need to support a "wait" operation which suspends the calling program and a "signal" operation which resumes execution of one of the suspended programs. The monitor supports these needs with special *condition* variables. A condition does not have any value associated with it. Practically, it is represented by a queue of processes which is initially empty. The monitor follows the following structure:

```
monitor_name: monitor ;
    begin
        var cond_var: condition ;
        ...
        procedure proc_name_1 (...parameters...)
            begin
                ...
                cond_var.wait ;
                ...
            end
        procedure proc_name_2 (...parameters...)
            begin
                ...
                cond_var.signal ;
                ...
            end
        ...
        init: ... all variables except conditions
    end
```

The invocation of a specific procedure in a monitor requires the identification of both the monitor and the procedure, as follows.

monitor_name.proc_name(...parameters...)

We can query the existence of processes waiting on a condition with a predicate *cond_name.queue*: true if anyone waits, false otherwise.

```
var cond_var: condition ;
    begin
        ...
        if cond_var.queue then
            ... someone is waiting on cond_var ...
        ...
    end
```

The monitor allows an application programmer to associate an invariant $\mathcal{I}(\text{monitor\_name})$ based on the monitor's local data, which holds before and after executing a procedure in this monitor. The following rules allow us to analyze programs which use monitors:

1. $\{$ true $\}$ **init** $\{\ \mathcal{I}(\text{monitor\_name})\ \}$
2. $\{\ \mathcal{I}(\text{monitor\_name})\ \}$ **monitor_name.proc_name** $\{\ \mathcal{I}(\text{monitor\_name})\ \}$
3. $\{\ \mathcal{I}(\text{monitor\_name})\ \}$ **cond_var.wait** $\{\ \mathcal{I}(\text{monitor\_name}) \wedge B\ \}$
4. $\{\ \mathcal{I}(\text{monitor\_name}) \wedge B\ \}$ **cond_var.signal** $\{\ \mathcal{I}(\text{monitor\_name})\ \}$

$B$ is an assertion which describes the condition under which the program resumes after waiting on cond_var. The last two rules, the expression $\mathcal{I}(\text{monitor\_name}) \wedge B$ can be replaced by an assertion $C$ [67] which is not necessarily equal, and in particular $C \neq \mathcal{I}$.

## 3.3 Resources

Another way of synchronizing concurrent programs is using the resource variables [109]. A resource is a set of logically connected variables shared between processes. Resource variables obey syntax restrictions to ensure conflicts protection within critical sections.

1. Critical section statements can occur only in parallel processes.
2. If a variable $x$ belongs to a resource $r$, it can appear in a parallel process only inside a critical section statement for $r$.
3. If a variable $x$ is changed in process $S_i$, it can appear in $S_j$ $(j \neq i)$ only within a resource (even for reading).

Language and Tools  53

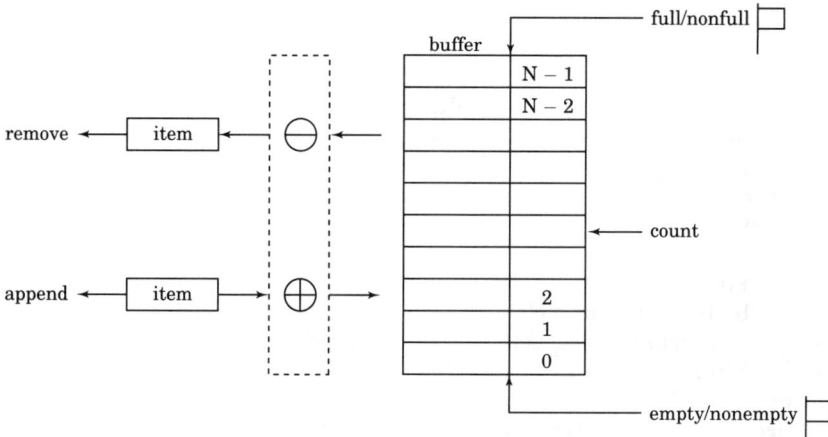

**Figure 3.1** The bounded buffer problem.

The parallel execution statement is expresses as

**resource** $r_1$(variable list),...,$r_n$(variable list) ;
**cobegin** $S_1 \| \ldots \| S_m$ **coend**

The statements $S_i$ between the **cobegin** and **coend** declarations are executed concurrently, in other words, they are parallel processes.

The critical section statement

**with** $r_i$ **when** $C$ **do** $S$ ;

where $r_i$ is a resource, $C$ is a boolean condition, and $S$ is a statement. When this statement is executed, the execution of $S$ is delayed until $C$ is true and no other process is using $r_i$. When $C$ holds and $r_i$ is available, $S$ is executed while declaring the unavailability of $r_i$ to all other processes. Upon termination of $S$, $r_i$ is released.

## 3.4 Examples

### 3.4.1 The bounded buffer problem

Let us consider some solutions to the bounded buffer problem. In this problem we have $N$ buffers, each capable of holding one item. Items are put in the buffers by the *append* operation ($\oplus$) and removed by the *remove* operation ($\ominus$). A first-in first-out discipline has to be maintained for the appending and removal of items from the buffers. Clearly, there can be 0 to $N$ items in the buffers at any moment. When there are 0 items, a removal operation cannot be carried out, and when there are $N$ items, an appending cannot be performed. These requirements form the synchronization constraints for this problem.

```
type item: ... /* with known size */
begin
    var
        mutex, empty, full: semaphore ;
        buffer: array 0..N − 1 of item ;
        nextp, nextc: item ;
    procedure append(element: item) ;
    begin
        P(empty) ;
        P(mutex);
        buffer ← buffer ⊕ element ;
        V(mutex) ;
        V(full) ;
    end
    procedure remove(result element: item) ;
    begin
        P(full) ;
        P(mutex) ;
        buffer ← buffer ⊖ element ;
        V(mutex) ;
        V(empty) ;
    end
init:
    buffer ← empty ;
    full ← 0 ;
    empty ← N ;
    mutex ← 1 ;
cobegin
    while true do
        [ produce nextp ; append(nextp) ∥ consume remove(nextc) ]
coend
end
```

**Program 3.1** The bounded buffer problem: a semaphore solution.

Figure 3.1 describes the ordered buffers, and the two operations to which they are subjected. Detection mechanisms for both a full and an empty buffer set are shown. These detection mechanisms are used to satisfy the synchronization requirements in the following solutions.

The first solution, from [112, page 323] is presented in Program 3.1. This program uses semaphore operations to ensure mutual exclusion of the access to the buffer set. It uses the semaphores *full* and *empty* to keep track of the constraints and to block the removal and appending appropriately.

The second solution, from [65, page 553] is presented in Program 3.2. This program uses a monitor to enforce mutual exclusion. Append and

```
bounded buffer: monitor
    type item: ... /* with known size */
begin
    var
        nonempty, nonfull: condition ;
        buffer: array 0..N − 1 of item ;
        last: 0..N − 1 ;
        count: 0..N ;
    procedure append(element: item) ;
    begin
        if count=N then nonfull.wait ;
            /* this keeps 0 ≤ count < N */
        buffer[last] ← element ;
        last ← last + size ;
        count ← count + 1 ;
        nonempy.signal
    end
    procedure remove(result element: item) ;
    begin
        if count=0 then nonempty.wait ;
            /* this keeps 0 < count ≤ N */
        element ← buffer[last - count×size] ;
        buffer ← buffer ⊖ buffer[last - count×size] ;
        nonfull.signal
    end
init:
        buffer ← empty ;
        count ← 0 ;
        last ← 0
end
```

**Program 3.2**  The bounded buffer problem: a monitor solution.

remove are procedures in this monitor, and they use the nonempty and nonfull monitor conditions to detect the buffers set state.

### 3.4.2 The readers writers problem

In the readers-writers problem there are a number of reader and writer processes which share a device, as shown in Figure 3.2. Each writer must have a nutually exclusive access to this device, while any number of readers can access it concurrently. A solution to this synchronization problem is presented in Program 3.3 (from [112, page 326]) which uses two semaphores *mutex* and *wrt* and a global variable *readcount*. Each reader executes procedure *startread*, and reads and executes procedure *endread*. A similar sequence is followed by the writers. As the *readcount* is changed by *startread* and *endread*, and

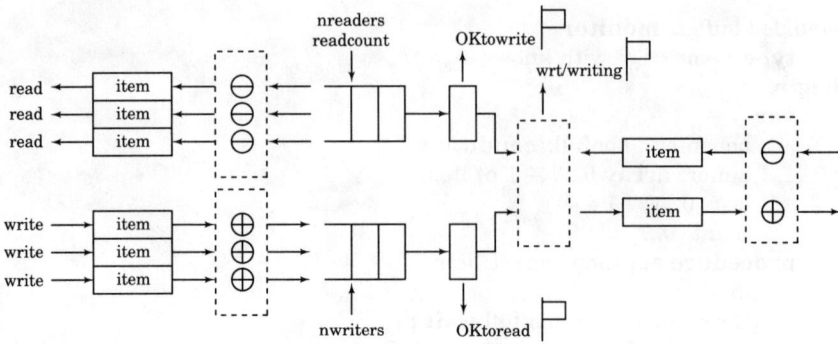

**Figure 3.2** The readers-writers problem.

many executions of these can proceed concurrently, protection is provided through the use of the *mutex* semaphore. When the readcount is incremented from 0 to 1 by the startread, the writers are blocked by the $P(wrt)$. They are released by the $V(wrt)$ executed by the endread on a zero readcount. In this solution the readers have priority in that once readers get started, new readers can continue without being blocked.

A monitor based solution, based on [65, page 556], is presented in Program 3.4. The third solution [112, page 362] uses resource variables [109] and is presented in Program 3.5.

### 3.4.3 Monitors and semaphores

Let us now examine how we can construct semaphores and monitors from each other.

**Building semaphores with monitors.** Program 3.6 describes the way we build a semaphore with monitors.

**Building monitors with semaphores.** Let us consider the following monitor based procedure.

```
monitor_name: monitor
begin
      var
          /* conditions */
              cond: condition
          /* non-conditions */
      procedure proc_name ;
          begin
                  /* section A */
              cond.wait ;
                  /* section B */
```

```
begin
    var
        mutex, wrt: semaphore ;
        readcount: integer ;
    procedure startread ;
    begin
        P(mutex) ;
        readcount ← readcount + 1 ;
        if readcount=1 then P(wrt) ;
        V(mutex) ;
    end
    procedure endread ;
    begin
        P(mutex) ;
        readcount ← readcount - 1 ;
        if readcount=0 then V(wrt) ;
        V(mutex) ;
    end
    procedure startwrite ;
    begin
        P(wrt) ;
    end
    procedure endwrite ;
    begin
        V(wrt) ;
    end
init:
        readcount ← 0 ;
        wrt ← 1 ;
        mutex ← 1 ;
cobegin
    while true do
        [ startread; (read) endread || startread; (read) endread || ...
            ... || startwrite; (write) endwrite ]
coend
end
```
**Program 3.3** The readers writers problem: a semaphore solution.


            cond.signal ;
                /* section C */
        end
    end


Program 3.7 describes the way we build the above monitor based procedure with semaphores.

readers writers: **monitor**
**begin**
    **var**
        OKtoread, OKtowrite: **condition** ;
        readcount: **integer** ;
        writing: **boolean** ;
    **procedure** startread ;
    **begin**
        **if** writing ∨ OKtowrite.queue **then** OKtoread.**wait** ;
        readcount ← readcount + 1 ;
        OKtoread.**signal** ;
            /* when one starts reading: every reader can */
    **end**
    **procedure** endread ;
    **begin**
        readcount ← readcount - 1 ;
        **if** readcount=0 **then** OKtowrite.**signal** ;
    **end**
    **procedure** startwrite ;
    **begin**
        **if** readcount≠0 ∨ writing **then** OKtowrite.**wait** ;
        writing ← true ;
    **end**
    **procedure** endwrite ;
    **begin**
        writing ← false ;
        **if** OKtoread.queue
            **then** OKtoread.**signal**
            **else** OKtowrite.**signal**
    **end**
**init**:
        readcount ← 0 ;
        writing ← false ;
**end**

**Program 3.4** The readers writers problem: a monitor solution.

## 3.5 Communicating Sequential Processes

A formal language that describes communicating sequential processes (CSP) has been introduced by Hoare [66]. A proof system that supports it is presented in [26].

### 3.5.1 CSP concepts and primitives

In [66] the CSP language is presented emphasizing the parallel composition of communicating sequential processes as a fundamental programming structure. The major concepts of the CSP language are as follows.

```
begin
    resource r (
        writing: boolean ;
        nreaders, nwriters: integer ) ;
    procedure startread ;
    begin
        with r when nwriters=0 do
            nreaders ← nreaders + 1 ;
        od
    end
    procedure endread ;
    begin
        with r do
            nreaders ← nreaders - 1 ;
        od
    end
    procedure startwrite ;
    begin
        with r when ¬writing ∧ nreaders=0 do
            nwriters ← nwriters + 1 ;
            writing ← true ;
        od
    end
    procedure endwrite ;
    begin
        with r do
            nwriters ← nwriters - 1 ;
            writing ← false ;
        od
    end
init:
        nreaders ← 0 ;
        nwriters ← 0 ;
        writing ← false ;
cobegin
    while true do
        [ startread; (read) endread || startread; (read) endread || ...
            ... || startwrite; (write) endwrite ]
coend
end
```

**Program 3.5** The readers writers problem: resource solution.

1. Dijkstra's concept of *guarded commands* is used as the only non-deterministic controller.

2. *Boolean* expressions can be used as guards, thus providing also for the sequential execution control.

Binary Semaphore: **monitor**
**begin**
    **var**
        s_eq_1: **condition** ;
        $s$: 0..1 ;
    **procedure** P ;
    **begin**
        **if** $s = 0$ **then** s_eq_1.**wait** ;
        $s \leftarrow 0$ ;
    **end**
    **procedure** V ;
    **begin**
        $s \leftarrow 1$ ;
        s_eq_1.**signal** ;
    **end**
    **init** $s \leftarrow 1$
**end**

**Program 3.6** A semaphore built with monitors.

3. A *parallel command* activates a simultaneous start of concurrent sequential processes.
4. Simple *I/O commands* are based on totally synchronous communication between the processes (rendezvous type).
5. The language supports *input commands in guards*, thus allowing conditional execution which depends on message arrival.
6. The language supports *arbitrary selection between alternatives*, out of which *only one* is selected for execution. The others remain dormant.
7. *Input guard on repetition* commands are allowed and supported in CSP.

Communication serves as the only data transfer mechanism, and guarded input commands synchronize the transfer.

Let us review the major notations and conventions of the language. Sequential commands are delimited with a semicolon ";" or by the *parbegin-parend* brackets "[" and "]," respectively. Type assignment is defined by a colon ":" and the process label is separated from the process body (command list) by a double-colon "::". A command list can contain declarations and comments. Variables have both types and bounds as attributes.

**Assignment commands.** An assignment command assigns value of an expression to a variable. For example, consider the following assignments:

    $x := 76.2$ ;    $y := (y+31)/2 * x$ ;    $s := s \cup \{ A \}$ ;

**begin**
    **var**
        /* conditions */
            cond_sem: **semaphore** ;
            cond_count: **integer** ;
        /* non-conditions */
        /* additional */
            mutex: **semaphore** ;
            urgent: **semaphore** ;
            urgent_count: **integer** ;
**procedure** proc_name ;
    **begin**
        P(mutex) ;
            /* section A */
        cond_count ← cond_count +1 ;
        **if** urgent_count > 0
            **then** V(urgent)
            **else** V(mutex) ;
        P(cond_sem) ;
        cond_count ← cond_count −1 ;
            /* section B */
        urgent_count ← urgent_count +1 ;
        **if** cond_count > 0
            **then begin**
                V(cond_sem) ;
                P(urgent)
            **end** ;
        urgent_count ← urgent_count −1 ;
            /* section C */
        **if** urgent_count > 0
            **then** V(urgent)
            **else** V(mutex)
    **end**
**init**:
    cond_sem ← 0 ;
    mutex ← 1 ;
    urgent ← 0 ;
    cond_count ← 0 ;
    urgent_count ← 0 ;
**end**

**Program 3.7** A monitor based procedure built with semaphores.


**Parallel commands.** We denote parallel execution of processes as follows. Let us consider three processes: $p_1$ to $p_3$. We want $p_1$ to be a process defined by a command list $A$, $p_2$ by $B$, and $p_3$ by $C$. The parallel execution of $p_1$ to $p_3$ is written as

[ $p_1$::$A$ || $p_2$::$B$ || $p_3$::$C$ ]

Another example demonstrates the execution of an "array" of processes. Say the above $p_3$ is a vector of three identical processes ($p_3^{(1)}, p_3^{(2)}, p_3^{(3)}$). We can then write the above example as follows.

[ $p_1$ || $p_2$ || $p_3^{(i:1...3)}$ ]

**I/O commands.** An *input* command $P_1?x$ executed in process $P_2$ is interpreted as a request to process $P_1$ to assign a value to the $P_2$'s local variable $x$. An *output* command $P_2!a$ which is executed in process $P_1$ is a request from $P_1$ to $P_2$ to receive the value of $P_1$'s local variable $a$. Since the communication is totally synchronous, $P_2$ waits on $P_1?x$ for $P_1$ to send a value, and $P_1$ waits on $P_2!a$ for $P_2$ to receive. When synchronization of the above example takes place, $P_2$'s $x$ is assigned with the value of $P_1$'s $a$.

**Guards.** Dikstra's guarded command is presented in the form

*guard* $\rightarrow$ *command-list*.

The *guard* is a boolean expression or/and an input command. The guard functions as a sequential control structure: if it fails, the process which contains it aborts and if it does not fail, *command-list* is executed. A failure of the guard is a consequence of evaluating a false value of the boolean expression.

For example, the guard

$a,b$: boolean; $c$: integer; $a \wedge b \wedge P?c \rightarrow$

fails if either $a$ or $b$ is false or if process $P$ is not ready with a value to assign to $c$.

**Alternations.** An alternative command is written as follows

[ $g_1 \rightarrow C_1$ □ $g_2 \rightarrow C_2$ □ ... □ $g_n \rightarrow C_n$ ]

where $g_1 \ldots g_n$ are the guards of the $n$ alternatives, and $C_1 \ldots C_n$ are the repective command lists of the alternatives.

An example of an IF-THEN-ELSE statement is written as

[ $a = b \rightarrow$ n:=1 □ $a \neq b \rightarrow$ n:=0 ]

**Repetitions.** An iterative execution of a command list has a guard on its execution.

*[ guard → command list ]

An iterative command can include alternative commands in it. An example of such a case is given below.

*[ client = dead → skip
□ client ≠ dead → cashier:= cashier+10 ]

**Subroutines and functions.** We can build a subroutine in CSP as a process which we execute in parallel to the user program. The subroutine has a guard on its input parameters, and it waits on this guard for receiving them from the user program.

SUBROUTINE = *[ user?(param_list) → ... ]
USER = ... ; subroutine!(param_list) ...
[ subroutine::SUBROUTINE ∥ user::USER ]

We note that once a rendezvous has been made for exchanging the parameters, the user program and the subroutine are completely asynchronous. They can execute side by side in a manner that resembles an object invocation more than a procedure call. Functions, on the other hand, return result parameters and users must wait for the results on input guards.

FUNCTION = *[ user?(param_list) → ... ; user!(results) ]
USER = ... ; function!(param_list) ; function?results) ...
[ function::FUNCTION ∥ user::USER ]

**Recursion.** A recursive invocation of a subroutine is supported by an array of processes, whose size is the recursion maximal depth.

USER = ... subroutine(1)!(params) ... subroutine(1)?(results) ...
SUBROUTINE = *[ subroutine(i-1)?(params) → ...
    subroutine(i+1)!(params) ... subroutine(i+1)?(results) ...
    subroutine(i-1)!(results) ]
[ subroutine(i:1...recdepth-1)::SUBROUTINE ∥ subroutine(0)::USER ]

We note how weak this recursion method is and the enormous number of problems it raises when the subroutines have shared variables which must be protected for mutual exclusion.

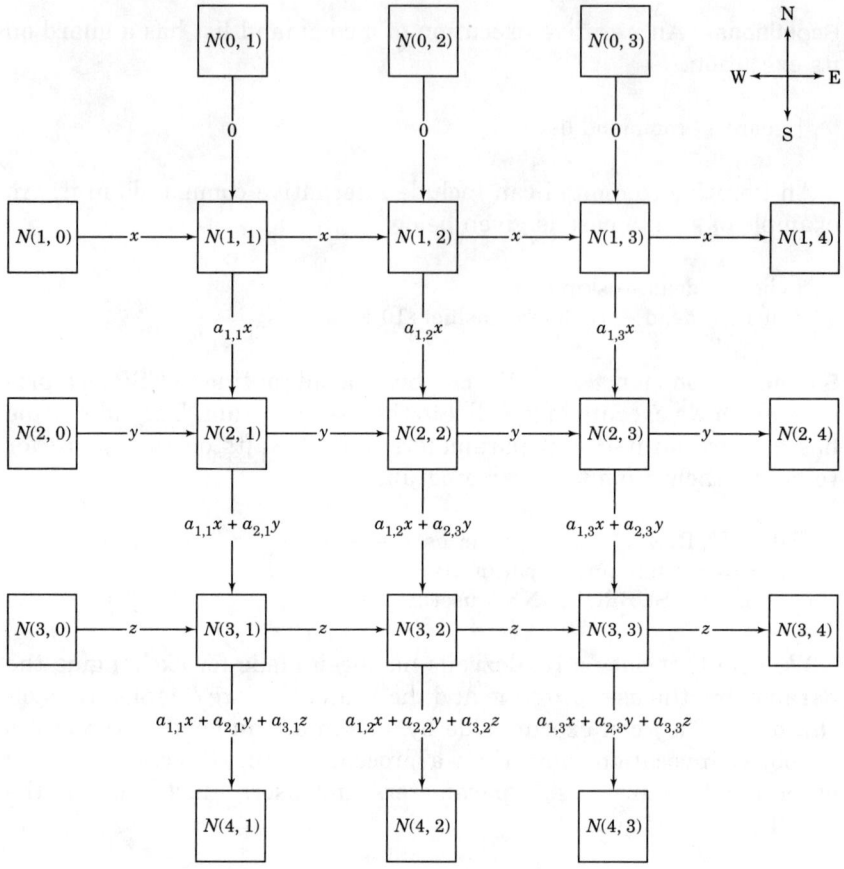

**Figure 3.3** Matrix multiplication in CSP.

### 3.5.2 Examples in CSP design

**Matrix multiplication.** Let us examine a solution to a problem with an iterative array: a multiplication of a 3 element vector by a $3 \times 3$ matrix. Program 3.8 [66] demonstrates the way such processes are constructed in CSP. Figure 3.3 illustrates the data flow and the 21 concurrent processes in this solution.

The program has three user inputs: the three WEST nodes denoted as $N(1,0)$ to $N(3,0)$. Three user result ports, the SOUTH nodes $N(4,1)$ to $N(4,3)$, receive the results of the multiplication.

The multiplication consists of nine similar nodes, denoted $N(i,j)$, $1 \leq i \leq 3$  $1 \leq j \leq 3$, each of which executes the calculation

$$a_{i,j} \times \phi + \theta \tag{3.3}$$

```
NORTH = *[ j : 1...3; true → N(1,j)!0 ] ;
        comment auxiliary zeros
EAST = *[ φ:real; i : 1...3; N(i,3)?φ → skip ] ;
       comment auxiliary sink

CENTER = *[ φ:real; i,j : 1...3; N(i, j − 1)?φ →
           N(i, j + 1)!φ; θ:real; N(i − 1, j)?θ;
           N(i + 1, j)!(a_{i,j} × φ + θ)
          ]

[ N(i: 1...3,0) :: WEST
∥ N(4,j:1...3) :: SOUTH
∥ N(0,j:1...3) :: NORTH
∥ N(i:1...3,4) :: EAST
∥ N(i:1...3,j:1...3) :: CENTER
]
```
**Program 3.8** Matrix multiplication in CSP.

```
b_buffer::
    buffer(0 ... N − 1) item;
    deposits, withdraws: integer; deposits:=0; withdraws:=0 ;
            comment 0 ≤ withdraws ≤ deposits ≤ withdraws+10 ;
    *[ deposits < withdraws+N ; consumer?buffer(deposits mod N) →
        deposits:=deposits+1 ;
    □ withdraws < deposits ; consumer?more() →
        consumer!buffer(withdraws mod N) ; withdraws:=withdraws+1 ;
    ]
```
**Program 3.9** The bounded buffer problem: CSP solution.

where $a_{i,j}$ is the matrix element, $\phi$ is a real number received from WEST side, and $\theta$ is a real number received from NORTH side.

Six auxiliary nodes assist the program: the NORTH nodes, $N(0, 1)$ to $N(0, 3)$, which send zeros to the first multipliers row and allow the nine multipliers to be identical, and the EAST nodes, $N(1, 4)$ to $N(3, 4)$, which serve as sink nodes for the multiplied vector.

**The bounded buffer problem.** Let us recall the bounded buffer problem which has been discussed in Section 3.4.1. A buffer of at most $N$ items is accessed both by a producer and by a consumer concurrently. The solution must ensure that no overflow in deposits and no underflow in withdraws occur.

We assume that the producer produces items and sends each to the buffer with a b_buffer!item command. Accordingly, the consumer verifies a nonempty buffer with a b_buffer?more() request, which upon a positive case is followed by a b_buffer?item command. Program 3.9 [66] demonstrates a CSP program for this problem.

```
cookie_type=(goody, poisonous) ;
client_type=(healty, ill, dead) ;
client(0 ... N − 1): client_type ;
WAREHOUSE::
    store: cookie(0 ... K − 1) cookie_type ; top: 0 ... K − 1 ; top:= 0 ;
    *[ baker?cookie(top+1) →
        cookie(top+1):= goody ; store:= store⊕cookie(top+1) ;
    ☐ mouse?more() ∧ ∃cookie∈store →
        store:= store⊖cookie(top) ; mouse!cookie(top) ;
    ☐ mouse?cookie(top+1) →
        cookie(top+1):= poisonous ; store:= store⊕cookie(top+1) ;
    ☐ (i: 0 ... N − 1) [ client$_i$?more() ∧ ∃cookie∈store →
            store:= store⊖cookie(top) ; client$_i$!cookie(top) ;
        ]
    ]
BAKER::
    cookie: cookie_type ;
    *[ terminated baking → warehouse!cookie ;
    ]
CLIENTELE::
    (i: 0 ... N − 1) [ client$_i$:= healty;
        *[ client$_i$=dead → skip ;
        ☐ client$_i$ ≠dead →
            [ warehose!more() ; warehouse?(cookie: cookie_type) ;
            cookie=poisonous →
                [ client$_i$=healthy → client$_i$:= ill ;
                ☐ client$_i$=ill → client$_i$:= dead
                ]
            ]
        ]
    ]
MOUSE::
    *[ warehouse!more() ; warehouse?(cookie: cookie_type) ; warehouse!cookie ]

[ warehouse:WAREHOUSE ‖ baker:BAKER ‖ mouse:MOUSE ‖ clientele:CLIENTELE ]
```

**Program 3.10** The cookie shop problem: CSP solution.

**The cookie shop problem.** Let us now consider the following problem. There is a cookie warehouse in which a baker stores his produced cookies. This baker has a regular clientele of $N$ clients. The clients come to the warehouse, buy the cookies and eat them. Unfortunately, there is a mouse in the warehouse. The mouse is dirty and infectious, and therefore, each cookie that it bites becomes poisonous. A client which eats a poisonous cookie becomes ill.

Program 3.10 describes this deadly process of the cookie warehouse. We note that this solution uses a WAREHOUSE process to maintain

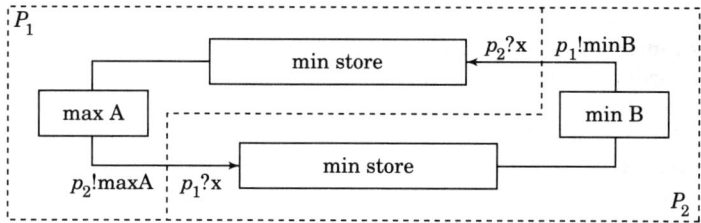

**Figure 3.4** Ordered set partitioning.

the cookies storage and to enforce mutual exclusion on clients and mouse access.

**Ordered set partitioning.** A well known problem frequently discussed in the literature is the problem of ordered set partitioning. Given two disjoint sets $A$ and $B$, we are allowed to exchange members between the sets, preserving their size. These exchanges produce new sets $A'$ and $B'$, accordingly. The exchange process continues until every element of $A'$ is smaller than any element of $B'$. Figure 3.4 describes this mechanism.

Program 3.11 [26] shows a formally proven CSP solution to this problem. There are two concurrent processes: $p_1$ and $p_2$. The two processes iterate while $p_1$ exchanges the current maximum of $A'$ (sending maxA) for the current minimum of $B'$ with $p_2$ (who sends minB). The iteration continues until maxA in $p_1$ equals the last element received from $p_2$.

## 3.6 Concluding Remarks

General programming problems often require the enforcement of synchronization for correct execution. In this chapter we considered several examples of synchronization problems and their solutions. Clearly the solutions are often based on primitive operations supported by the hardware. Note that a common attribute of synchronization problems is that while we can express the temporal relationship among events, these relationships and conditions are defined in relative terms but no absolute temporal relations are defined.

Synchronization, particularly in heterogeneous systems, requires standardization of services provided by the mechanisms. For example, services of semaphore management [69] have been standardized, including the semaphore object model and its entry object. These services include the definition and deletion of a semaphore, taking and relinquishing control, as well as the reporting of the status of the semaphore and its entry object.

$P_1$::

      **comment** The min store ;
  x,maxA: integer ; A: set of integet;
  maxA:= max(A) ;
  $p_2$!maxA ; A:= A−{ maxA} ;
  $p_2$?x ; A:= A∪{ x} ;
  maxA:= max(A) ;
  *[ maxA>x →
    $p_2$!maxA ; A:= A−{ maxA} ;
    $p_2$?x ; A:= A∪{ x} ;
    maxA:= max(A)
  ]

$P_2$::

      **comment** The max store ;
  x,minB: integer ; B: set of integer;
  $p_1$?x ;
  B:= B∪{ x} ; minB:= min(B) ;
  $p_1$!minB ; B:= B−minB ;
  *[ $p_1$?x →
    B:= B∪{ x} ; minB:= min(B) ;
    $p_1$!minB ; B:= B−minB ;
  ]

      **comment** The parition ;
[ $p_2$ : $P_2$ ∥ $p_2$ : $P_2$ ]

**Program 3.11** The ordered set partitioning problem: CSP solution.

CSP has often been used to construct examples of concurrently operating processes which must interact. Several examples of synchronization have been presented using CSP. In many cases the use of CSP makes the problem simpler.

# Chapter 4

# Networks

This chapter introduces issues related to a very important resource in a distributed system: the communication subsystem. We start with a brief explanation of architectural concepts, and then turn to various communication protocol models and characteristics.

A distributed system is generally physically separated and its components demonstrate various degrees of heterogeneity. In order to analyze and to construct such systems, a layered approach is well suited which allows local abstractions and isolation. The layered architecture, introduced in Section 1.2.2, included the hierarchical composition of the system components. The layers are constructed on top of the physical definitions of the system, such that separated peer layers interconnect through the physical layer. Therefore, each component of the system generally contains a service layer built above a kernel layer, which is constructed on the physical layer [134]. Application computations as well as many types of services are built on top of these layers, in a layer that is denoted as the application layer.

The layered architutecture is strongly reflected in communication standards [128] [129]. In this chapter we present basic concepts in the context of the OSI standards first. Each communication activity occurs in a particular layer. Since there are at least two partners for this activity (the sender and the receiver), the sets of steps at both sides must correspond to each other properly. This set of regulations and practices is called a *protocol*. Layers are defined according to services they provide to the upper level layer. On the other hand, each layer relies on the services it is given by its lower level layer. The boundary between two adjacent layers is called the *interface*.

Communication primitives can be represented by a logical-messages model as depicted in Figure 4.1. An entity sends a request to another

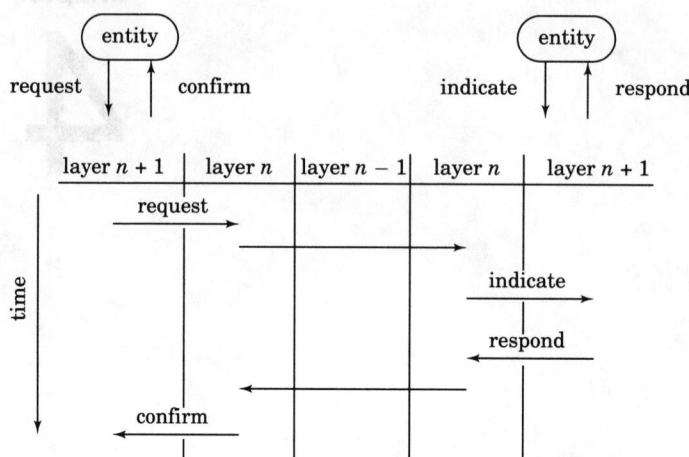

**Figure 4.1**  Logical messages model of communication primitives.

entity. The request is transferred through a lower layer to the required destination. It is accepted as an request at the destination entity, which responds accordingly. Its response is also transferred back through a lower layer and is received by the request initiator (from its lower layer) as a confirmation. The entities which communicate at a specific layer are called *peers*.

In the following sections we consider the details of the layered architecture of communication protocols as formulated by International Standards Organization in the framework of Open Systems Interconnection (OSI) reference model and protocols.

## 4.1  The Layered Architecture

A network of interconnected machines can be viewed as just that. Such a view partitions the network vertically into a number of distinct machines that are interconnected using physical transmission media. The OSI Reference Model views the network in its totality, but partitions it as a series of horizontal layers. Here, a layer cuts across the vertical boundaries of systems. Such a view is helpful in more than one way:

- it allows for a discussion on exchange of information between peer objects within a layer, independent of other layers,
- it allows for gradual and modular development of functionality of each layer, and
- it simultaneously allows an open system to be viewed as a succession of sub-systems, thereby permitting a modular implementation of the open system itself.

### 4.1.1 Layers, services and functions

For simplicity, a given layer is referred to as an (N)-layer, the one below it as (N-1)-layer and the one above as (N+1)-layer. At the bottom we have the physical transmission media forming (0)-layer.

This succession of layers not only partitions the entire network, but it also partitions each system into a succession of *subsystems* – a subsystem being identified (or formed) by the intersection of a system and a layer. Subsystems within a layer are said to be of the same *rank*, while subsystems belonging to adjacent layers within an open system are said to be *adjacent*. Adjacent subsystems communicate through their common boundary. Communication between subsystems of the same rank is more complex. A subsystem is logically viewed as consisting of a number of *entities*. An entity is a representation of a process (or processes) within a computer system. In the OSI environment the entities and their inter-relationships are well structured.

One concept central to the layered architecture of OSI is that of *service*. Each layer provides a different set of services to the layer above. As one moves up the layers, the set of services provided by a layer is either enhanced or improved in quality. In other words, a layer provides services to the layer above it, and also uses services provided by the lower layer, and those below. The layer hierarchy can be looked upon recursively as:

- a layer provides services;
- part of the services are implemented as functions within the layer, while the rest are derived from (N-1)-services provided by the (N-1)-layer and those below;
- the (N-1)-services are partly implemented as (N-1)-functions in the (N-1)-layer, while others are derived from (N-2)-services, and those below. And so on.

Figure 4.2 illustrates the layers and their interactions. The concept of layered architecture allows identification of different functions for implementation within various layers.

Services made available by a layer are implemented in the form of functions in that layer and those below. Entities of the layer are responsible for implementing its functions. Note that it is the (N+1)-entities that are the users of (N)-services. As a consequence, a service provided by an (N)-layer may be accessed by an (N+1)-entity whenever it interacts with an (N)-entity. A service can only be accessed through a *service-access-points*. Formally, an (N)-service-access-point, (N)-SAP, is a point at which services are provided by an (N)-entity to an (N+1)-entity. A service access point is similar to an interface

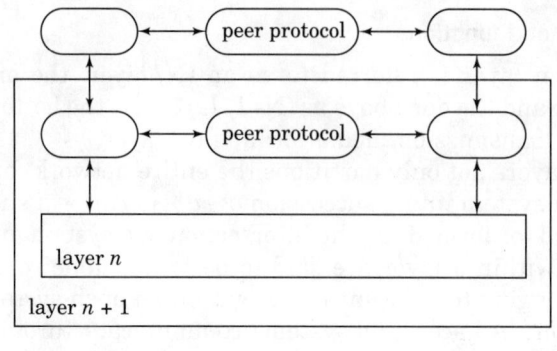

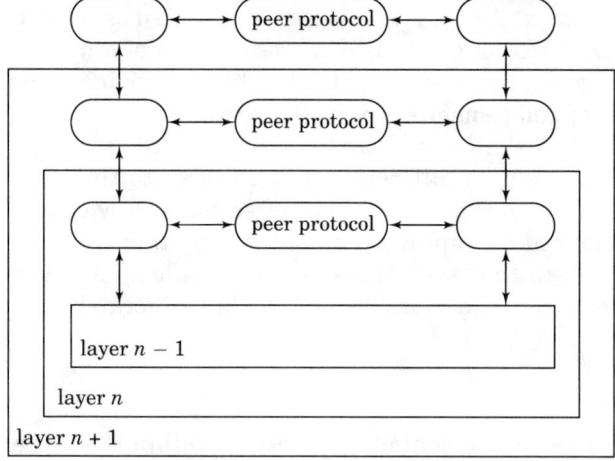

**Figure 4.2** OSI layer transparency principle.

through which two entities from adjacent layers can interact with each other.

To provide services to the layer above, the two (N)-entities themselves need to communicate with each other. Either the two (N)-entities communicate directly with each other, or they do so using (N-1)-services. In the latter case, communication between the peer entities is governed by a set of rules and formats that each entity must adhere to in order to provide services. This set of rules and formats (semantics, syntax, and timing) is referred to as *(N)-protocol*.

Two (N+1)-entities communicate with each other by accessing (N)-services made available at the corresponding (N)-SAPs. One of the important services made available by an (N)-layer relates to the logical *connections* that may be established by the supporting (N)-entities. Such connections are established between corresponding (N)-SAPs.

Exchange of data between peer entities over a connection is not the only such mechanism available within OSI. Information can be transferred by an (N+1)-entity to another (N+1)-entity without having to first formally establish an association. This is made possible by the use of *connection-less data transfer* services made available by the (N)-layer.

Exchange of information may take place either between two peer entities or between an (N+1)-entity and an (N)-entity that are attached to the same SAP. The nature of information exchanged between a pair of entities can be classified into two types:

- user data, and
- control information.

Transfer of *data* is the prime objective of all communication between entities. But, entities also need to exchange *control information* which enables them to coordinate their operations. Examples of control information include address of destination, sequence number associated with data being exchanged, and acknowledgement information.

Two (N+1)-entities can communicate with each other over an (N)-connection that is established and maintained on their behalf by the supporting (N)-entities between the corresponding (N)-SAPs. Such a connection is, in fact, an association between three parties, namely, the two (N+1)-entities and the (N)-layer. The establishment of this association enables the two (N+1)-entities to, first, express agreement on their willingness to communicate with each other. Further, while agreeing to do so, they also decide upon the syntax and semantics (viz., (N+1)-protocol) of all information exchanges that would take place over the connection. This process of establishing a connection also enables the communicating (N+1)-entities to initialize themselves to a mutually known global state so that subsequent exchanges of information can be interpreted and acted upon in accordance with the agreed (N+1)-protocol.

Connection-oriented interaction between (N+1)-entities proceeds through three distinct phases: connection establishment, data transfer, and connection release. Data transfer can only take place once a connection has been established. Preferably the connection is released once the data transfer is complete since committed resources can then be reallocated for use with other connections.

A major requirement of a layer that provides connections is that supporting entities should be able to communicate with each other. Either they are within the same open system and a direct (outside the OSI environment) interface exists between them, or they communicate over an (N-1)-connection. The OSI framework supports the idea of multiplexing in that an (N-1)-connection can be used by several (N)-connections. At the other end, then, de-multiplexing or splitting is supported.

Once a connection has been established, user data originating at an (N+1)-entity is made available to the supporting (N)-entity. The data units are then transferred to its corresponding peer entity which subsequently delivers them to the corresponding (N+1)-entity. A number of issues pertaining to the transfer of such a sequence include:

- regulating the rate of flow of user data over a connection;
- guaranteed delivery of data in the proper sequence;
- confirming the delivery of user data to the destined (N+1)-entity;
- detection of errors and loss of data, and recovery; and
- re-initializing the connection.

In flow control the major concern is with limiting the rate at which user data is made available to that which can conveniently be supported over a connection. Flow control between peer entities limits the rate at which user-data is exchanged between them. This peer flow control is defined as part of the protocol.

Issues relating to preserving the sequence of data across a connection and of acknowledgements are part of the larger issue of reliable data transfer. The reliability of data transfer refers to the requirement that data be communicated without any error, loss of data, or duplication, and (possibly) in the same sequence with an acceptably high probability.

The OSI architecture and its protocols are concerned not so much with the impairments, but with detecting the occurrence of errors and of failures. Generally, if a layer detects an error, it makes every effort to recover from it using, for example, error detecting or correcting codes and, possibly, a positive acknowledgement with re-transmission scheme. Normally, such attempts succeed, but when errors persist with great frequency, a reinitialization of the connection may be undertaken in the hope that recovery can still take place. There is a finite probability that some errors may go unnoticed leading to data loss or duplication. If, in spite of all efforts, the layer is unable to recover, it simply *signals* a failure of the connection to the user (N+1)-entities. It is then the responsibility of the user (N+1)-entities to attempt a recovery or to abandon communication altogether.

Procedures to detect errors or failures and to recover from them are specified as part of the protocol. The procedures to reset a connection are also specified. A method by which a layer signals failure to user entities is specified as part of services. Typically, the layer must also provide a reason for the failure, if it is known, and whether such a condition is temporary or permanent.

An alternate approach to data transfer without first establishing a connection is called Connectionless Data Transfer. Connectionless data transfer is the transmission of independent, unrelated data units from one (N)-SAP to another in the absence of a connection. To support such data transfer an (N)-layer can offer connectionless (N)-service. The transfer is, typically, carried out in three steps:

1. The (N+1)-entity A passes the data unit across the local (N)-SAP C, to the supporting (N)-entity E, together with the (N)-address of the (N)-SAP to which the destination (N+1)-entity B is attached.

2. The supporting (N)-entity E transfers the user-data to the corresponding (N)-entity F which supports the (N)-SAP D, together with the addresses of the source and destination (N)-SAPs.

3. The (N)-entity F passes the data unit across the (N)-SAP D to the attached (N+1)-entity B, together with the address of the (N)-SAP C to which the sending (N+1)-entity A is attached.

The three step procedure ends with the delivery of the data to the destination (N+1)-entity. It is up to the receiving (N+1)-entity to act upon the received data or to simply ignore it.

Connectionless data transfer exhibits the following characteristics:

- Only a single interaction between a user (N+1)-entity and the supporting (N)-entity is required to initiate transmission of data. Once a request for data transfer has been made (or an (N)-data unit is delivered to the destination (N+1)-entity), no further interaction takes place between the user (N+1)-entity and the supporting (N)-entity at its (N)-SAP.

- Since a connection is not established prior to data transfer, data transfer is based on an *a priori* knowledge shared between the two communicating (N+1)-entities. Similarly, at each end, there is an *a priori* agreement between a user (N+1)-entity and its supporting (N)-entity regarding (N)-services available at the (N)-SAP. Further, since negotiation is not performed, this *a priori* knowledge or agreement is not altered.

- Each data-unit is considered to be self-contained, in that the required address information is communicated together with the data. Independence of data-units from others implies that a sequence of data units handed over to the (N)-layer at one end may not be delivered to the destination (N+1)-entity without loss or duplication or even in the same sequence.

## 4.2 Service Definition and Protocol Specification

As noted above, *services* are central to the concept of layering. Each layer in an OSI environment provides a step-by-step enhancement of the services available to entities within the next higher layer. Interaction between an (N)-service-user and the (N)-service-provider across an (N)-SAP can be viewed as a series of certain basic or primitive interactions. During each primitive interaction either a user passes control information or data, or both, to the service-provider, or *vice versa*. A service primitive is an abstract, implementation-independent element of interaction between a service-user and the service-provider. It is an indivisible *atomic action* assumed to occur instantaneously.

For instance, from an implementation viewpoint, the issuing of a service primitive by a service-user, may involve a number of physical interactions over a finite time duration, each consisting of an exchange of control information or only a part of the data. But the service-provider receiving the primitive, processes these as a single primitive interaction, in that the service-provider does not change state *while the primitive is being issued*, or at least such change is not visible from the standpoint of the architectural description of OSI. Simple examples of such primitives include a *request primitive* and an *indication primitive*. Associated with each service-primitive is a collection of zero or more *parameters*. These parameters provide a mechanism specifying the nature of control information or data, or both, that can be passed by the service-user to the service-provider when it issues a service-primitive.

Services made available by a service-provider are a collection of several individual services, called *service-elements*, and associated with each service-element is a procedure (or a function) implemented by the service-provider. The examples of service elements include:

- connection establishment,
- connection release,
- expedited data transfer.

Type of services are well defined for each layer in the OSI structure. In addition, quality of service parameters are also specified. The term quality of service refers to certain characteristics of (N)-services as observed by user (N+1)-entities. Performance related QOS parameters measure performance during the three different phases of connection establishment, data transfer, and connection release. Figure 4.3 classifies some of the parameters commonly encountered in the specification of QOS.

| Phase | Performance Criterion | |
|---|---|---|
| | Speed | Accuracy/Reliability |
| Connection Establishment | Establishment Delay | Establishment Failure Probability |
| Data Transfer | Transfer Delay Throughput | Residual Error Rate Transfer Failure Probability Resilence |
| Connection Release | Release Delay | Release Failure Probability |

**Figure 4.3** Quality of Service parameters related to performance over a connection.

## 4.3 Layer Functions

The goal of a protocol specification is to describe the form and content of all collaboration between the (N)-entities in their attempts to provide (N)-services. Such a specification would, of course, assume availability of an (N-1)-service, and its defined properties.

A protocol specification is in the form of

1. a collection of (N)-functions to be implemented,
2. the procedure used by (N)-entities to implement an (N)-function, and
3. the rules that govern all communication between the (N)-entities.

The collection of (N)-functions are those needed to bridge the gap between services available and those offered, and in some way correspond to the (N)-service-elements that the layer supports. This specification is collectively referred to as an (N)-protocol.

For connectionless data transfer services, the only set of functions needed are the ones that correspond to data transfer. These are

- transfer of data, and
- treatment of protocol errors.

But in the context of connection-oriented data transfers, the (N)-protocol must specify procedures to implement a wide variety of additional functions including:

- connection establishment,
- data transfer, and
- connection release.

Their implementation is considered mandatory, since these form the basis for connection-oriented services. Since a connection request may be refused by the responding service-user, a procedure corresponding to

- connection refusal

is also, typically, part of protocol specification.

Protocols may also specify procedures for the following optional functions:

- expedited data transfer,
- flow control,
- sequencing,
- acknowledgement, and
- error detection, reporting, and recovery.

Regarding error detection and recovery, a protocol may, in the simplest of cases, require connection release, namely,

- error release,

or provide for elaborate recovery mechanisms. More specifically, if a supporting (N-1)-connection were to fail, then the protocol may require reestablishment of the (N-1)-connection and subsequent

- reassignment after failure.

## 4.4 Protocol Specifications

A major portion of the specification of an (N)-protocol relates to procedures required to implement the specified (N)-functions. These procedures are described in terms of the form and content of all messages exchanged between the cooperating (N)-entities. The corresponding rules include specification of

- syntax,
- semantics, and
- timing

of all messages exchanged. On some occasions, however, a function may not require exchange of any messages at all. Instead, the protocol specifies the actions to be initiated by a protocol machine. These actions take the form of the issuing of primitives at the service boundary with either the (N)-layer or the (N-1)-layer.

Syntactical issues pertain to the encoding of messages exchanged between (N)-entities in the form of (N)-protocol-data-units ((N)-PDUs), each of which carries (N)-Protocol-Control-Information ((N)-PCI) and possibly (N)-user-data. The syntax specifies the encoding of each PDU parameter. Thus a PDU may contain, for instance,

- the type of PDU,
- the type of parameter,
- the length of each parameter value, and
- the parameter values.

Syntactical issues are extremely important in the context of protocol specification since the communicating (N)-entities may reside in different open systems. This is in contrast to service definitions where issues relating to the representation of service primitives and their parameters are considered to be implementation dependent.

Semantic issues, on the other hand, relate to the content of messages communicated between (N)-entities. A specification may thus include the different types of PDUs that can be sent, and the interpretation that a receiving entity can associate with it in terms of changes in its state. Since implementation of certain functions is optional, only those PDUs that correspond to functions that have been implemented can be transmitted or received.

Reception by a protocol machine of a PDU

- which is outside the repertoire of the selected protocol,
- which includes an inadmissible parameter,
- which contains an invalid parameter value, or
- which cannot be decoded

is considered to be erroneous. As such the receiving protocol machine may initiate error recovery procedures that can require

- re-transmission of the erroneous message,
- reporting of an error to the user (N+1)-entities,
- re-initialization of the (N)-connection, or
- connection release.

Generally, the protocol requires transmission of one or more PDUs and/or initiation of certain related actions when an external *event* occurs. Three distinct types of events are recognized. These are:

1. reception of a (valid or invalid) PDU from a corresponding (N)-entity,
2. issuing of a (request or response) primitive by a user (N+1)-entity,

3. issuing of an (indication or confirm) service primitive by the (N-1)-service-provider, and
4. time-out.

The actions taken by an (N)-entity, upon detecting the occurrence of an external event, can include one or more of the following

1. transmission of a PDU,
2. issuing of an indication or a confirm primitive at the (N)-service boundary,
3. issuing of a request or a response primitive at the (N-1)-service boundary, or
4. starting or stopping a timer.

## 4.5  OSI Reference Model

The OSI architecture is centered around the concept of layers and has been used extensively in developing the OSI Reference Model. The Model consists of the following seven layers (see Figure 4.4):

- the Application layer,
- the Presentation layer,

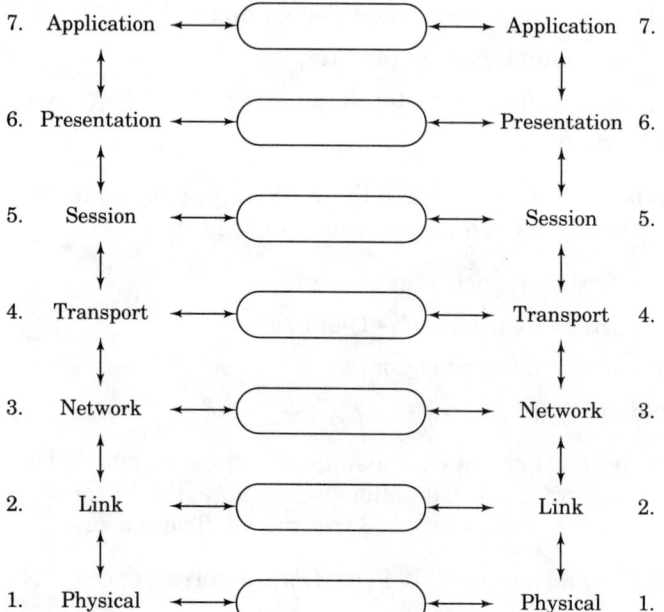

**Figure 4.4**  OSI interlayer relations.

- the Session layer,
- the Transport layer,
- the Network layer,
- the Data Link layer, and
- the Physical layer.

The highest layer is the Application layer. It consists of Application entities that cooperate with each other to provide application related services in an OSI environment. The lower layers, Physical through Presentation layers, provide services which make it possible for Application entities to communicate with each other. At the bottom, the Physical layer uses the communication media to exchange encoded bits of information.

The Application entities are the final source and destination of all data. Some of the open systems, however, simply perform the functions of relaying information from one open system to another. Such a system, therefore, implements functions included in the three lower layers only.

Although the OSI Basic Reference Model prescribes the use of seven layers, the same capability can, in principle, be provided by fewer than seven layers, or by using more than seven layers. Further, the Reference Model defines, for each of the seven layers, the service that it provides to the next higher layer. In doing so, it implicitly specifies the collection of functions to be included in each layer.

As the data package descends from the upper layer, each layer appends additional information, usually in the form of a header or trailer. Figure 4.5 illustrates the additional information each layer appends. For example, the link layer receives the application data already headed by five headers, marked as H7 to H3. It adds to it a header H2 and a tail T2, and then transfers it to the physical layer for transmission.

The main purpose of the OSI is to permit users to implement distributed applications across a network of open systems. The general structure of the OSI layers is shown in Figure 4.2.

We next present a brief description of the nature of services provided by each layer and the functions required to be implemented to support the services. We distinguish those functions that are optional from those that are mandatory.

### 4.5.1 The Application layer

The Application layer is the highest layer of the OSI architecture, and permits application processes to access OSI capabilities. The purpose

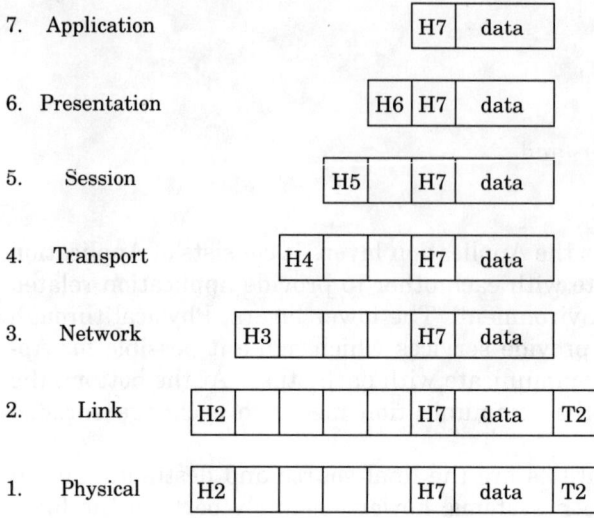

Figure 4.5  Layers append head-tail data.

of the layer is to serve as a window between corresponding application processes so that they may exchange information in the open environment. The description of the Application layer makes use of three definitions.

An *Application entity* is a model of those aspects of an application process that are significant from the viewpoint of accessing OSI capabilities. It consists of one *user element* and a set of *Application service elements*. Each Application service element uses the underlying OSI communication services to provide a specific application-level service, *reliable transfer* or *message handing*, for instance. Unlike services provided by the lower layers, application related services are not provided to any higher layer and, therefore, do not have access points attached to them. Application service elements themselves use services provided by each other and by the next level, the Presentation layer.

A user element is that part of an application process which models a user's application program, but only to the extent that it uses services provided by the Presentation layer and the required Application service elements.

Application layer services and related protocols are classified into two groups, *Common Application Service Elements* (*CASE*) and *Specific Application Service Elements* (*SASE*). CASE elements are commonly required by user and SASE elements, whereas a SASE element is included as part of an Application entity only when the application specifically requires the corresponding service. Some examples of the latter are message handling, *file transfer*, and *virtual terminal* access. On the other hand, *association control*, *reliable transfer*, and *remote operations* are common application services which are typically used by SASE elements. Association control, for instance, enables its users to negotiate and establish the communication environment between Application entities. Once that is done, protocol-data-units concerning user elements and Application service elements can be exchanged.

Functions implemented within the Application layer are very much dependent upon the service provided by each service element, but there are a number of functions that are commonly found in most Application layer protocols. These include

1. identification of communicating Application entities,
2. determination of their access rights and user authentication,
3. negotiation of the "abstract syntax" of Application protocol and user data,
4. the use of lower layer services, and
5. error detection and notification.

### 4.5.2  The Presentation layer

The Presentation layer is responsible for the appropriate representation of all information communicated between Application entities. It covers two aspects, the structure of user data, and its representation during transfer in the form of a sequence of bits or bytes. Note that the Presentation layer is only concerned with the syntax and its logical structure and not with the meaning given to it by Application entities.

A notation, called *Abstract Syntax Notation (ASN)*, for defining the structure of Application protocol-data-units and of the associated user information is available. It enables a sending entity to represent information using a syntax that is local to the open system. The main functionality of the Presentation layer is the transformation of information from its local representation to the one used during transfer, or vice versa. It thereby relieves Application entities from issues related to representation of information.

To support this, the Presentation layer implements the following functions,

1. connection establishment, and its termination,
2. negotiation and possibly re-negotiation of the abstract syntax of Application protocol-data-units,
3. syntax transformation including data compression, if required, and
4. data transfer.

### 4.5.3 The Session layer

The main functionality of the Session layer is to provide Presentation layer entities with the means to organize the exchange of data over a connection either in the full-duplex or half-duplex mode of communication. That is, depending upon the application, user entities can decide to take turns to transfer data. It also enables users to release a connection so that there is no loss of data during a release operation. In fact, the connection release can even be negotiated, in which case a user entity retains the option to reject a connection release.

*Synchronization points*, when established in the data stream exchange, enable the two users to structure their communication in the form of *dialogue units*. It thereby enables them to resynchronize data exchange to an earlier synchronization point. Resynchronization may be useful in case of errors or, more generally, to reset the connection to an earlier defined environment. The Session layer also allows users to define an *activity*. Activities are another way of providing structure to data exchange between users. Aside from starting or ending an activity, a user may interrupt the activity in the midst of communication and later resume it.

In order to support the above services, the Session layer implements the following functions:

1. connection establishment and its maintenance,
2. orderly connection release, that can optionally be negotiated,
3. normal data transfer, that can be half-duplex or full-duplex,
4. typed data transfer, which is not subject to restrictions imposed by the half-duplex mode of communication,
5. expedited data transfer, which is not subject to flow control restrictions,
6. establishment of synchronization points and resynchronization,
7. activity management, and
8. reporting of exceptional conditions.

### 4.5.4 The Transport layer

While the Network layer, and those below it, provide a path for data transfer between host computers, the Transport layer provides a facility to transfer data between Session entities in a transparent, reliable, and cost effective manner. It is the responsibility of this layer to optimize the use of Network services and ensure that the quality of Transport services is at least as good as that requested by the Session entities.

The Transport layer protocol has end-to-end significance, and is therefore implemented in host computers only. The protocol makes use of the available Network services and is not concerned with issues of routing, etc. In view of the fact that the characteristics and performance of the Network service may vary substantially, a variety of Transport protocols are available to ensure that the service it provides is largely independent of the underlying communication network. At one extreme, whenever the Network layer provides a reliable service (that is, acceptably low error rates and low failure probability), the functions implemented within the Transport layer are limited to:

1. connection establishment and its maintenance,
2. normal and expedited data transfer, and
3. error detection and reporting.

But, if the Network service is such that user data may be corrupted, lost, duplicated, or delivered out of sequence, then the Transport layer protocol must detect these these errors and recover from them. Functions that are additionally implemented are:

1. error detection and recovery, and
2. end-to-end sequence control of protocol-data-units.

In order to transfer data in a cost-effective manner and to match user requirements in terms of the quality of Transport service, the Transport layer uses one or more of the following functions:

1. multiplexing or splitting of Transport connections onto Network connections,
2. end-to-end flow control, and
3. segmentation, blocking and/or concatenation.

Typical features provided by the transport layer can be demonstrated by the transmission control protocol (TCP) of the U.S. Department of Defense[1]. The TCP is intended for reliable interprocess communication between processes which reside on pairs of host computers. Note that TCP is not a part of the OSI protocol suite and is being used here only as an example of a transport protocol in extensive use. The TCP assumes lower layer connectionless (datagram)

---

[1]MIL-STD-1778

service of internet protocol (IP), and includes operational tools in the following areas:

- *Basic data transfer.* The TCP transfer continuity is managed by the protocol. If a user needs to assure a transfer completion in cases where transfer has been asynchronously blocked by the protocol, appropriate functions are defined to complete the delayed transaction.
- *Reliability.* TCP supports recovery from network layer errors such as damaged, lost, or duplicated data as well as out-of-sequence data fragments.
- *Flow control.* The receiver governs the amount of data sent by the sender by means of acknowledgement of a window of acceptable fragments that can be sent without further permission.
- *Multiplexing.* TCP supports a simultaneous use of ports within each host. Namely, each socket can be used in multiple connections, where each connection is uniquely defined by each pair of sockets.
- *Connection management.* Each connection is uniquely defined by the pair of sockets which define its two ends. Further, connection status information includes the sequence numbers and the window sizes. Upon connection establishment, TCP initializes the status information on both sockets with a handshake mechanism with clock-based sequence numbers.
- *Precedence and security.* The precedence and security of the communication is user defined. Default values are installed by the protocol.

### 4.5.5 The Network layer

The basic purpose of the Network layer, is to provide data transfer capability across the communication subnetwork. The required functions are specific to the communication subnetwork and must be implemented by each open system in the subnetwork, including *intermediate systems* capable of routing and relaying information between possibly dissimilar communication subnetworks. Thus, the Network layer relieves the Transport layer entities from all concerns regarding subnetwork topologies and their interconnection as to the routing and relaying through one or more of the subnetworks.

The Network layer provides the means to establish, maintain, and terminate Network connections between open systems. It specifies the functional and procedural means to transfer user data between Transport entities over a Network connection. A Network connection may involve messages to be stored and later forwarded through several communication subnetworks. In order to suitably relay user data from the source host to the destination computer through one or more subnetworks, a route must be determined either centrally or in a distributed manner. Messages must also be routed within each subnetwork through which the connection is established.

The major set of functions required to be implemented by a connection-oriented Network layer protocol includes:

1. connection establishment and its maintenance,
2. multiplexing and possibly splitting,
3. re-initialization, or *reset*, of connection,
4. addressing, routing, and relaying,
5. normal and expedited data transfer,
6. sequencing and flow control,
7. error detection, notification, and possibly recovery.

Alternatively, the Network layer can provide connectionless data transfer service, in which case the relevant set of functions built into the Network layer are the data transfer, routing, and relaying functions. Segmentation can also be used to ensure that Network protocol-data-units be accommodated within buffers maintained by the Data Link layer.

Examples of Network layer protocols include the internetworking protocol (IP) of the U.S. Department of Defense[2] and the X.25 of the International Consultive Committee on Telegraphy and Telephony (CCITT).

The IP provides pure datagram[3] services for packet-switched networks with no provisions for flow control, sequencing, or reliable operation. Therefore its major functions are two: addressing and fragmantation.

The X.25 protocol has a frame level above the physical level, and a packet level above the frame level. At the packet level of the X.25 protocol we find three major functions:

- logical channels management
- packet structure composition/decomposition
- datagram service

---

[2]MIL-STD-1777

[3]A packet-switching network uses fixed length logical elements in its message fragmantation. In a datagaram service, each packet is treated independently of the others, a fact which may cause out of sequence reception of logical elements.

### 4.5.6 The Link layer

The purpose of the Data Link layer is to provide functional and procedural means to establish, maintain, and release connections between Network entities and to transfer user data. This layer is also responsible for the detection and possible correction of errors occurring over the Physical connection. Connection-oriented Data Link services are supported by the following functions:

1. connection establishment and release,
2. splitting of Data Link connections,
3. delimiting and synchronization of protocol-data-units,
4. error detection and recovery, and
5. flow control and sequenced delivery.

Alternatively, a Data Link layer may simply support connectionless data transfer capability. In that case each service-data-unit is transferred independently of all other service-data-units. Such a Data Link layer requires a minimal set of functions to be implemented.

The data-link layer attempts to make the physical layer reliable and support initiation, maintenance, and release of the physical layer. We can classify Data Link layer protocols into three classes:

1. *Start/stop* protocol transfers one character at a time. It is an old asynchronous protocol, mostly used within devices with no buffer capacity.

   | start bit | character | stop bit |
   |---|---|---|

2. In a *character oriented synchronous* transmission, the text part of the transmission is provided in bytes. An example can be seen in the BISYNC protocol (by IBM):

   | SYNC | SYNC | SOH | header | STX | text | ETX | CRC |
   |---|---|---|---|---|---|---|---|
   | | | start of header | (user defined) | start of text | (variable length optional) | end of text | |

3. *Bit oriented synchronous* protocols belong to the type mostly used in local area networks. As it is bit oriented, it is code independent. The high-level data-link control (HDLC) is an example of such a protocol. Its frame structure follows.

| 8bit | 8bit | 8bit | ≥ 0bit | 16bit | 8bit |
|---|---|---|---|---|---|
| FLAG | DEST | CNTRL | DATA | CRC | FLAG |
| 01111110 | dest. address | | | 01111110 | |

It is important to notice that the synchronization FLAG pattern 01111110 cannot be a part of the data, and that it requires bit manipulation (e.g., compressing) when this data item exists. The CNTRL field supports both flow control and error control. Based on this field, the HDLC supports a "sliding window" control [129] in the following way. The sender maintains a store of messages for which a positive acknowledge has not been received. Therefore there are two queues of messages: those which wait for transmission and those which wait for acknowledgement. The queue of the unacknowledged messages acts as a window which slides over the transmitted messages queue.

### 4.5.7 The Physical layer

The Physical layer is the lowest layer in the communication subsystem. It defines the rules governing the way in which bits are passed from one physical entity (device) to another. The layer is characterized by four types of properties.

1. Electrical information defines operating voltage or current, voltage referencing (differential or absolute), and timing of pulses.
2. Mechanical data delineates the connector that one must use along with pin size and location in the connectors.
3. Functional properties are assigned to each pin in the connector.
4. Procedural properties are defined, such as the sequences and the event ordering in the physical level.

An example of a physical layer is the RS-422 standard. The RS422 connector and its pin assignment are described in Figure 4.6. All the electrical signals are differential, and thus each signal requires two pins. Signals are specified (EIA) up to wire length of less than 5000 feet and at bit rate lower than 1 Mbps, with the current drive limited to 100 mA on short circuit and receive the input higher than 100 mV differential. The data transfer in one direction is accomplished through transmission from the pair TX(+):TX(−) to the pair RX(+):RX(−). The request-to-send (REQ) triggers the clear-to-send (CLR) and the data set ready (SET RDY) and data terminal ready (TRM RDY) signals are used for hand shake and synchronization of the devices.

```
13°    °25      pin assignment
12°    °24      11. RING IND (−)
11°    °23                              22. RING IND (+)
10°    °22      9. TRM RDY (−)
 9°    °21      8. DCD (+)              20. TRM RDY (+)
 8°    °20      7. GND                  19. DCD (−)
 7°    °19      6. SET RDY (+)          18. SET RDY (−)
 6°    °18      5. CLR (+)              17. CLR (−)
 5°    °17      4. REQ (+)              16. REQ (−)
 4°    °16      3. RX (+)               15. RX (−)
 3°    °15      2. TX (+)               14. TX (−)
 2°    °14      1. GND
 1°
```

**Figure 4.6** RS422 connector and pin assignment.

## 4.6 The IEEE 802 Standard

The OSI structure has been defined for the most general cases of interconnection of arbitrary open systems. In simpler environments significant simplification of the structure is not only possible but highly desirable since it may lead to much higher efficiencies. One such example is the IEEE 802 standard developed for local networks.

The IEEE 802 standards address the protocols for use in local area networks. These standards define the protocols for the lower three layers of the OSI reference model. In particular they address the following functions:

- manage communication over the physical link
- for transmission construct frames with address and error-detection fields added
- perform address recognition and error detection, and disassemble the frame on receiving a frame
- support the interfaces with the higher layer by providing the SAPs

In IEEE 802 standards the first three functions are grouped and used to define a separate layer, *Medium Access Control* (MAC). The last function is supported by *Logical Link Control* (LLC). Some distinct advantages result from this separation of functions. Typical Data-Link layer protocols do not define the logic for managing access to a multiple-source, multiple-destination link which are common for local area networks. Further, several MAC layer options can be supported by a single LLC implementation.

The IEEE standard 802.1 correlates several protocols into an open system interconnected architecture. Figure 4.7 portrays the IEEE layers architecture. The lower layer is the Physical layer, above it we find the Medium Access Control (MAC) layer, and above the latter we find the Logical Link Control (LLC) layer.

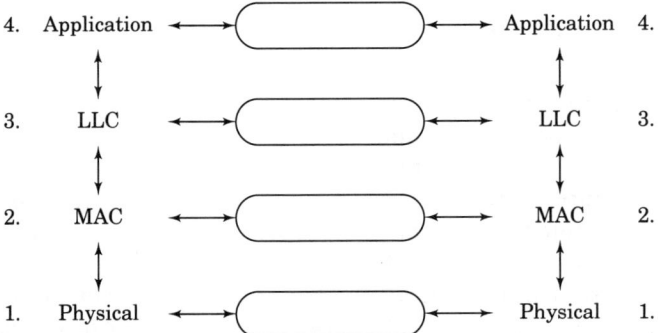

**Figure 4.7** IEEE layers architecture.

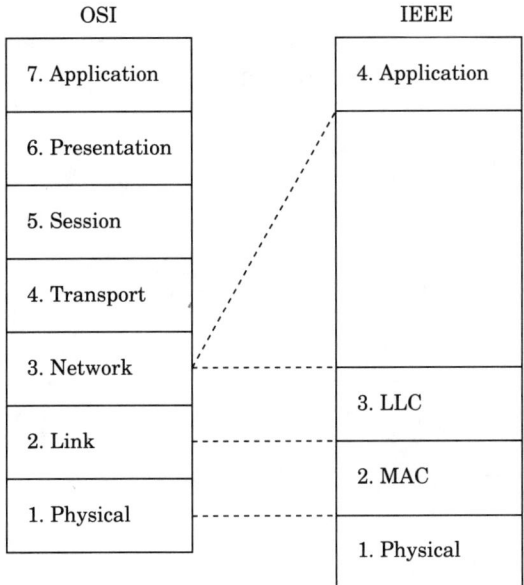

**Figure 4.8** OSI versus IEEE layers model.

Figure 4.8 compares the functional coverage of the IEEE standard 802.1 and those of the seven layer OSI model.

### 4.6.1 The Physical layer

The physical layer, as in the OSI model, transmits an unstructured stream over a physical link. It defines the mechanical, electrical, and procedural characteristics of the hardware along with the control circuits and signals. Encoding/decoding of signals, preamble generation/removal (for synchronization), and bit transmission/reception are the primary functions for this layer.

## 4.6.2 The Medium Access Control layer

A common characteristic of most local area networks is that they use shared transmission medium. For any two devices connected to this medium, their usage of it for communication requires that they be able to access the medium without interference by any other device.

Medium access control interfaces the Logical Link Control layer and the Physical layer. Figure 4.9 demonstrates how medium access can be categorized into various classes. The medium can be time-division multiplexed (time sharing) or frequency-division multiplexed (as in radio). Time sharing media can further be divided into those which are synchronously shared and those which are asynchronously shared. The asynchronously accessed media can be randomly accessed or support controlled access.

Three medium access control protocols are defined in IEEE standards 802.3 to 802.5: a carrier-sense, multiple-access, collision-

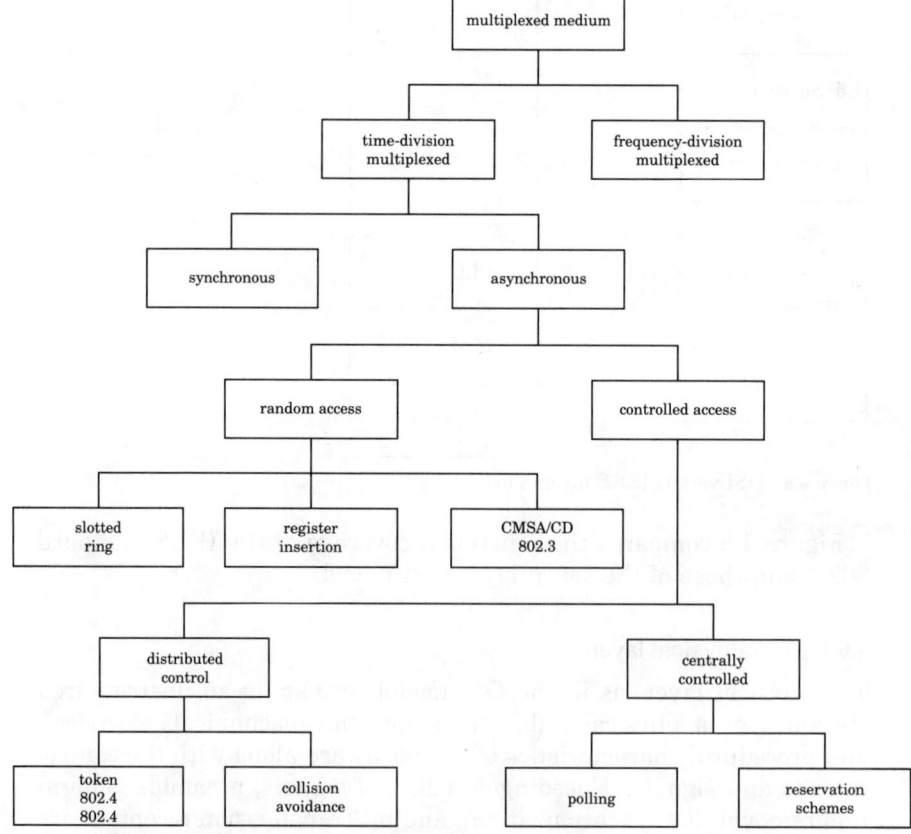

**Figure 4.9** IEEE medium access control classification.

detection protocol (CSMA/CD) and two token ring protocols. These protocols are also indicated in Figure 4.9 according to their classification.

### 4.6.3 The Logical Link Control layer

The primary concern of this layer is the tranmission of a frame of data between two devices directly connected to the medium with no intermediate store and forward functions being carried out. Unlike the traditional Data Link layer, LLC must support multi-access nature of the medium and provides some Network layer functions. Due to the lack of switching nodes in a local area network, many of the functions of the Network layer are not needed and functions including multiplexing and the support of connectionless service as well as connection-oriented service are provided by LLC. In addition, the LLC provides the multicasting and broadcasting functions.

The LLC standards are specified in IEEE 802.2 standards.

## 4.7 Network Concepts and Characteristics for Embedded Systems

Let us recall our system distribution model. In this model a distributed system contains a distributed program running on a distributed computer. A distributed program is a program with multiple threads and multiple address spaces which do not intersect. A distributed computer consists of a set of multiple processors. Each of these processors is allocated with a distinct address space. The processors are interconnected by a communication subsystem, commonly called a *network*. The hardware architecture requires two special means, one for supporting internal access to the network and another for allowing two (or more) networks to cooperate.

### 4.7.1 The network interface

Each of the nodes in the system is connected to the communication network through a network interface unit (NIU). The network interface interconnects three items: the logical link protocol that executes on the host, the devices to which (or from which) the data should be delivered (or obtained), and the communication medium. We describe a general network interface unit in Figure 4.10.

The network interface contains servers and drivers. The device servers *get* data from the output devices, while the drivers *put* data to the input devices.

Data segments that are transferred between the devices and the communication medium are *buffered* in input and output buffers. The

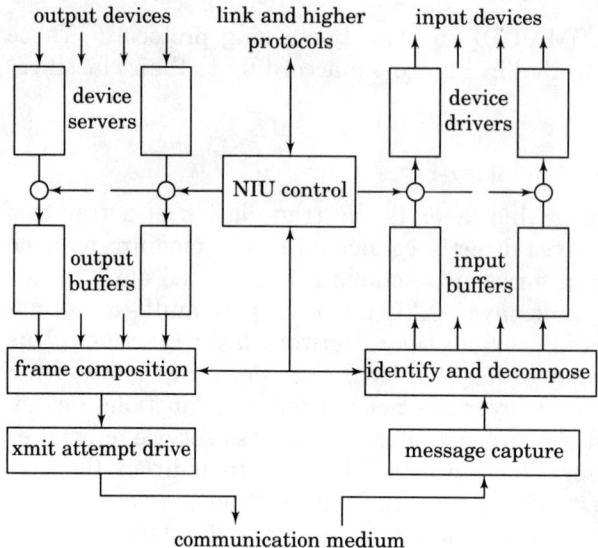

**Figure 4.10** Network interface unit (NIU)

NIU-control handles the buffering, allowing buffer usage to be at the convenience of the protocol, as it carries out functions such as acknowledging, flow control, etc.

The NIU also *composes* the frames for transmission, by dividing a long message into fragments and appending the proper head/tail to the message. Similarly, when a message is received, it must be identified ("is it for me?") and *decomposed* by removing the head/tail.

The transmission of a message involves gaining control of the communication medium. In a CSMA/CD system, *attempts* to transmit may result in collisions, followed by a number of retransmissions depending on the persistence of the unit. In a token network the unit *waits* for a free token before attempting transmission. In some other architectures the unit may wait for a time slot to transmit.

The economical benefits in servicing a number of devices with one NIU raise some timing uncertainties. We cannot know the state of NIU-control at the time a message arrives, either from a device or from the communication medium. Hence, it is difficult to predict how long its response time is going to be. Furthermore, the fact that the logical link protocol is not synchronized with NIU-control introduces additional time uncertainties :

- upon sending: from initiation to xmit/attempt
- upon receipt: from capture to link protocol interruption

Note that there is a time delay from identification ("it is for me!") to the interruption of the logical link protocol. This time involves input buffer and input driver delays of data transfer. But in cases where a message is purely a control function, the NIU-control can inform the link protocol in a shorter way, as observed in Figure 4.10.

### 4.7.2 Internetwork communication

Let us now turn to issues of higher level layers in the communication layered structure. On top of the Local Network (LN) layers we find the *transport* layer. Most communication protocols support the error-free delivery of data using checks and retransmissions. Note that a "retry" may be useless if there is a deadline associated with an execution. Therefore, the Transport layer, just like an allocation mechanism, must provide means for physical redundancy in the absence of temporal redundancy. We note that when we employ physical redundancy, we transmit message duplicates in parallel. These messages must use different paths, to assure improved reliability. In such a case, we expect duplication even in nonfaulty cases. There are very simple solutions to this problem, some of which are indexing, timestamps, and other message descriptors that support idempotency. We present later in this chapter, another way to support the physical redundancy.

The *session* layer, usually built on top of the Transport layer, supports connection establishment and disconnection for a session. We consider connection establishment an allocation of a virtual circuit. Note that for real-time environments we must base this allocation on time, as in the case of local resources. In addition, we note that when we connect nodes at different networks, the name spaces must associate unique end points for this connection. Furthermore, heterogeneous networks may have different conventions and practices.

Most modern computer embedded systems utilize microprocessors at various control points of the plant they control. This approach allows specialization at the low level of the system architecture, while maintaining generality at higher levels. Computer embedded systems of this nature include a heterogeneous set of computing devices that interact with each other. Any specialization according to local metrics can lead to heterogeneity of the computing devices. For example, an output device that deals with serial (one bit) data streams, does not necessarily need a 32 bit word computer. We must make the decision on the word size of this computer according to the tasks and expected load of this device. Therefore, we require from our operating system, the support of such heterogeneous environments. The *presentation* layer, in the open system interface definition, is one answer to this requirement. However, its cost in time is high. Another solution is the use of *typed messages*.

Communication via typed messages requires an *a priori* agreement on the structure of the message. Each typed message contains fields in an order specified *a priori*. The internal fields of the message have specified sizes and meanings, and the source and destination of the message share this interpretation.

### 4.7.3 The internetwork interface

The internetwork interface carries out the tasks just defined for internetwork communication. The principle of bridging networks is depicted in Figure 4.11. Although the figure describes the case of two networks, the *bridge* concept can be extended to connect $n$ networks. Note that a bridge Physical layer includes two NIUs, each serving a local network.

We expect the bridge to transfer the communication data between the networks, through the Physical and Logical Link layers which correspond appropriately to the corresponding layers of the networks. We definitely expect the bridge to filter the communication and to transfer only relevant data, otherwise the networks will be overloaded with irrelevant traffic. Other desirable properties of a bridge include duplication avoidance, maintenance of sequentiality, and minimization of unnecessary overhead. Data delivery can be either passing through for similar networks on both sides of the bridge, or passing after translation otherwise. Translation may require frame reconstruction due to different conventions as well as resegmentation due to different frame sizes. The translation device that connects two different protocols is called *gateway*.

Note that as we connect the bridge, the address spaces of both networks are increased. Actually, three issues must be answered:

- Naming: who is it for?
- Addressing: where is it?
- Routing: how to get it there?

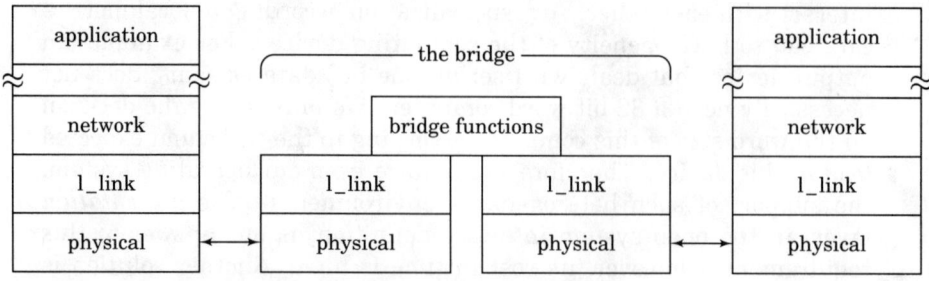

**Figure 4.11** The principle of bridging networks.

The selected solution must always support unique names and addresses. Routing uniqueness in unnecessary, but loops must be avoided in order to increase efficiency and decrease delays.

Internetwork communication creates important issues which must be resolved in an error-free environment:

- How is a packet loss in one network (say the destination net) recovered in another (say the sender's net)?
- How does the solution deal with orphans (messages arriving at a failing node)?
- What time-outs should be taken? Should there be any local knowledge on end-to-end link length?
- What buffering strategy should we adopt in cases of different local strategies?

We do not intend to deal with these issues in this book, but we encourage the reader to further investigate these topics since they significantly influence the system robustness and performance parameters.

## 4.8 Concurrency, Commitment and Recovery

In systems involving permanent (or even semi-permanent) data storage, consistency of stored information is of prime importance. For instance, if two or more related operations are to be carried out on a file, then either each operation executes successfully and the resulting changes are made permanent, or the file is left unaltered in case any operation fails. The last step, in which a processor makes the change permanent, is referred to as a *commit* operation. An operation can fail either because it is invalid, or because the resources required for its execution are unavailable, or because the processor itself fails. The need for consistency of data storage requires that the sequence of operations be viewed as an *atomic action* which either succeeds or fails. In case an operation fails, the processor aborts the execution. Such an abort is also referred to as a *rollback*.

### 4.8.1 Commitment

The situation is more complex in distributed applications where two or more Application entities, possibly residing in two different systems, cooperate to manipulate a shared or distributed data base. At the end of processing (and necessary communication), not only must the cooperating entities be sure that the data base is consistent, but all entities must be aware of the final state of the data base. This requires the use of atomic actions.

Formally, an atomic action is a sequence of operations performed by a distributed application with the property that it be controlled by a unique Application entity, and that parts of it can be executed by two or more communicating entities. Further, the atomic action either completes successfully, or terminates without any change in the state of the data controlled by the participating entities.

The atomic action may be modeled as a rooted tree with two or more nodes. Each node of the tree models a participating Application entity. The root node is also called the *master* entity, and determines whether the atomic action is subsequently committed or aborted. Each branch of the tree models the superior-subordinate relationship between entities, discussed earlier. Such a relationship is also called a *CCR (or Commitment, Concurrency, and Recovery) relationship*. Successful execution of the entire atomic action is dependent upon successful execution of each subtask, or equivalently of each subatomic action.

Before discussing the protocol to handle commitment or rollback, we describe the nature of failures that can be encountered in a distributed application. Two types of failures are recognized. The first type, due to failure of an Application entity, can result in loss of *unsecured* stored data. Subsequently, after execution of a local recovery procedure, the failed Application entity can be restored. However, secured data, typically on a reliable disk, is assumed to be available. When applying commit procedures, it is assumed that the data to be manipulated by a subordinate as part of an atomic action is stored on a secure medium. Such data is called *bound data*, since it is *bound* to the atomic action. Changes, if any, to the bound data are also stored on a secure medium together with *atomic action data*. The latter includes a description of the Application association, the atomic action, concurrency controls, and pointers to bound data. Subsequently, when a commit operation is performed, the old copy is released and the update is made permanent. In case a rollback to the original state of bound data is performed, additional resources acquired during the execution of the atomic action are released.

Communication breakdown is the second type of failure. Such failure causes loss of data in transit only. The Application association may be reestablished and communication restarted. A restart procedure enables the Application entities to resynchronize communication to the beginning of the atomic action (or to an intermediate checkpoint). The restart procedure can be used in case of failure of an Application entity.

### 4.8.2 Concurrency

It is quite conceivable that two or more atomic actions may concurrently attempt to manipulate the same data. In such a situation, commitment of an atomic action becomes more complex, since not only do their actions modify the data simultaneously, but either of them can decide to rollback their computations. Concurrency controls, imposed by an implementation, must therefore ensure that

1. Data bound to an atomic action is not changed other than by the atomic action, itself.
2. If an atomic action, to begin with, uses data that is subject to rollback (as part of another atomic action), then it is not committed until and unless the other atomic action commits.

### 4.8.3 CCR services

Commitment, concurrency, and recovery procedures are optionally supported by an implementation in the form of a set of common Application service elements, called CCR elements. CCR services, when available, can be used to ensure atomicity of interactions in a distributed application. A user, or another Application service element included within the Application entity, which requires CCR services, must include CCR elements in the Application context within which it negotiates at the time of establishing an association with its peer user. Each CCR relationship requires a *distinct* Application association to be established between the superior and its subordinate entities. Further, over an Application association, there can not be more than one atomic action in progress at any given time. In this respect, the nesting of atomic actions is not supported.

The collection of CCR service primitive available to a user are listed in Figure 4.12, together with an indication of whether a service element is confirmed or not.

The execution of an atomic action consists of two phases, the first phase ending with the subordinate indicating its readiness to commit (using *C-READY* primitives) or its inability to progress further (using *C-REFUSE* primitives). The second phase consists of a commit (or rollback) operation. Further, the superior can optionally issue a *C-PREPARE* request primitive to solicit a response in the form of either a C-READY indication or C-REFUSE indication primitive. Where a C-PREPARE is not issued, the Application determines the event that prompts the subordinate to indicate its ability to commit, or otherwise.

Rollback of the atomic action is not the only option available to cooperating entities in case failures occurs. Either Application entity can attempt a *restart* of the atomic action, in the hope that execution of the atomic action may still progress. Whether the restart is to

| Service | Primitive | Type of Primitive | Parameter(s) |
|---|---|---|---|
| Initiation | C-BEGIN | Unconfirmed | atomic action identifier, branch identifier, atomic action timer, user data |
| Indication of Termination | C-PREPARE | Unconfirmed | User data |
| Offer of Commitment | C-READY | Unconfirmed | User data |
| Refusal of Commitment | C-REFUSE | Unconfirmed | User data |
| Commitment | C-COMMIT | Confirmed | - |
| Rollback | C-ROLLBACK | Confirmed | - |
| Restart | C-RESTART | Confirmed | resumption point, atomic action identifier, branch identifier, restart timer, user data |

**Figure 4.12** CCR service primitives.

resynchronize the state to that at the beginning of the atomic action, or to an intermediate point is specified as the parameter *resumption point* of the primitive *C-RESTART* request.

In addition to the restart procedure, an application can establish checkpoints (or minor synchronization points) and resynchronize as necessary. In fact, issuing of *C-BEGIN* and C-COMMIT primitives establishes a *major* synchronization point. A C-ROLLBACK service initiates a resynchronization to the major synchronization point established at the time of establishing the association.

Thus far, we have discussed the use of CCR services by Application entities in the commitment of atomic actions, and in the recovery from failures. The CCR protocol is implemented in the form of two protocol machines which interpret, and respond to, service primitives issued by users. To enable users to interact with each other, the protocol machines exchange information between themselves. This they do in the form of *CCR Application protocol-data-units* (*CCR APDUs*). Exchange of CCR APDUs between the protocol machines enables the semantics of CCR service elements to be conveyed between the users. The CCR protocol specifies, in unambiguous terms, the actions that a protocol machine takes when it receives:

1. a CCR service request or response primitive, or

2. a CCR APDU from the corresponding protocol machine (contained in a Presentation service indication or confirm primitive).

In response to any of the above incoming events, the action taken by a protocol machine is to issue (or transfer)

1. a CCR service indication or confirm primitives, or
2. a CCR APDUs to the corresponding CCR protocol machine (contained in a Presentation service request or response primitive).

Of course, the specific action or the sequence of actions is dependent upon the current state of the protocol machine, as well as upon the parameters of the service primitive or APDU received. The specification is in the form of a state transition table, which also specifies the state to which the protocol machine transitions. We shall not specifically discuss the table. The CCR APDUs are listed in Figure 4.13.

| Service | Protocol data unit | Parameter(s) of APDUs | Presentation service which carries APDUs |
|---|---|---|---|
| Initiation | C-BEGIN-RI | atomic action identifier, branch identifier, atomic action timer, user data | P-SYNC-MAJOR request/indication |
|  | C-BEGIN-RC | - | P-SYNC-MAJOR response/confirm |
| Termination | C-PREPARE-RI | user data | P-TYPED-DATA request/indication |
| Offer to commit | C-READY-RI | user data | P-TYPED-DATA request/indication |
| Refusal to commit | C-REFUSE-RI | user data | P-RESYNC (abandon) request/indication |
|  | C-REFUSE-RI | - | P-RESYNC (abandon) response/confirm |
| Commit | C-COMMIT-RI | - | P-SYNC-MAJOR request/indication |
|  | C-COMMIT-RC |  | P-SYNC-MAJOR response/confirm |
| Rollback | C-ROLLBACK-RI | - | P-RESYNC (abandon) request/indication |
|  | C-ROLLBACK-RC | - | P-RESYNC (abandon) response/confirm |
| Restart | C-RESTART-RI | atomic action identifier, branch identifier, restart time, resumption point, user data | P-RESYNC (restart) request/indication |
|  | C-RESTART-RC | resumption point, user data | P-RESYNC (restart) response/confirm |

Figure 4.13  CCR APDUs and their parameters.

Collisions can occur at any time during the execution of an atomic action. As an example, consider the situation where a superior issues a C-ROLLBACK request primitive. At about the same time, the subordinate issues a C-REFUSE request primitive. The CCR protocol specifies that in such an event, the protocol machine corresponding to the superior entity should ignore the incoming *C-REFUSE-RI* APDU and await a response to the *C-ROLLBACK-RI* APDU that it transmitted earlier. More generally, the protocol specifies as to which of the two—the superior or the subordinate—is the winner, and also specifies the actions to be initiated by the winner.

## 4.9 Concluding Remarks

Clearly, the implementation of any distributed system requires the implementation of a communication capability which permits the machines to transfer data and control signals. The capability is provided through the implementation of a network. As the communication requires a cooperation among two or more autonomous entities, protocols must be defined to permit systematic communication. In this chapter we reviewed the major directions in the design and implementation of computer networks. We observed that many standards have been developed for the protocols. The OSI reference model provides the primary framework for the definition and specification of protocols.

The OSI reference model defines the seven layer model for the protocols and standards that have been developed for protocols for most of the layers. Note that the primary emphasis of the standardization is to permit the interconnection of diverse systems. When a designer controls all parts of a system, the overheads of the multiple layer implementations can be significantly reduced, but at the cost of not being able to communicate with other systems directly.

The issues of error detection and recovery are considered as a part of the design of all protocols. The primary idea is that each layer should detect and recover from errors as far as possible. If it can not, it must notify the layer above. The system level error detection is not considered part of the design of networks except in some cases. An example of system level handling of errors is the protocol defined for Concurrency, Commitment, and Recovery and discussed in this chapter.

Communication technology is developing very rapidly and is requiring a revision of the old protocol standards as well as the formation of new standards. For example, significant changes are likely as a consequence of the availability of very high speed networking capability.

# Part 2

# Distribution

The second part of the book discusses distributed system mechanisms which are required to make the tolerating of faults feasible. In this part we discuss several topics including the algorithms and techniques for mutual exclusion, election, deadlock detection and prevention, and termination detection. Each topic reviews the various approaches in a manner which allows the reader to compare performance parameters of the proposed solutions. This part starts with the definition of several concepts of distribution, constructing a distributed system model and analyzing its sensitivity to control and environment, thereby establishing criteria for the comparison of solutions. Then, the above topics and problems are developed and analyzed for various solutions. A particularly important topic discussed in this part is the reaching of agreement between distinct computations. This topic is discussed under various fault environments, providing tools for reaching approximate agreement, verifying consensus, and then committing and voting. The concepts, algorithms, and tools are examined from various points of view. Some are examined formally, including the axiomatic definitions and their properties, while others are introduced to the level of detailed implementation in the pseudo code.

**Chapter**

# 5

# Concepts and Measures

This chapter begins the second part of the book, whose main concern are the various approaches used in the distribution of subsystems. Hence, we start this part with a chapter that explains what distribution is and provides some measures to compare different approaches and solutions.

## 5.1 Terminology, Definitions, and Conventions

### 5.1.1 Programs and their images

Sequential and concurrent programs are used to carry out the design of computerized systems. Historically, sequential programs were the first programs to be designed, and only later did concurrent programs appear, providing an additional dimension to programming. In a sequential program, each event has a unique predecessor which is always completed before the successor event begins its execution. This is not the case in concurrent programs, where events do not always assure the certainty of completion of their predecessors, and thus provide the foundation for potential parallel execution.

Since the main difference between concurrent and sequential systems is in the traces of their events, let us start with some definitions that explain our terminology and their various properties of concern.

> DEFINITION 10. A *thread* of program control is a sequential trace of events caused by the program execution, in which any two events in the trace maintain a total order independent of the resource allocation.
> □

The executing image of a sequential program is a *process* whose implementation requires a single-thread processor and a set of resources. However, if our system contains multiple processes which are executed in parallel, access control becomes essential.

> DEFINITION 11. The *address space* of a program is the space into which the program is given access for its execution image. □

Therefore, a process can be described as an execution image of a program with a single thread and a single address space[1]. But since a process is a single image of a sequential program, we must understand what happens during the concurrent execution of multiple images of the same program. Two replicas of a program allocated with separate and distinct resources for execution must have different address spaces, or else the asynchronous nature of the execution will result in chaos. These address spaces can intersect, but they cannot be totally identical.

When we advance from sequential to concurrent programs, we must take care of additional characteristics which have not been relevant in the sequential scheme:

- synchronization of corresponding and correlating events
- communication between concurrent processes

Concurrency of distinct processes may introduce address space intersection, which is commonly referred to as shared memory. In order to properly distinguish between programs which employ shared memories and programs which do not, we define distributed program as follows.

> DEFINITION 12. A *distributed program* is a program with multiple threads and multiple address spaces which do not intersect. □

### 5.1.2 Computers and systems

The hardware with which we build the computers is discussed in detail in Chapter 2. A computer essentially contains the following components:

- Computation nodes (often called processing nodes) which carry out actions (see Section 1.2.1). These nodes consist of:
  - Processors, which may or may not be of the same type.

---

[1] See definitions in in Chapter 1

- Address spaces, each associated with a processor at a particular time interval, which may or may not overlap.

- Communication subsystem connects the computation nodes through arcs which are organized in a defined topology. Some examples of topologies are illustrated in Figure 5.1.

A single computation node computer consists of a single processor, with a single address space always accessed by this processor, and no communication subsystem. On the other hand, a multiple computation node computer, includes processors which have an address space allocated for each and that can share memory for multiple access. The multiple computation nodes are interconnected by a communication network which defines a particular topology.

Figure 5.2 describes the general scheme of a distributed computer, which includes the various components of such a scheme.

> DEFINITION 13. A *distributed computer* consists of a set of multiple processors, each allocated with a distinct address space at a given period of time, and interconnected by a communication subsystem. □

The interconnection topology of the communication subsystem can carry many forms:

- star topology, in which all nodes surround the communication subsystem on its perimeter,

- ring topology, in which all nodes are serially connected,

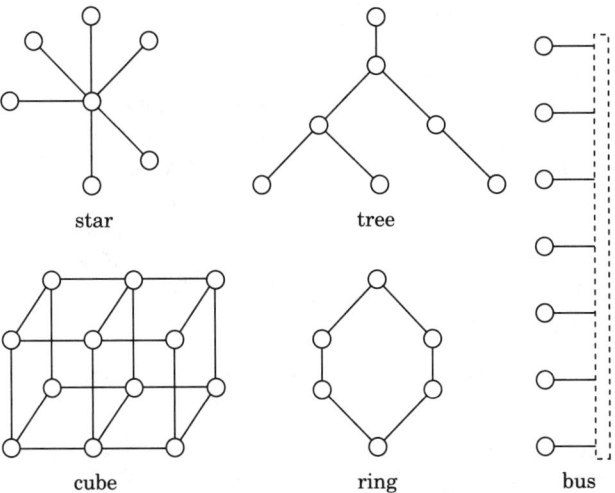

**Figure 5.1** Network topology examples.

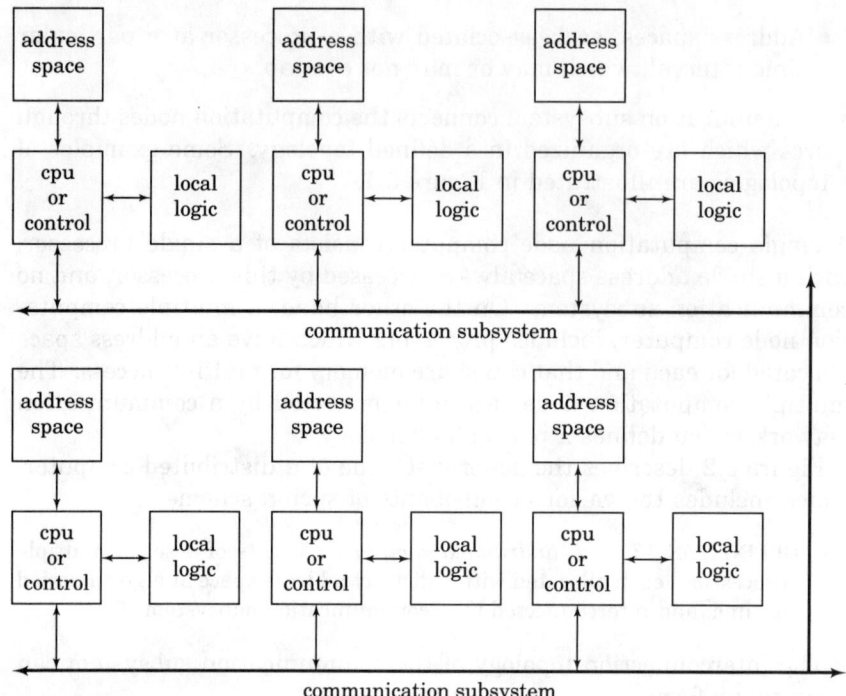

**Figure 5.2** Distributed computer.

- tree topology, with hierarchical structure,
- cube structure, in which every node interconnects with its neighbors organized in a cube formation, or
- other topologies, including irregular ones.

We note that all the above topologies fit the description given in Figure 5.2, and thus, may be considered as orthogonal to other characteristics.

In order to construct a system, we must introduce with it the computers, their peripherals, and the application specific devices. In addition, the system depends entirely on the programs which run on these devices. In Section 1.2.1 Definition 1, we have defined a system as an identifiable mechanism that maintains an observable behavior at its interface with its environment. Based on this definition, let us define what a distributed system is.

DEFINITION 14. A *distributed system* is a system which contains a distributed program running on a distributed computer. □

According to this definition, it is evident that neither a distributed program which runs on a single processor nor a sequential program which runs on a distributed computer can constitute a distributed sys-

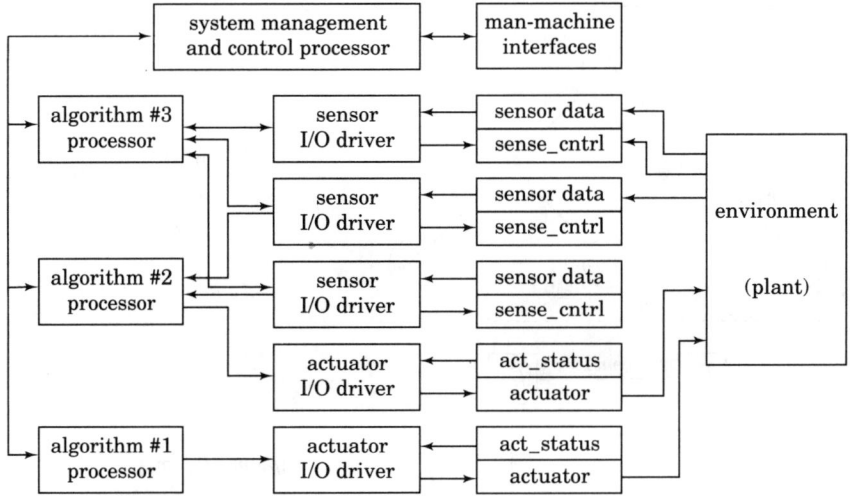

**Figure 5.3** Distributed computer-embedded system.

tem. Furthermore, neither a parallel program with processes whose address spaces intersect each other nor a program with processes in which the thread of control does not impose a total order can be a part of a distributed system.

Figure 5.3 describes an example of a distributed computer-embedded system. In this example we can see that one can have a distributed computation, namely algorithms #1, #2 and #3, running on identical interconnected processors. On the other hand, the sensor I/O drivers can be fully device specific.

## 5.2 Distributed System Model

The computerized part of a distributed system consists of sets of processes and resources defined as follows. A set of processes $\mathcal{V}_p = \{p_1, \ldots, p_n\}$ are related to each other through a set of logical links $\mathcal{E}_p$, to form a graph

$$\mathcal{G}_p = (\mathcal{V}_p, \mathcal{E}_p). \tag{5.1}$$

A set of resources $\mathcal{V}_P = \{P_1, \ldots, P_m\}$ are related to each other through a set of physical links $\mathcal{E}_P$, to form a graph

$$\mathcal{G}_P = (\mathcal{V}_P, \mathcal{E}_P). \tag{5.2}$$

Resources $\mathcal{V}_P$ are associated with computations $\mathcal{V}_p$, forming the mapping $\mathcal{G}_P \mapsto \mathcal{G}_p$, as described in Figure 5.4 in conjunction with the

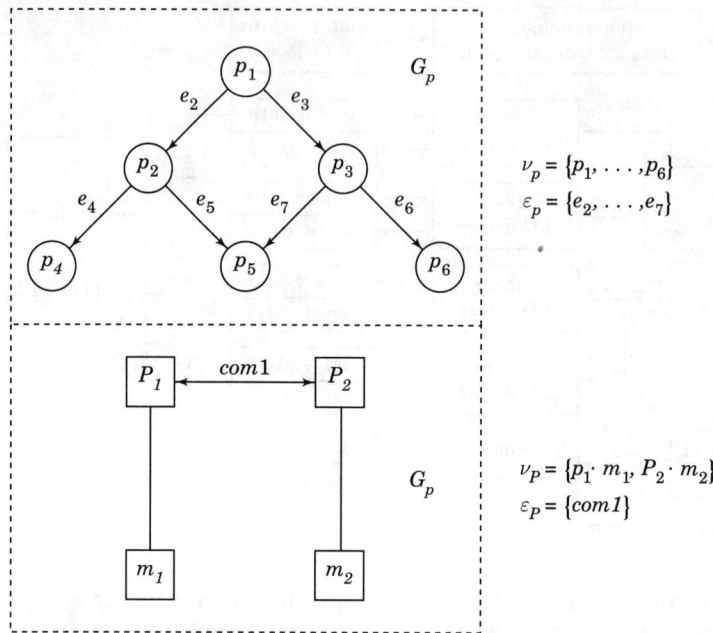

**Figure 5.4** Distributed system model.

following table of $\mathcal{G}_P$ mapping:

| $P_1 \cdot m_1$ | $P_2 \cdot m_2$ | $com_1$ |
|---|---|---|
| $P_1 \cdot m_1 \mapsto p_1$ | $P_2 \cdot m_2 \mapsto p_3$ | $com_1 \mapsto e_3$ |
| $P_1 \cdot m_1 \mapsto p_2$ | $P_2 \cdot m_2 \mapsto p_5$ | $com_1 \mapsto e_5$ |
| $P_1 \cdot m_1 \mapsto p_4$ | $P_2 \cdot m_2 \mapsto p_6$ | |
| $m_1 \mapsto e_2$ | $m_2 \mapsto e_6$ | |
| $m_1 \mapsto e_4$ | $m_2 \mapsto e_7$ | |

The inter-node relations are defined through the arcs through message based relations. Thus, a relation between objects can be considered an invocation, that triggers an object operation and is syntax and semantics restricted. This type of relation also imposes a total order between the sending event and the receiving event, and thus synchronizes two remote sites. However, an arc can contain more than one communication activity, and therefore, it can represent a set of send-receive events.

Various properties of concurrent programs, and particularly in distributed systems, can be defined and proven formally in an axiomatic approach. The axiomatic approach we use here is based on [109]. This approach uses *assertions* which describe the state of the variables in the system. In this approach the notation $\{P\}\ S\ \{Q\}$ stands for: if $P$

is true before executing $S$, and $S$ halts then $Q$ is true after executing $S$. In this notation, $S$ is a statement and $P$ and $Q$ are logical assertions. These assertions can be regarded as a precondition $P$ and a postcondition $Q$. For a set of parallel statements $\{S_i\}$ there is a set of preconditions $\{P_i\}$ and respectively, a set of postconditions $\{Q_i\}$.

Critical variables, which are logically connected, are especially emphasized as a resource. Variables which do not belong to any resource can be regarded as free variables. The usage of a variable of a resource $r$ in a statement $S$ is always guarded with a logical condition $C$ in a **with**[2] statement of the form

**with** $r$ **when** $C$ **do** $S$

The parallel program is therefore of the form

**resource** $r_1, r_2$; **cobegin** $S_1 \| S_2 \| \ldots \| S_n$ **coend**

where $r_1$ and $r_2$ are resources of guarded variables and $S_1 \ldots S_n$ are the parallel statements executed between the cobegin and coend. We note a special assertion $I(r)$, the invariant for resource $r$, which is true when parallel execution begins and remains true whenever execution is observed outside the critical sections of $r$. The parallel execution axiom can be defined as follows:

DEFINITION 15. **Parallel execution axiom:** If

1. $\forall i$: $\{P_i\}S_i\{Q_i\}$ is proved, and
2. No free variables in $\{P_i\}$, $\{Q_i\}$, are changed by $S_j$, for $j \neq i$, and
3. All the variables in $I(r)$ belong to $r$,

then

$\{P_1 \wedge P_2 \wedge \ldots \wedge I(r)\}$
  **resource** $r$; **cobegin** $S_1 \| S_2 \| \ldots \| S_n$ **coend**
$\{Q_1 \wedge Q_2 \wedge \ldots \wedge I(r)\}$

□

We use this approach later in the book to define formally the various properties of a distributed system.

---

[2] Sometimes referred to as **region**.

## 5.3 Distribution Measures

Distributed architectures simplify the solutions, as each node deals with its local activities and the nodes with which it has direct relations. However, in these architectures, we are unable to obtain an immediate global view of the system, as only a neighborhood is accessible.

The purpose of this section is to summarize a set of quantified criteria with which we can compare and examine the distributed behavior of an algorithm. We first consider the aspects of the sensitivity of system control such as synchronization, resource management, concurrency, and data integrity. We now turn to aspects of sensitivity to environment, such as event ordering and failure.

### 5.3.1 System control sensitivity

System control schemes use various means to synchronize distributed activities, in particular, those which share some parts of their address space. It is therefore only natural to verify that a proposed solution to some problem affects neither system control nor depends on a particular synchronization scheme. Furthermore, as system resources are generally limited, effects related to resource management must be examined as well. The way resources are acquired and released influences the overall system performance. Therefore, it is essential to detect computations' termination as well as blocked situations.

Concurrency improves system resiliency and performance. However, it raises issues of data integrity which are not evident in serial systems. Questions of database partitioning and consistency of replicated data must be examined carefully. Most of today's processors use replicated data in their cache mechanisms even in a single locality, and thus emphasize even more, the sensitivity of this aspect.

**Synchronization.** One of the most sensitive aspects of concurrent programming is synchronization. We expand the discussion on this topic in Chapter 6. Two programs which execute concurrently can require the use of the same data item, whether for reading it or for modifying it, and thus raise contention on accessing it. The sequential program segment which contains this shared item access is called a *critical section* of the program. Based on the axiomatic approach previously introduced, we can define the critical section axiom:

DEFINITION 16. **Critical section axiom:** If

1. $I(r)$ is invariant for resource $r$ from the cobegin statement, and
2. $\{I(r) \wedge P \wedge C\}\ S\ \{I(r) \wedge Q\}$, and
3. No free variables in $\{P\}$ and $\{Q\}$ are changed by another process

then

$\{P\}$ **with** $r$ **when** $C$ **do** $S$ $\{Q\}$

□

In Chapter 3 we introduced various synchronization devices, such as semaphores, monitors and resources. These tools help us synchronize the access to a shared item, assuring the required mutual exclusion of the execution of critical sections. The implementation of such devices must demonstrate the following criteria:

- Independence: the solution does not depend on hardware or instruction speed.
- Fairness: a process which intends to enter its critical section when no other process is in a critical section will eventually be able to do that.

In centralized systems, where a synchronization device is controlled by an appointed node, it is essential to support a recovery from this node's failure. As such a failure can be permanent, we may need election procedures in which another node is elected to replace the faulty one.

**Resource management.** The major significance of resource availability in distributed systems, calls for resiliency with respect to resource blocking (deadlock) and for means of termination detection. Deadlock situations can be either avoided (prevented) or detected upon occurrence. Aspects of resource availability in deadlock and termination situations are discussed in Chapter 8.

Figure 5.5 describes resource deadlock, where each of the two processes holds a partial set of the required resources, and awaits resources held by the other. Situations of this type may be caused by a design fault, but such effects can also be a result of a processing node failure. Hence, detection of such situations along with avoidance of their effects is a significant property which must be carefully verified.

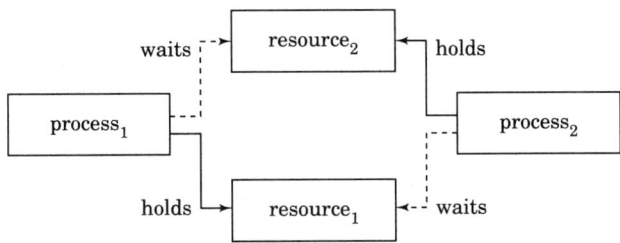

**Figure 5.5**  Resource deadlock.

**Data integrity and management.** Nodes are unable to have an immediate global view of the system. In fact, only neighboring nodes are generally accessible, limiting the nodes view to its locality. However, distributed architectures promote the use of independent alternatives to achieve a higher resiliency in the presence of faults. The use of alternatives and replicas raises the significance of data integrity, because data items are distributed between them to form a view of the system state in various localities. A consistent view of the system is mandatory.

The issue of managing replicated objects is discussed in detail in Section 12.3. A set of replications of an executable object, called a *troupe* [41], is a set of instances of the same objects all executing concurrently. A set of replicated non-executable objects, called a *pack* [92], is a set of instances of the same passive object upon which the same set of operations are carried out. The issue of maintaining replicas consistent with each other is discussed further in Chapter 14.

Figure 5.6 shows a troupe of objects, all of which are replicas of the same object and all carrying out operations on a pack of replicated passive objects. We note that our presentation of a single entity operating on another single entity requires all members of each entity to be consistent with each other. This issue becomes more complicated when troupe or pack members reside in different localities, and thus increase their vulnerability to fragmentation of the system into isolated subsystems, in other words partitioning. System resiliency to uncorrelated and uncoordinated updates of replicated objects is therefore critical.

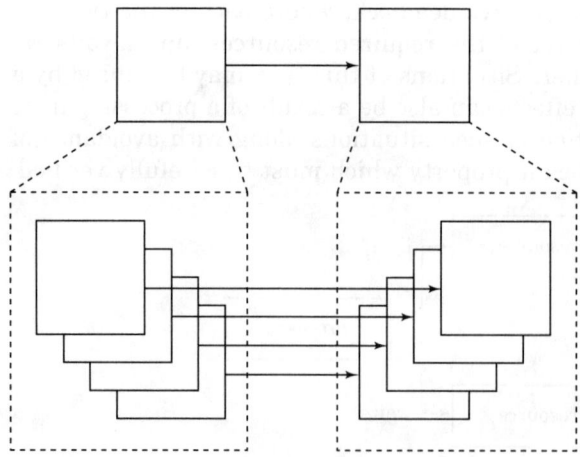

**Figure 5.6** Object replication.

### 5.3.2 Sensitivity to environment

The environment of a processing node is comprised of its neighboring nodes and the arcs which connect them. Recall that a neighborhood is not necessarily physical, but rather, it is logical in nature. Neighboring nodes relate to each other during execution, and thereby depend heavily on the quality of these relations. As discussed in Chapters 2 and 3, each node which executes a process (and thereby demonstrate no concurrency nature) is assumed to satisfy the following restrictions:

- no memory is shared between two processes
- only message based communication constructs the arcs which interconnect the nodes

The environmental effects appear in the form of two issues. The first relates to the concurrent execution of communicating nodes and raises the issue of ordering events which occur in different localities. The second relates to the communication quality and raises a series of related issues on the message communication structure.

**Events ordering.** Event ordering is extremely important not only for archiving purposes but also for system timing and progress control. Precedence relation between events contains the knowledge of which event occurred before the other. The principle of ordering events is based on two properties. The first property is the local total order of events, since a process is a single image of a sequential program. Thus, a local logical clock, incremented on each event occurrence, maintains a local ordering. The second property is the association of particular events between the nodes. This property is based on the axiomatic fact that the event of receiving a message always occurs after the event of sending it.

Figure 5.7 illustrates this principle. A logical clock on each of the nodes tags the local events and maintains a total order between them. A message sent from node $B$ (event $B_0 = e_s$) to node $A$ is received there causing event $A_0 = e_r$. Since $e_s < e_r$ we can conclude $B_0 < A_0$. The above principle achieves only a partial ordering [79] of the system events. An example of this inconclusive ordering is seen if we add another transmission from $A$ ($A_{-1} = e'_s$) to $B$ ($B_1 = e'_r$). Knowing that $B_0 < A_0$, $A_{-1} < B_1$, $A_{-1} < A_0$ and $B_0 < B_1$ gives us no clue on the precedence relation between $B_0$ and $A_{-1}$ or between $B_1$ and $A_0$.

Therefore, algorithms which rely on event ordering must be carefully examined to the extent that they require this knowledge. Algorithms in which this knowledge is established are compared in two parameters: the cost involved in acquiring the knowledge and the accuracy and granularity in which the knowledge is obtained. Moreover,

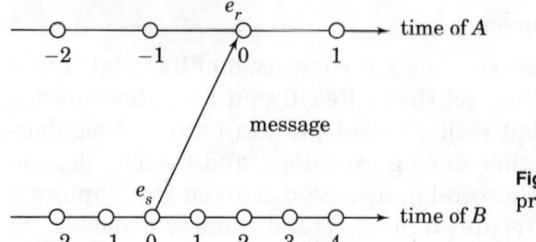

Figure 5.7 Event ordering principle.

the use of multiple routes to increase arc resiliency can produce inconsistencies as we may have multiple receive events for every send event.

**Robustness to link failures.** The arcs which interconnect the computation nodes are implemented as logical links, each mapped into a set of physical links. Since the internode relations are realized by these links, failure modes of the links can become a major potential source of errors.

1. *Message loss effects.* The effect of a lost message can be quite harmful if no precautions are taken. For example, let us consider a state transition in a system which depends on a message arrival. If this message is lost and there is no alternative indication on its being sent, the transition can never occur. Another example is a message which transfers control of a resource from one node to another. If this message is lost, the whole system may become blocked.

2. *Message duplication effects.* In order to prevent message-loss catastrophes, a message can be sent on several different routes to increase its robustness with respect to link failures. However, if this message increments a counter upon arrival, the algorithm will not function properly if no precautions are taken to eliminate the duplicates. In order to be able to deal with message duplication, algorithms must be idempotent. In idempotent algorithms, the execution of a sequence of actions result is the same as if it was executed only once.

3. *Message desequencing effects.* The use of multiple routes, such as described in the datagram approach, can cause message desequencing effects. An algorithm's resiliency to such phenomena releases restrictions and increases the network's dynamic control possibilities.

4. *Modification of the message control.* Network dynamic control can shift the network from one administration type to another. For ex-

ample, an urgent mode can change the priorities of pending messages in some buffer, and thus damage the previously predictable behavior of the network.

5. *Bounded transmission time.* Various algorithms assume that the transmission time of each message is bounded. This assumption is extremely important and occurs quite frequently in hard real-time systems. However, the less this assumption is made, the more robust the system becomes.

### 5.3.3 Comparison criteria

We can now summarize the criteria to use for comparing algorithms proposed as solutions for a particular problem.

1. The first parameter is the algorithm symmetry. An asymmetric algorithm is one in which all nodes do not execute the same program text. Text symmetry means that all nodes execute the same program text, while each node may have a distinct process identifier. In a strong text symmetry even the process identifier does not intervene.

   Another criterion for comparing algorithms is the necessity of the availability of global system information to each node in order to execute successfully. Sometimes this information is a restricted communication system topology, for which the algorithm is adequate.

   | Class | Symmetric/Asymmetric | Remarks |
   |-------|----------------------|---------|
   | S1    | +/−                  | symmetry degree |
   | G1    | +/−                  | global info necessity |
   | G2    | +/−                  | topology dependence |

2. Robustness with respect to node failures raises sensitivity metrics and the failure consequences are application dependent. Recalling that we need these parameters for comparison criteria, allows us to define application dependent criteria. The symmetry parameters appear here with a strong emphasis, as both the sensitivity and the consequences differ as the case of a coordinator node failure differs from that of a peer node failure. The metrics must represent the system resiliency: how many nodes can fail and the algorithm still be properly executed.

   | Class | Sensitivity | Effect | Remarks |
   |-------|-------------|--------|---------|
   | N1    |             |        | coordinator node failure |
   | N2    |             |        | peer node failure |

3. Robustness with respect to link failures again raise sensitivity metrics and failure consequences are application dependent. Recalling that we need these parameters for comparison criteria, allows us to define application dependent criteria. The failure modes considered here have been discussed previously.

| Class | Sensitivity | Effect | Remarks |
|-------|-------------|--------|---------|
| L1 | | | message loss |
| L2 | | | message duplication |
| L3 | | | message desequencing |
| L4 | | | modified message control |
| L5 | | | bounded transmit time |

4. Communication traffic parameters represent the load imposed by the specific solution on the global network, since network load affects other tasks in the system. It is important to classify these parameters in both average and peak terms, as well as peak correlation to other activities. An algorithm with a peak load on a lightly loaded network disturbs less than one with a lower peak but with a higher average consumption.

| Class | Metric | Remarks |
|-------|--------|---------|
| T1 | | average load |
| T2 | | peak load |
| T3 | | peak frequency |
| T4 | | peak correlation |

5. Performance criteria assesses how well the system functions. These criteria are generally application dependent, but they all associate the cost functions of response time, throughput, and resource management.

| Class | Metric | Remarks |
|-------|--------|---------|
| P1 | | response time |
| P2 | | resources required |
| P3 | | blocking probability |
| P4 | | resource efficient use |

These criteria can be summarized to evaluate each proposed solution as follows:

| Class | Metrics | Effects | Remarks |
|---|---|---|---|
| S1 | | | symmetry degree |
| G1 | | | global info necessity |
| G2 | | | topology dependence |
| N1 | | | coordinator node failure |
| N2 | | | peer node failure |
| L1 | | | message loss |
| L2 | | | message duplication |
| L3 | | | message desequencing |
| L4 | | | modified message control |
| L5 | | | bounded transmit time |
| T1 | | | average load |
| T2 | | | peak load |
| T3 | | | peak frequency |
| T4 | | | peak correlation |
| P1 | | | response time |
| P2 | | | resources required |
| P3 | | | blocking probability |
| P4 | | | resource efficient use |

### 5.3.4 Network topology dependence

Solution approaches generally depend on the network topology that they are intended for. Although most solutions can be topology independent, their performance and sensitivity metrics demonstrate a drastic change from one topology to another. The following characteristics are often used:

- Ring networks usually employ token based algorithms, where messages marked as tokens are transferred around the ring from one node to another in an agreed upon sequence.
- Tree structured networks usually use diffusion algorithms. In this scheme a parent node passes messages to its children until a message meets a leaf node. The leaf node delays the message and answers back. Each parent collects answers from all its children, and only then answers.
- Fully connected graph networks use timestamps which impose the proper event ordering as required by the problem definition.

Clearly these characteristics are not essential. A token based algorithm does not have to be used in a ring topology although it may be used in other topologies also. As we discuss distributed algorithms in the following chapters, we will visit these topologies again.

# Chapter 6

# Mutual Exclusion

The concurrent execution of programs that may share resources requires that the access to the resources be controlled to avoid undesirable effects. For example, an uncontrolled sharing of a printer by two processes can result in printed output which may be unreadable. The value of a variable shared by two processes and which can be set and accessed by either, can be set incorrectly if the operations from the two processes are interleaved inappropriately. A way to assure a correct operation is to define the region of a process using shared variables or resources as a *critical section* and then use a mechanism to assure a mutually exclusive execution of the critical section. The mechanism used cannot depend on special hardware support, number of processors, their relative speed of execution, etc.

In addition, in order to avoid deadlocks and starvation, we force noncritical sections of our programs not to intervene in the execution of any critical section. Furthermore, the selection of a process that is allowed to enter a critical section is not allowed to be postponed indefinitely.

We start this chapter with a definition of the mutual exclusion property and a description of the critical section problems. We then continue with various types of algorithms that achieve this property.

## 6.1 Problem Definition

A distributed computation can use and modify variables that are shared by its concurrently executing objects. The access to these shared variables must therefore be conditional and dependent on the state of the variable with regard to its access points. This shared access capability must be managed as a critical section, and guarded to enforce the mutual exclusion property, a property in which no two critical sections of the same resource are allowed to overlap in their execution time.

### 6.1.1 The axiomatic approach model

Based on the axiomatic approach introduced in [109], with the guarded variables gathered in the resources ($r$) used in the parallel execution of statements $S_i$, we can formulate the mutual exclusion condition. Let us recall that the invariant $I(r)$ is true when parallel execution begins and remains true whenever execution is observed outside the critical sections of $r$. In addition, recall that a statement $S_{cr}$ in a critical section is protected by a condition $C$ **with** statement of the form

**with** $r$ **when** $C$ **do** $S_{cr}$.

Therefore, the mutual exclusion condition follows.

CONDITION 1. If

1. $S_1$ and $S_2$ are statements in different parallel processes of a program $S$, and
2. $S_1$ and $S_2$ do not belong to any critical section of a resource $r$, and
3. $P_1$ and $P_2$ are the precondition assertions of $S_1$ and $S_2$, respectively, from the proof of $\{P\}S\{Q\}$, and
4. $P$ is true when execution of $S$ begins, and
5. $P_1 \wedge P_2 \wedge I(r) \Longrightarrow$ false

then

$S_1$ and $S_2$ are mutually exclusive.

□

### 6.1.2 The dining philosophers problem

Figure 6.1 depicts Hoare's solution of the classical synchronization problem of the dining philosophers due to E. W. Dijkstra. Five philosophers surround a table. Each pair shares a fork which can be used by one of them at a time for eating. However, eating requires the use of two forks, as they need to get the spaghetti from the bowl in the center of the table. It is evident that a simple lock mechanism is insufficient, because if each of them locks his left fork (and they may do it simultaneously) all of them will starve. Moreover, it is clear that

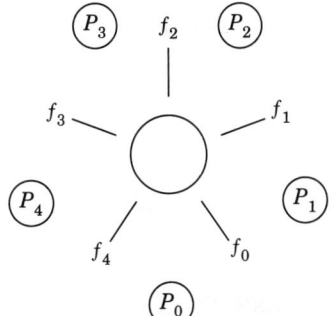

**Figure 6.1** Dining philosophers problem.

**begin** *Dining Philosophers*
    $af := 2$ ; eating := false
        **comment** $af[i]$ shows the available forks for phil$[i]$
        **comment** eating$[i]$ is 1 (true) when phil$[i]$ is eating, 0 (false) otherwise
    **resource** forks (af):
        **cobegin** PHIL$_0$ ∥ ... ∥ PHIL$_4$ **coend**
**end**
PHIL$_i$::
  **while** true **do**
  *get_forks_i*: **with** forks **when** $af[i] = 2$ **do**
    **begin**
      $af[i \ominus 1] := af[i \ominus 1] - 1$ ; $af[i \oplus 1] := af[i \oplus 1] - 1$ ;
      eating$[i] := 1$
    **end**
  *eat_i*: time-bounded eating ;
  *release_forks_i*: **with** forks **when** true **do**
    **begin**
      $af[i \ominus 1] := af[i \ominus 1] + 1$ ; $af[i \oplus 1] := af[i \oplus 1] + 1$ ;
      eating$[i] := 0$
    **end**
  *think_i*: time-bounded thinking
  **endo**

**Program 6.1** Solution to dining philosophers problem.

the forks must be protected as a resource, with a mutually exclusive access and a conditional locking mechanism.

**Resource driven solution.** Program 6.1 [109] brings a resource protecting solution to the dining philosophers problem. It uses two variable types:

- The array $af$: $af[i]$ indicates the amount of available forks for philosopher $[i]$.

- The array eating: eating[i] is 1 when philosopher [i] is eating, 0 otherwise.

All philosophers start not eating, and therefore, each has two forks available. The array of available forks is protected in a resource "forks" to which access is permitted only through a **with-when** statement. Due to the cyclic nature of the solution, we use modulo 5 addition and subtraction ($\oplus$ and $\ominus$, respectively).

**Axiomatic analysis of the solution.** Let us annotate Program 6.1 [109] with precondition and postcondition assertions, to try to show that no two adjacent philosophers are eating at the same time.

The special assertion $I(forks)$, the invariant of this resource, must be true when parallel execution begins and remain true whenever observed outside the critical sections of forks. We therefore choose:

$$I(forks) \equiv \{\forall 0 \leq i \leq 4 : (0 \leq eating[i] \leq 1) \wedge (eating[i] = 1 \Rightarrow$$
$$af[i] = 2) \wedge (af[i] = 2 - eating[i \ominus 1] - eating[i \oplus 1])\}. \quad (6.1)$$

The annotated program follows.

$\{$ true $\}$
**begin** *Dining Philosophers*
$\quad af := 2$ ; eating := false
$\quad\quad \{ I(forks) \wedge (\forall i : eating[i] = 0) \}$
$\quad$ **resource** forks (af):
$\quad\quad$ **cobegin** PHIL$_0$ $\| \ldots \|$ PHIL$_4$ **coend**
$\quad\quad\quad \{ I(forks) \wedge (\forall i : eating[i] = 0) \}$
**end**
PHIL$_i$::
$\quad\quad\quad \{ eating[i] = 0 \}$
**while** true **do**
*get_forks_i*: **with** forks **when** $af[i] = 2$ **do**
$\quad\quad\quad \{ I(forks) \wedge eating[i] = 0 \wedge af[i] = 2 \}$
$\quad$ **begin**
$\quad\quad af[i \ominus 1] := af[i \ominus 1] - 1$ ; $af[i \oplus 1] := af[i \oplus 1] - 1$ ;
$\quad\quad eating[i] := 1$
$\quad$ **end**
$\quad\quad\quad \{ I(forks) \wedge eating[i] = 1 \}$
$\quad\quad\quad \{ eating[i] = 1 \}$
*eat_i*: time-bounded eating ;
*release_forks_i*: **with** forks **when** true **do**
$\quad\quad\quad \{ I(forks) \wedge eating[i] = 1 \}$
$\quad$ **begin**
$\quad\quad af[i \ominus 1] := af[i \ominus 1] + 1$ ; $af[i \oplus 1] := af[i \oplus 1] + 1$ ;
$\quad\quad eating[i] := 0$

               **end**
                        { $I(forks) \land eating[i] = 0$ }
                        { $eating[i] = 0$ }
      *think_i*: time-bounded thinking
               **endo**
                        { $eating[i] = 0$ }

Now let us examine how Condition 1 holds. $I(forks)$ is true when parallel execution begins and remains true whenever observed outside the critical sections of forks. In addition,

$$S_1 \equiv \text{``}eat\_i\text{''}\ ,\ P_1 \equiv \{eating[i] = 1\}, \qquad (6.2)$$

$$S_2 \equiv \text{``}eat\_i \oplus 1\text{''}\ ,\ P_2 \equiv \{eating[i \oplus 1] = 1\}. \qquad (6.3)$$

Hence,

$$P_1 \land P_2 \land I(r) \Longrightarrow false. \qquad (6.4)$$

## 6.2  Token Ring Algorithms

Having defined the mutual exclusion property, we can now become acquainted with algorithms that achieve it. There are various types of algorithms that allow concurrent execution of programs while mutual exclusion is ensured in critical sections. We describe some representative examples, classified according to the principles on which they are based, starting with token ring algorithms.

### 6.2.1  The generic token ring solution

The generic token ring algorithm passes a token for the critical section from one process to another, where the processes are ordered in a unidirectional ring. Each process, $P_i$, executes the following rule:

$P_i$ ::
    **var** waiting: **boolean**
**begin**
    **wait** "token-has-arrived" from $P_{i-1}[n]$ ;
    **if** waiting **then**
        **do** critical section ;
        waiting := **false**
        /* else $P_i$ doesn't need the resource */
    **fi**
    **send** "token-has-arrived" to $P_{i+1}[n]$
**end**

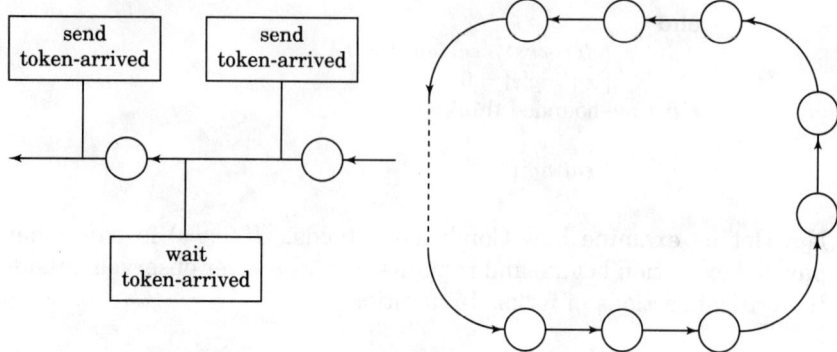

**Figure 6.2** Token-ring mutual-exclusion algorithm structure.

The topology of the processes is shown in Figure 6.2, in which the nodes represent the processes, and the arcs mark the path followed by the token. When a process wants to enter a critical section, it waits until it has received the token. Then it enters the critical section. Upon exit, it forwards the token to the next process.

Note that each node executes the same code and it only has to know the ID of the successor node in the ring topology. Clearly, the communications topology does not have to be a ring for this approach to be used, since a logical ring topology can be defined.

This approach is critically dependent on the availability of one unique token in the system. A loss of a token can result in an indefinite wait for all the nodes. A way out is to incorporate a time-out mechanism which activates the algorithm to generate a new token (see Chapter 7). Note that the failure of a node may result in an open ring, which can also result in an idefinite wait. Therefore, when a time out is used, and before a new token is generated, the connectivity of the ring has to be ascertained. Different approach is required for a broken ring.

### 6.2.2 Two tokens solution

A solution to the mutual exclusion problem which is based on a bidirectional ring has been introduced in [16]. In this algorithm a token "ping" is sent to the right and a token "pong" is sent to the left at the same time. Both tokens continue to circulate. The main advantage of having two tokens is that if one of the tokens is lost, the loss is detected and corrected quickly. A count is sent with the token also. Each process maintains a local counter to indicate the value of the count it saw last time it processed the token ping. Each time the tokens meet (i.e., cross each other's way) at the same process, the counts for both are adjusted by incrementing the count for ping and decrementing the count for pong which maintains a count whose value is the negative of the value of the count for ping. If one of the tokens was lost, the two consecutive visits of the token will have the same count. This is used to detect the loss of the token and a new one (ping or pong) is generated by the process detecting the loss.

A program describing this algorithm is presented in Program 6.2. The implementation assumes that the tokens do not overtake each other on a link or that the transit time of a token on a link is zero. Some additional precautions will be required if a finite communication time has to be taken into account.

Note that for this implementation the counters can have a finite range. While there are no losses the counters will maintain the same value. When a loss occurs the counters can differ by one. Therefore, a rather small range of values for the counts is sufficient. Further, while this algorithm can detect the loss of one token, it does not handle the case when both the tokens are lost.

### 6.2.3 Logical tokens

A disadvantage of having a physical token is that it can be lost. The detection of such a loss and the recovery from it can take time and resources. An alternate approach is to use a logical token which is a condition that circulates and can be detected by any process by keeping a local variable and comparing it with the value of the similar variable of its neighbor.

A program for this algorithm is described in Program 6.3 The conditions which are detected and used as logical tokens are shown in Figure 6.3.

Note that the node $i - 1$ can only send a message when it changes the value of its $s(i-1)$ for better traffic efficiency. This algorithm is not symmetric in that it checks the value of its left neighbor only. The algorithm does not address the problem of site failure, which leads to a low robustness for this algorithm. Clearly, a node can ask the neighbor again or use a timeout procedure to detect site failures. Appropriate recovery mechanisms in which a node is assigned a new neighbor has to be defined.

An alternate way to having all sites in the same state is to use a unit-step initialization. This approach is described in Program 6.4.

The token transfer is described in Figure 6.4, for the case of node-0 critical section.

This algorithm can be extended to solve the problem of mutual exclusion of $m$ resources for $n$ processes. We circulate a gap of size $m$ instead of size 1. Each process "clips" its needs and changes its variables accordingly. The gap the process sees is the number of resources it can use. We leave this solution as an exercise to the reader.

**var**
    ping, pong: token ;
    numping, numpong: integer ;
**local**
    $v(i)$: integer ; /* the last value of numping seen at $P_i$ */
    *pinghere*, *ponghere*: boolean;
upon REC(ping,numping)::
**begin**
    *pinghere* := *true* ;
    **if** *ponghere* **then**
        numping := numping + 1 ;
        numpong := numpong - 1 ;
        v(i) := numping
    **elseif** $v(i)$=numping **then** /* pong is lost: generate a new one */
        numping := $v(i)+1$ ; numpong := -numping ;
        send(pong,numpong)
    **else** $v(i)$ := numping
    **fi**
    **fi**
    **do** critical section (if required) ;
    *pinghere* := *false* ;
    send(ping,numping)
**end**
upon REC(pong,numpong)::
**begin**
    *ponghere* := *true* ;
    **if** *pinghere* **then**
    **while** *pinghere* **then** wait **endwhile**
    **elseif** $v(i)$=-numpong **then** /* ping is lost: generate a new one */
        numping := $v(i)+1$ ; numpong := -numping ;
        send(ping,numping)
        **fi**
    **fi**
    *ponghere* := *false* ;
    send(pong,numpong)
**end**
    init::
    **begin**
        numping := 1 ;
        numpong := -1 ;
        ping and pong at the same site $i$ ;
        $v(i) := 1$;
        $(\forall j \neq i): v(j) := 0$
    **end**

**Program 6.2** Two-token mutual-exclusion algorithm.

**var**
    $s[i]$: binary integer ;
$P_0$::
**begin**
    **wait** $s(n-1) = s(0)$ ;
    **do** critical section (if required) ;
    $s(0) := (s(n-1) + 1) \bmod 2$
**end**
$P_i$:: $i = 1 \ldots -1$
**begin**
    **wait** $s(i) \neq s(i-1)$ ;
    **do** critical section (if required) ;
    $s(i) := s(i-1)$
**end**
init::
    $(\forall i : 0 \ldots n-1)\, s(i) := 1$ ;

**Program 6.3** Logical-token mutual-exclusion algorithm.

```
            ↓ init
1-  0   1   2   ⋯   n-1
0-

            ↓ first used
1-  0   1   2   ⋯   n-1
0-  —

            ⇵
1-  0   1   2   ⋯   n-1
0-  —   —   —

            ⇵ all used
1-  0   1   2   ⋯   n-1
0-  —   —   —   —   —   —

            ↓ first used again
1-  0   1   2   ⋯   n-1
0-      —   —   —   —

            ⇵
1-  0   1   2   ⋯   n-1
0-          —   —   —
```

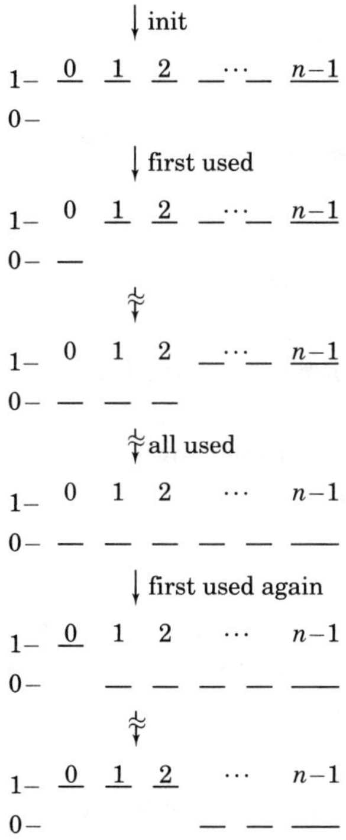

**Figure 6.3** Logical-token mutual-exclusion algorithm.

**var**
  $s[i]$: binary integer ;
$P_i$:: $i = 0 \ldots n-1$
**begin**
  **wait** $s(ni) = s(i-1)$ ;
  **do** critical section (if required) ;
  $s(i) := (s(i-1) + 1) \bmod n$
**end**
init::
**begin**
  $(\forall i : 1 \ldots n-1)\ s(i) := i$ ;
  $s(0) := 1$
**end**

**Program 6.4** Another logical-token mutual-exclusion algorithm.

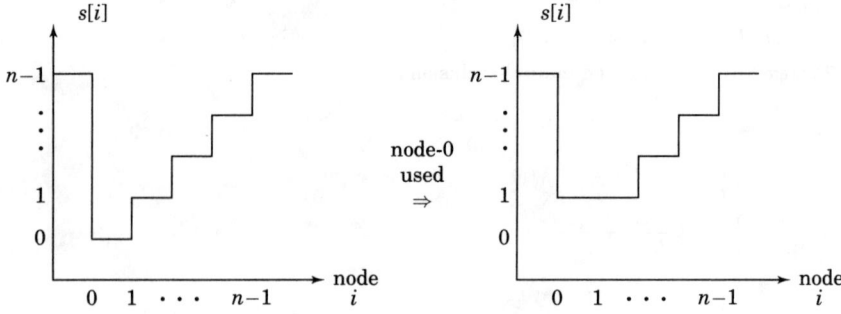

**Figure 6.4** Another logical-token mutual-exclusion algorithm.

## 6.3 Time Ordering Algorithm

Let us review some properties we impose on our system model:

1. The system is constructed from a set of communicating processes, organized in a fully connected network.

2. Each process has access to a local clock $C_i$.

3. The processes communicate with each other only via messages, where each message has the form (msg,$C_i$,$i$). The state defined by REC(msg,$C_i$,$i$) on $j$, is the state where process $j$ detects reception of a message of type "msg" from process $i$.

4. Upon each REC(msg,$C_i$,$i$) on $j$, it advances its clock as follows

$$C_j := \max(C_i, C_j) + 1. \tag{6.5}$$

5. There are certain local events for which $i$ advances its clock

$$C_i := C_i + 1. \tag{6.6}$$

6. Total ordering is obtained by

$$(C_i, i) \succ (C_j, j) \; iff \; C_i > C_j \; \vee \; C_i = C_j \wedge i > j \tag{6.7}$$

### 6.3.1 Lamport's timestamp ordering

A solution based on a distributed FIFO organized according to a total ordering of timestamps has been introduced in [79]. The processes uses three message types to maintain the required total ordering.

1. A process $i$ which needs a resource that it shares with other processes, broadcasts ($REQ,C_i,i$) to these processes.
2. A process $j$ which REC($REQ,C_i,i$) answers $i$ with ($ACK,C_j,j$).
3. A process $i$ which completes the use of the resource, broadcasts to the processes with which it shares this resource ($REL,C_i,i$).

When we calculate the overhead associated with this algorithm, we find it requires $3(n-1)$ messages for each critical section execution, where $n$ is the number of processes that share the resource. The first and the last broadcasts made by $i$ are sent to $n-1$ processes, and each of these $n-1$ processes answers the request with an acknowledge.

The algorithm is described in Program 6.5. Each process maintains an array $q_i$ of size $n$ in which it stores the messages received from the other processes. Along with the execution of the clock update rules, a process $i$ updates its $q_i$ array according to the following rules.

1. If $i$ receives ($REQ,\ldots,j$) or ($REL,\ldots,j$), it stores this message in $q_i[j]$.
2. If $i$ receives ($ACK,\ldots,j$) it updates $q_i[j]$ only if the current value type of $q_i[j] \neq REQ$.

We let $i$ update $q_i[i]$ respectively. Therefore, $i$ can enter a critical section if the message stored in $q_i[i]$ has the oldest timestamp in $q_i$.

This algorithm is an important milestone in the development of mutual exclusion solutions. However, we note some weak points in this algorithm. It is definitely wasteful to send an ($ACK,\ldots,i$) message if $i$ has a pending request which is not yet fulfilled. Another weak point is its robustness: the system can be blocked by either a site (and thus a process) failure or a loss of a message. On the other hand, we note that the algorithm is insensitive to message duplications, recalling that messages do not overtake each other in the links.

```
var
    C_i: clock ;
    msg: (REQ, ACK, REL) ;
    q_i: array [0...n - 1] of (msg,C_j,j) ;
init::
    ∀i, ∀j: q_i[j] := (REL,0,j) ;
when i invokes a critical section::
begin
    ∀j: send (REQ,C_i,i) ;
    q_i[i] := (REQ,C_i,i) ;
    C_i := C_i + 1 ;
    (∀j ≠ i) wait timestamp(q_i[j])>timestamp(q_i[i]) ;
    do critical section ;
    ∀j: send (REL,C_i,i) ;
    q_i[i] := (REL,C_i,i) ;
    C_i := C_i + 1 ;
end
        upon REC(msg,C_j,j) by i ::
        begin
            C_i := max(C_j, C_i) + 1 ;
            if msg=REQ then
                q_i[j] := (REQ,C_j,j) ;
                send (ACK,C_i,i) to j ;
            fi
            if msg=REL then
                q_i[j] := (REL,C_j,j) ;
            fi
            if msg=ACK then
                if type(q_i[j])≠ REQ then
                    q_i[j] := (ACK,C_j,j) ;
                fi
            fi
        end
```

**Program 6.5** Lamport's time-stamp mutual-exclusion algorithm.

An apparent recovery scheme from a process failure can be attained by distributing the knowledge of its failure when it is detected. Say that some mechanism has detected the failure of process $i$. Upon such detection it broadcasts to all the rest of the processes a message of a new type, $(FAIL,...,i)$. Each of the receiving processes, say $j$, reacts to that message with a nullification of $q_j[i]$.

### 6.3.2 Ricart and Agrawala

An improvement of the above algorithm has been introduced in [121]. The wasteful use of the release and acknowledge messages is reduced to a single acknowledgment type. We make use of the following variables:

**var**
    $C_i$: current clock ;
    $oC_i$: old clock ;
    msg: $(REQ, ACK)$ ;
    nb_answer: integer $(1 \ldots n-1)$ ;
    *using, waiting*: boolean ;
    $deferred_i$: **array** $[0 \ldots n-1]$ of boolean ;

init::
    $\forall i, \forall j: deferred_i[j] :=$ False ;

The algorithm is described in Program 6.6. The algorithm performance is enhanced; the overhead associated with this algorithm is $2(n-1)$ messages for each critical section execution. The first broadcast made by $i$ is sent to $n-1$ processes, and each of these $n-1$ processes answers the request with an acknowledgment. A process of higher priority cannot send an acknowledgment until it has finished using the critical section.

The fairness is maintained through timestamping for total order (FIFO). However, we note some weak points in this algorithm. The system can be blocked by either a site (and thus a process) failure or a loss of a message. The robustness can also be improved by replacing the counters which are very sensitive to message duplications with boolean arrays that are nullified properly. We leave such an improvement as an exercise[1] to the reader.

### 6.3.3 Carvalho and Roucairol

Communication traffic can be further reduced by using global knowledge which exists locally [3]. For example, if a process completes its critical section without receiving any request for the resource, there is no need to release it and request it all over again. In fact, for the extreme case of acquiring a resource nobody else needs, there is no overhead at all. On the other hand, in the case all of the processes requested a resource, we come back to Ricart and Agrawala's solution.

The above principle is described in Program 6.7 by using an array of *authorization* given to each process by all the other processes. However, using this method, the FIFO fairness is lost. Furthermore, a "larceny" can start each time a process has only one authorization ab-

---

[1]Note that a simple array is insufficient. One needs an array of timestamps dealt only if the $ACK$'s timestamp is older than that of the $REQ$'s. In addition, one should consider $REQ$ duplication as well.

**when** $i$ invokes a critical section::
  **begin**
    $oC_i := C_i := C_i + 1$ ;
    nb_answer := 0 ; $waiting$ := True ;
    $\forall j$: **send** $(REQ,C_i,i)$ ;
    **wait** nb_answer = $n - 1$ ;
    $using$ := True ; $waiting$ := False ;
    **do** critical section ;
    $using$ := False ;
    **for** $(j \neq i) := 0$ **to** $n - 1$ **do**
      **if** $deferred_i[j]$ **then**
        $deferred_i[j]$ := False ;
        **send** $(ACK,C_i,i)$ to $j$ ;
    **od fi**
  **end**
**upon** REC($REQ,C_j,j$) by $i$ ::
  **begin**
    $C_i := \max(C_j, C_i) + 1$ ;
    **if** $using$ **then**
      $deferred_i[j]$ := True ;
    **fi**
    **if** $waiting \wedge C_j \succ oC_i$ **then**
      $deferred_i[j]$ := True ;
    **fi**
    **if** $waiting \wedge \neg(C_j \succ oC_i)$
        $\vee \neg(waiting \vee using)$ **then**
      **send** $(ACK,C_i,i)$ to $j$ ;
    **fi**
  **end**
**upon** REC($ACK,C_j,j$) by $i$ ::
  **begin**
    $C_i := \max(C_j, C_i) + 1$ ;
    nb_answer := nb_answer + 1 ;
  **end**

**Program 6.6** Ricart and Agrawala's mutual-exclusion algorithm.

sent. While waiting for that missing authorization, the process can lose the authorizations it has already acquired.

This algorithm uses the following variables:

        **var**
          $C_i$: current clock ;
          $oC_i$: old clock ;
          msg: $(REQ, ACK)$ ;
          $auth_i$: **array** $[0 \ldots n - 1]$ of boolean ;
          $using, waiting$: boolean ;
          $deferred_i$: **array** $[0 \ldots n - 1]$ of boolean ;

**when** $i$ invokes a critical section::
  **begin**
       $oC_i := C_i := C_i + 1$ ; $waiting :=$ True ;
       **for** $(j \neq i) := 0$ **to** $n - 1$ **do**
           **if** $\neg auth_i[j]$ **then**
               **send** $(REQ, C_i, i)$ ;
       **od fi**
       **wait** $\forall j$: $auth_i[j] =$ True ;
       $using :=$ True ; $waiting :=$ False ;
       **do** critical section ;
       $using :=$ False ;
       **for** $(j \neq i) := 0$ **to** $n - 1$ **do**
           **if** $deferred_i[j]$ **then**
               $deferred_i[j] :=$ False ; $auth_i[j] :=$ False ;
               **send** $(ACK, C_i, i)$ to $j$ ;
       **od fi**
  **end**
**upon** REC($ACK, C_j, j$) by $i$ ::
  **begin**
       $C_i := \max(C_j, C_i) + 1$ ;
       $auth_i[j] :=$ True ;
  **end**
**upon** REC($REQ, C_j, j$) by $i$ ::
  **begin**
       $C_i := \max(C_j, C_i) + 1$ ;
       **if** $using$ **then**
           $deferred_i[j] :=$ True ;
       **fi**
       **if** $waiting \wedge C_j \succ oC_i$ **then**
           $deferred_i[j] :=$ True ;
       **fi**
       **if** $waiting \wedge \neg(C_j \succ oC_i)$ **then**
           **if** $auth_i[j]$ **then**
               $auth_i[j] :=$ False ;
               **send** $(REQ, C_i, i)$ to $j$ ;
       **fi fi**
       **if** $\neg(waiting \vee using)$ **then**
           $auth_i[j] :=$ False ;
           **send** $(ACK, C_i, i)$ to $j$ ;
       **fi**
  **end**

**Program 6.7**   Carvalho and Roucairol's algorithm.

init::
$\forall i, \forall j: deferred_i[j] :=$ False ;
$\forall i, \forall j: auth_i[j \neq i] :=$ False ; $auth_i[i] :=$ True ;

Let us now consider the robustness of the algorithm. There is no support in this algorithm for lost messages. In case such losses occur, it can result in a blocked waiting process. Errors in messages or addresses can have the same result. On the other hand, using an array instead of Ricart and Agrawala's counter gives better results for $ACK$ message duplications. The reader (who has already solved the exercise suggested in the Ricart Agrawala algorithm) can clearly see that this algorithm has the same problem with $REQ$ message duplication as the previous one.

### 6.3.4 Suzuki and Kasami

A further reduction in communication traffic is achieved by having the requests sent only to neighbors, and using a token instead of acknowledgements [21]. The token travels with a global variable, in which the time-count of the last "visit" to each process is stored. Again, if there has been no request for the resource while it is used in a process, the token stays with that process.

This algorithm is described in Program 6.8. It requires $n-1$ request messages, and a single $O(n)$ step of the token. We note that the algorithm does not support token loss. The double token solution does not fit this algorithm since token traffic is not limited to a ring. However, robustness with respect to token duplication can be obtained at the wait statement by adding a comparison to the last token received. If such protection is not added, the consequences of token duplication are fatal to mutual exclusion which is violated.

Duplication of $REQ$ messages are ignored due to the maximum function with which log_site is updated. The algorithm is insensitive to receiving the $REQ$ messages out of sequence. It is also somewhat sensitive to a loss of $REQ$ messages. Let us consider the case in which only a single $REQ$ message, directed to $j$, was lost. A process $i \neq j$ can still get the token when coming to $k \neq j$. Nevertheless, when $j = i-1$ and the whole set of processes is extremely active, $i$ will not be able to receive priority.

Excluding losses, starvation is prevented and each process will eventually receive the resource. Nevertheless, this fairness is much weaker than the FIFO types we have seen so far. Furthermore, the unfairness of the scheme raises difficulties in implementing any timeout mechanism.

```
        global var
            log_token: array [0...n - 1] of timestamp ;
        local vars
            C_i: current clock ;
            present, waiting: boolean ;
            log_site: array [0...n - 1] of timestamp ;
            msg: (REQ, token) ;
when i invokes a critical section::
    begin
        C_i := C_i + 1 ;
        ∀j: send (REQ,C_i,i) ;
        wait REC(token,log_token) ;
        using := True ; present := True ;
        do critical section ;
        log_token[i] := C_i ;
        using := False ;
        cycle k := i + 1 to i - 1 do
            if (log_token[j] < log_site[j]) ∧ present then
                present := False ;
                send (token,log_token) to k ;
            fi
        od
    end
upon REC(REQ,C_j,j) by i ::
    begin
        log_site[j] := max(log_site[j], C_j) ;
        if present ∧ ¬using then
            present := False ;
            send (token,log_token) to j ;
        fi
    end
```
**Program 6.8** Suzuki and Kasami's mutual-exclusion algorithm.

### 6.3.5 Mamoru Maekawa

A reduction of communication traffic based on a group representatives that respond on behalf of members of their group has been suggested by Mamoru Maekawa [14]. The communication overhead associated with this algorithm is $3\sqrt{n} \div 5\sqrt{n}$, where $n$ is the number of processes that share the resource.

This approach is based on the fact that it is not necessary to transmit requests to all the nodes, and to receive grants from all of them. We can divide the system into subsets which intersect each other, and approach only one representative of each group within our own subset. This principle is described in Figure 6.5. We notice that node $i$ requests the resource from its neighbors in $S_i$ $a$, $b$, and $c$. However,

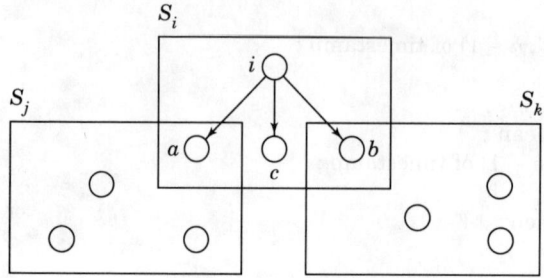

**Figure 6.5** Intersecting subsets of processes.

since $a$ and $b$ are intersecting members with $S_j$ and $S_k$, they respond with information on all the members of $S_j$ and $S_k$.

The algorithm which implements this approach is described in Program 6.9. It requires four properties to hold.

1. The *non-null intersection property*. There must be an intersecting member with the other subsets.

$$\forall i, j : 0 \leq i, j \leq n-1, \ S_i \cap S_j \neq \emptyset \qquad (6.8)$$

2. The *automatic self permission*. No request has to be made to itself. Hence,

$$\forall i : 0 \leq i \leq n-1, \ i \in S_i \qquad (6.9)$$

3. The *balanced overhead*. Each node sends and receives the same number of messages to acquire a resource. Hence, the sizes of all subsets are equal, say $K$.

$$\forall i : 0 \leq i \leq n-1, \ |S_i| = K \qquad (6.10)$$

4. The *balanced arbitration*. Each node serves as an arbitrator for the same number of subsets. Hence, every node is contained in the same number of subsets, say $D$. In other words, each node serves in $D-1$ subsets as an arbitrator.

$$\exists (D-1) j's \ j \neq i, \ 0 \leq i \leq n-1 \ : \ i \in S_j \qquad (6.11)$$

The reader is encouraged to verify that for a system in which the four properties hold,

$$n = (D-1)K + 1 \qquad (6.12)$$

and $K = D$, therefore, with a fractional error, $K = \sqrt{n}$. The way we cast the $S_i$'s is not unique, and some methods are suggested in

when $i$ invokes a critical section::
   **begin**
      $C_i := C_i + 1$ ; $done_i :=$ False ; $waiting_i :=$ True ;
      $\forall j \in S_i \; (j \neq i)$ **do**
         **send** $(REQ,C_i,i)$ ;
         $locked_i[j] :=$ False ;
      **od**
   **end**
upon REC($INQ,j$) by $i$ ::
   **begin**
      **if** $\neg done_i \wedge \neg waiting_i$ **then**
         **send** $(CLR,i)$ to $j$ ;
         $locked_i[j] :=$ False ;
      **else if** $\neg done_i \wedge waiting_i$ **then**
         $deferred_i[j] :=$ True ;
      **else**   /\* **if** $done_i$ **then** \*/
         **send** $(REL,i)$ to $j$ ;
      **fi**
   **end**
upon REC($CLR,j$) by $i$ ::
   **begin**
      $temp_i := top_i$ ;
      $top_i :=$ **remove** $(C_k, k)_{min}$ **from** $queue_i$ ;
      **insert** $temp_i$ **to** $queue_i$ ;
      **send** $(ACK,i)$ to $top_i$ ;
   **end**
upon REC($REL,j$) by $i$ ::
   **begin**
      **if** $queue_i \neq \emptyset$ **then**
         $top_i :=$ **remove** $(C_k, k)_{min}$ **from** $queue_i$ ;
         **send** $(ACK,i)$ to $top_i$ ;
      **else**
         $top_i :=$ nil ;
      **fi**
   **end**

**Program 6.9**   Mamoru Maekawa's mutual-exclusion algorithm.


[14]. When $n$ cannot be expressed as $K(K-1)+1$, the algorithm performance decreases a little, because the system is unbalanced and some nodes have to send extra messages.

The algorithm uses five types of messages: the $REQ$ message which conveys the timestamped request, the $INQ$ message which inquires a locking process, the $CLR$ message which relinquishes a locked commitment, the $ACK$ message which grants priority to the requesting process, and the $REL$ message which releases a request.

**upon** REC($REQ,C_j,j$) **by** $i$ ::
**begin**
    $C_i := \max(C_j, C_i) + 1$ ;
    **if** $top_i$ = nil **then**
        **send** ($ACK,i$) to $j$ ;
        $top_i := (C_j, j)$ ;
    **else**
        **insert** $(C_j, j)$ **to** $queue_i$ ;
        **if** $\exists (C_k, k) \in (queue_i \cup top_i) : (C_j, j) \succ (C_k, k)$ **then**
            **send** ($FAIL,i$) to $j$ ;
        **else**
            **send** ($INQ,i$) to $top_i$ ;
    **fi fi**
**end**
**upon** REC($ACK,j$) **by** $i$ ::
**begin**
    $locked_i[j] := $ True ;
    **if** $\forall j \in S_i : locked_i[j] = $ True **then**
        **do** critical section ;
        $done_i := $ True ;
        $\forall j \in S_i \ (j \neq i)$ **do**
            $locked_i[j] := $ False ;
            **send** ($REL,i$) to $j$ ;
            **if** $deferred_i[j]$ **then**
                $deferred_i[j] := $ False ;
    **fi od fi**
**end**
**upon** REC($FAIL,j$) **by** $i$ ::
**begin**
    $waiting_i := $ False ;
    **for** $\forall j \in S_i \ (j \neq i)$ **do**
        **if** $deferred_i[j]$ **then**
            $deferred_i[j] := $ False ;
            **send** ($CLR,i$) to $j$ ;
    **od fi**
**end**

**Program 6.9** Mamoru Maekawa's algorithm *(continued)*.

The algorithm uses the following variables.

    **var**
        $C_i$: current clock ;
        $top_i, temp_i$: (clock, integer) ;
        $queue_i$: **list** of (clock, integer) ;
        msg: ($REQ, INQ, CLR, ACK, FAIL, REL$) ;
        $waiting_i, done_i$: boolean ;
        $deferred_i$: **array** $[0 \ldots K-1]$ of boolean ;

$locked_i$: **array** $[0 \ldots K-1]$ of boolean ;

init::
$\forall i, \forall j: locked_i[j] := deferred_i[j] :=$ False ;
$\forall i: done_i :=$ True ; $queue_i := emptyset$ ; $top_i :=$ nil ;

Considering communication traffic in a lightly loaded system shows that we need $3\sqrt{n}$ messages for obtaining mutual exclusion:

- $K - 1$ $REQ$ messages
- $K - 1$ $ACK$ messages
- $K - 1$ $REL$ messages

In a heavily loaded system we need $5\sqrt{n}$ messages to obtain mutual exclusion:

- $K - 1$ $REQ$ messages
- $K - 1$ $INQ$ messages
- $K - 1$ $CLR$ messages
- $K - 1$ $ACK$ messages
- $K - 1$ $REL$ messages

## 6.4 Path Reversal Based

One of the most efficient algorithms for mutual exclusion for distributed systems was proposed by Trehel and Naimi [106]. This algorithm uses path reversal techniques for a rooted tree to achieve mutual exclusion while retaining the average complexity $O(Log\ N)$. Our description of the algorithm follows the presentation in [58] which also presents a proof of correctness of the algorithm.

This algorithm uses two types of messages. (REQUEST, $n$) indicates that node $n$ wants to access the shared resource, and (PERMIT) represents the permission to access the resource. Node $n$ can be in *thinking, waiting,* or *using* the resource. In addition, node $n$ has the boolean variable $permit_n$, and pointers $last_n$ and $next_n$ that take values from $\{nil, 1, .., N\}$. $permit_n$ is true if and only if $n$ is allowed to access the resource. $last_n$ points to the last node that $n$ knows to have issued a request. $next_n$ points to the node that will use the resource next after $n$ completes using it, and is *nil* whenever $n$ is thinking. Both $last_n$ and $next_n$ are *nil* if $n$ knows of no node that issued a request after $n$'s last request.

In order to start the algorithm properly, there should be no messages in transit, all nodes must be thinking, all $next$ pointers should be *nil*, one node should have the permit, and the *last* pointers should form a tree over the nodes {1,...N}, rooted at the node with the permit.

When a thinking node $w$ wants to use the resource, it originates a request which is satisfied when $w$ completes using the resource. The request will either visit $w$ alone (if $w$ already has the permit), or it is sent to $last_w$ and travels via zero or more nodes until it comes to a node $v$ with $last_v = nil$. Each intermediate node forwards the request to the node pointed to by its *last* pointer and sets its *last* pointer to point to $w$. If the last node v is thinking, then v has the permit and sends it to w. Otherwise v sets both $next_v$ and $last_v$ to point to $w$. In either case, the request originated by $w$ is no longer in transit. In the latter case, when $v$ next completes using the resource, it sends the permit to the node pointed to by $next_v$ and resets it to *nil*.

One distinguishing feature of this mutual exclusion algorithm is that unlike other algorithms, a waiting node sends its request to a node whose identity varies with time.

Each node $n$ executes the following:

**begin**
    when *thinking* and "want to use the resource" do
        if $permit_n$ then become *using*
            else [ become *waiting*; Send (REQUEST,$n$) to $last_n$;
            $last_n := nil$]

    upon reception of (PERMIT) do [ become *using*; $permit_n := true$]

    when *using* and "want to stop using" do
        become *thinking*;
        if $next_n \neq nil$ then [Send(PERMIT) to $next_n$; $next_n := nil$;
        $permit_n := false$]

    upon reception of (REQUEST, $v$) do
        if $last_n \neq nil$ then [Send(REQUEST, $v$) to $last_n$; $last_n := v$];
        if $last_n = nil$ and not *thinking* then [$next_n := v; last_n := v$];
        if $last_n = nil$ and *thinking* then [Send(PERMIT) to $v$;
        $permit_n := false; last_n := v$]
**end**

This algorithm maintains a rooted tree of all nodes at all times with the root being the node which made the request last. When the messages are in transit the structure cannot look exactly like a rooted tree but it has been shown that the consistency of the view is still maintained. This is the most efficient distributed mutual exclusion algorithm that has been reported in the literature to date.

## 6.5 Concluding Remarks

Whenever a system has concurrent operations going on, a synchronization among them is often required. One of the fundamental synchronization problems is that of mutual exclusion in which when one critical activity is going on, no other critical activity can go on concurrently. A number of solutions to the problem of mutual exclusion have been developed for implementation in a distributed environment. In this chapter we presented the primary techniques to solve this problem. In order to select a technique which can be used, the designer has to consider the system resource environment and its constraints along with the robustness requirements. For example, if a distributed system is using a token ring for communication, ring-based algorithms are the most appropriate. Note that the algorithms presented in this chapter make some assumptions. For example, some algorithms assume that the communication of a message from one node to another node takes no time. In practice, such communication does take time. Therefore, the algorithms have to be extended to take into account such delays. As the extensions are straightforward they have been left as an excercise for the reader.

## 6.5 Concluding Remark

**Chapter**

# 7

# Election Algorithms

In many algorithms one of the nodes acts as a special node. In such cases this node can become the weakest point of the system. It can become a bottleneck of the system performance, and furthermore, it can paralyze the whole system when it fails. Robustness considerations call for the ability of the system to renominate such a node, in case there is a failure of the current distinguished node. Such a nomination procedure uses an election algorithm.

## 7.1 Problem Definition

The election algorithm requires that if multiple elections are started concurrently, a single node is elected. This calls for a property which is uniquely associated with each of the candidates which participate in the election. We can use process identifiers as such a property[1]. This can be implemented in a selective extinction technique, where not all messages are forwarded. This emphasizes another property required for carrying out an election, which is the connectivity of the candidates' forum. The candidates need some connectivity through which they can deliver their selection or block undesirable candidates.

We can summarize the properties we require in an election algorithm as follows:

- The algorithm must ensure that one and only one node is elected.

---

[1] One can also use the priority related to the processes as long as it is uniquely associated with each candidate.

- No node knows how many nodes participate in the algorithm. Specifically, a node knows only its own identifier.
- The elected node must be able to renew the control which was lost before the election.
- Election algorithms that are initiated simultaneously by some nodes, must result in the same conclusion as a single election initiated by a single node.

## 7.2 Election in Ring Architecture

### 7.2.1 Unidirectional ring

An election algorithm based on a unidirectional ring structure has been introduced in [4]. A node which "wants" to initiate an election sends a $(REQ,i)$ message on the unidirectional ring. If this message returns, passing through all the nodes on the ring, the initiator sends an $(ELECTED,i)$ message.

In this way we can employ a selective extinction technique, ceasing the transmission of a request that cannot become elected, as demonstrated in Figure 7.1. A node forwards only election requests with identifiers which are evaluated as higher than its own. If the identifier received in the message has a lower value than that of the receiver, it forwards a message with its own identifier[2].

Program 7.1 describes this algorithm. As mentioned above, it uses two message types:

1. $(REQ,i)$: which conveys the election request.
2. $(ELECTED,i)$: which declares the new nominee.

We note that the two cycles performed by the winner are required for cases of multiple initiations as well as for selective extinction.

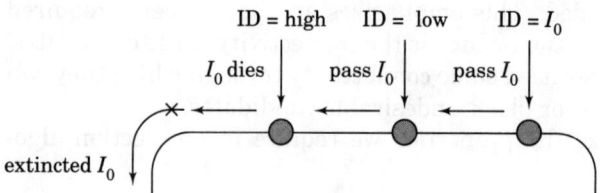

**Figure 7.1** Extinction of messages in unidirectional ring.

---

[2]The algorithm can clearly work with an adverse logic, passing lower identifiers and blocking higher ones.

**var**
    $me_i$: integer /* my identifier */
    $in\_game_i$: boolean /* initialized to False */
    $winner_i$: integer
when $i$ decides to invoke an election::
    **begin**
        $in\_game_i$ := True ;
        **send** $(REQ,i)$ ;
    **end**
upon REC($REQ,j$) by $i$ ::
    **begin**
        **if** $j > me$ **then**
            $in\_game_i$ := True ;
            **send** $(REQ,j)$ ;
        **fi**
        **if** $(j < me) \land \neg(in\_game)$ **then**
            $in\_game_i$ := True ;
            **send** $(REQ,me)$ ;
        **fi**
        **if** $j = me$ **then**
            **send** $(ELECTED,me)$ ;
        **fi**
    **end**
upon REC($ELECTED,j$) by $i$ ::
    **begin**
        $winner_i := j$ ;
        $in\_game_i$ := False ;
        **if** $j \neq me$ **then**
            **send** $(ELECTED,j)$ ;
        **fi**
    **end**

**Program 7.1** Unidirectional ring election algorithm.

The extinction is implicit in the rule of receiving $(REQ,\ldots)$. In case the rule finds $(j < me) \land (in\_game)$, it terminates the message cycle.

Let us analyze the message traffic of this algorithm. In case all nodes start simultaneously, we have: $n-1$ messages of the unfortunate initiators which will not be *winner*, $n$ winner's $(REQ,\ldots)$ messages, and $n$ winner's $(ELECTED,\ldots)$ messages. In case only the winner starts, we have $2n$ messages. The maximal number of messages results in the case of

$$\sum i = \frac{n(n+1)}{2} \qquad (7.1)$$

which is an $O(n^2)$ bound. Timewise, the first cycle can occur in parallel, resulting in an $O(n)$ maximal delay, and the second occurs serially in an $O(n)$ delay. Hence, we have an $O(n)$ delay.

Let us now consider the algorithm's robustness. The algorithm is indifferent to message duplication: double $(REQ,...)$ are ignored due to the variable $in\_game$, and double $(ELECTED,...)$ have no effect. The algorithm is also indifferent to message desequencing since it has no effect on the results. On the other hand, message loss is fatal to the algorithm. So is the case of a node failure, which is fatal to any unidirectional ring activity.

| Class | Metrics | Effects | Remarks |
|-------|---------|---------|---------|
| S1    |         | +       | symmetry degree |
| G2    |         | ring    | topology dependence |
| N2    |         | −       | peer node failure |
| L1    |         | −       | message loss |
| L2    |         | +       | message duplication |
| L3    |         | +       | message desequencing |
| L4    |         | +       | modified message control |
| L5    |         | +       | bounded transmit time |
| T1    | $O(n^2)$ |        | average load |
| T2    | $O(n^2)$ |        | peak load |
| P1    | $O(n)$  |         | response time |

### 7.2.2 Bidirectional ring

Performance can be enhanced if we carry out local elections, and then the local winner extends the electing set. This "oil-slick" behavior (see Figure 7.2) results in an $O(n \log n)$ performance. In the above principle the two intersecting sets, which concluded their local elections form together a larger set, out of which one of the winners can be chosen. We can continue this broadening process until we cover the entire population.

In [10] we find a method that combines the above $O(n \log n)$ performance with a bidirectional communication network. Program 7.2 delineates this solution.

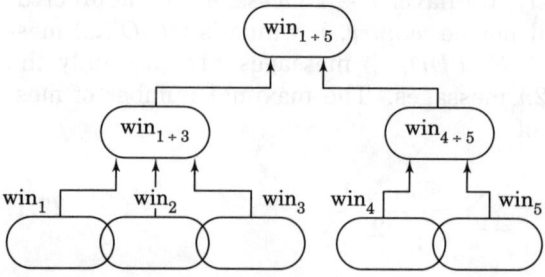

**Figure 7.2** Hierarchical elections: "oil-slick" behavior

**when** $me$ decides to invoke an election::
    **begin**
                     /* broadcast to both sides */
                     /* and wait for answers */
       state := candidate ; $max\_lg$ := 1 ;
       **while** state = candidate **do**
           $nb\_ans$ := 0 ; $ans\_ok$ := True ;
           **twofold send** $(CANDIDATE,me,0,max\_lg)$ ;
           **wait until** $nb\_ans$ = 2 ;
           **if** $\neg ans\_ok$ **then**
                  /* terminate iterations */
               state := looser ;
           **else**
                  /* continue and broaden the "slick" */
               $max\_lg$ := $2 \times max\_lg$
           **fi**    **od**
    **end**
**upon** REC$(ANSWER,bool,j)$ **by** $me$ ::
    **begin**
       **if** $j$ = me **then**
           $nb\_ans$ := $nb\_ans + 1$ ; $ans\_ok$ := $ans\_ok \wedge bool$ ;
       **else**      /* not for me */
           **forward** $(ANSWER,bool,j)$ ;
       **fi**
    **end**

**Program 7.2**  Bidirectional ring election algorithm.

This algorithm uses three message types:

1. $(CANDIDATE,i,lg,lg\_max)$ where $lg$ is a traveling counter and $lg\_max$ is its overflow limit, the size of the current "oil slick",

2. $(ANSWER,bool,\text{dest})$ where $bool$ is a boolean which defines the value of the answer, and

3. $(ELECTED,j)$.

Three primitives support the above communication:

1. Twofold send: is a broadcast to both directions.

2. Forward: uses the direction of the arriving message.

3. Backward: uses the direction opposite to the arriving message.

Messages which become irrelevant are not forwarded so that communication traffic is reduced.

Each node can be in one of four states with respect to the election process. It starts in an "unconcerned" node, which later becomes a

**upon** REC($CANDIDATE,j,lg,lg\_max$) **by** $me$ ::
   **begin**
      **if** $j < me$ **then**
         **backward** ($ANSWER$,False,$j$) ;
         **if** state = unconcerned **then**
            *decide to invoke an election* ;
      **fi fi**
      **if** $j > me$ **then**
         state := looser ; $lg := lg + 1$ ;
         **if** $lg < lg\_max$ **then**
            /* me is not the last */
            **forward** ($CANDIDATE,j,lg,lg\_max$) ;
         **else**    /* me is the last */
            **backward** ($ANSWER$,True,$j$) ;
      **fi fi**
      **if** $j = me$ **then**
         state := elected ; $winner := me$ ;
         **twofold send** ($ELECTED,winner$) ;
      **fi**
   **end**
**upon** REC($ELECTED,j$) **by** $me$ ::
   **begin**
      **if** $winner \neq j$ **then**
         **forward** ($ELECTED,j$)
         $winner := j$ ; state := unconcerned ;
      **fi**
   **end**

**Program 7.2** Bidirectional ring election (*continued*).

"candidate" and remains so throughout the election process, and then the election terminates in a "loser" or an "elected" node.

The algorithm uses the following variables:

**var**
      $me_i$: integer /* my identifier */
      $winner_i$: integer /* winner's identifier */
      $max\_lg$: integer /* max length of current "slick" */
      $nb\_ans$: integer /* number of answers */
      $ans\_ok$: boolean /* true when answer is ok */
      state: (candidate, unconcerned, elected, looser)

A node which has a winning chance is allowed to extend its election set. This chance is represented by the value of two variables: *ans_ok* and *state*.

- *ans_ok*=True: no False $ANSWER$ has been received.
- state=candidate: no better site's $CANDIDATE$ request has arrived.

Let us now examine the algorithm performance and show that it is an $O(n \log n)$. Let us consider the case of a simultaneous decision of all the nodes to invoke an election. Since each pair of sites compare their identifiers, after the first step at least one out of each pair is out of the election. Thus, if after $i$ steps, a node is known to be in the game, at least $2^{i-1}$ are not.

- In the first step $n$ nodes are in the game exchanging 4 messages.
- In the second step $\lfloor \frac{n}{2} \rfloor$ nodes are in the game exchanging $4 \cdot 2$ messages.
- In the $i$th step $\lfloor \frac{n}{2^{i-1}} \rfloor$ nodes are in the game exchanging $4 \cdot 2^{i-1}$ messages.
- When $i = \log n + 1$ all the sites are out of the game.

Thus, the total number of messages $T$ can be expressed as

$$T = 4 \cdot n + 4 \frac{n}{2} \cdot 2^1 + 4 \frac{n}{2^2} \cdot 2^2 +$$
$$\cdots + 4 \lfloor \frac{n}{2^{i-1}} \rfloor \cdot 2^{i-1}$$
$$= 4 \cdot n(1 + 1 + \cdots + 1) = 4 \cdot n \cdot \log n \qquad (7.2)$$

The reader may notice that the algorithm performance can be improved on its response to a $CANDIDATE$ message. An unconcerned node which receives such a message from a node with a higher identifier, replies a False $ANSWER$ and starts the election all over again. We leave it as an exercise to the reader to utilize the knowledge that the node has about the side with respect to which it is a winner.

| Class | Metrics | Effects | Remarks |
|---|---|---|---|
| S1 |  | + | symmetry degree |
| G2 |  | ring | topology dependence |
| N2 |  | − | peer node failure |
| L1 |  | + | message loss |
| L2 |  | + | message duplication |
| L3 |  | + | message desequencing |
| L4 |  | + | modified message control |
| L5 |  | + | bounded transmit time |
| T1 | $O(n \log n)$ |  | average load |
| T2 | $O(n \log n)$ |  | peak load |
| P1 | $O(n)$ |  | response time |

### 7.2.3 Bidirectional ring simulated by unidirectional ring

An election algorithm in which a bidirectional ring is simulated by a unidirectional one is given in [9]. Figure 7.3 demonstrates this prin-

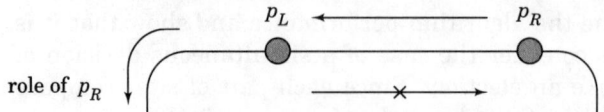

**Figure 7.3** Unidirectional-ring in election algorithm.

ciple. Say that $p_L$ and $p_R$ are adjacent nodes on a unidirectional ring as described. Here $p_L$ which receives messages from $p_R$ knows what the latter needs. Since it cannot return messages to $p_R$, $p_L$ represents its neighbor to its left side (to which it is allowed to send messages).

This algorithm, given in Program 7.3, includes two message types:

1. (1, identifier) which conveys the election of an identified candidate to the left.
2. (2, identifier) a message generated by a left neighbor which received a (1,...) message from its right side.

The algorithm uses the following local variables in each node $i$:

**var**
    $me_i$: integer /* my identifier */
    $max_i$: integer /* identifier of node whose role $me_i$ plays */
    $right\_neighbor_i$: integer /* identifier of my right active neighbor */
    $state_i$: (idle, unconcerned, active)

Program 7.3 provides a specific implementation of this approach. A node which starts election becomes active and sends (1,...) message to its left neighbor. If the left neighbor has not yet started an election itself, it starts it upon receiving this message. Thus, assuming no message overtaking, we assure that no node receives (2,...) message before (1,...). If the left neighbor is already out of the election (becoming idle), it forwards the requests to its left neighbor. As long as it is a candidate itself, it toggles the message type each time it receives another message, until $max$ is found.

Figure 7.4 illustrates the case of initiating an election simultaneously[3] by all the nodes. At the first step, we see the initial set up of the variables $max_i$ and $right\_neighbor_i$ with respect to the node's identifier and the transmission of the (1, $me_i$) message. Once the (1,...) messages have been conveyed to the proper left neighbors, the

---

[3] The simultaneous case has been chosen in order to focus on the election process, rather than on the transients of the "wake up".

**when** $me_i$ decides to invoke an election::
    **begin**
        **send_left** $(1, me_i)$ ;
        $state_i :=$ active
    **end**

**upon** REC$(1,j)$ **by** $i$ ::
    **begin**
        **if** $state_i$=unconcerned **then**
            **send_left** $(1, me_i)$ ;
            $state_i :=$ active
        **fi**
        **if** $state_i$=active **then**
            **if** $j \neq max_i$ **then**
                **send_left** $(2,j)$ ;
                $right\_neighbor_i := j$
            **else** /* $max_i$ is found */
            **fi**
        **else** /* $state_i$=idle */
            **send_left** $(1,j)$ ;
        **fi**
    **end**

**upon** REC$(2,j)$ **by** $i$ ::
    **begin**
        **if** $state_i$=active **then**
            **if** $(right\_neighbor_i > max_i) \wedge (right\_neighbor_i > j)$ **then**
                $max_i := right\_neighbor_i$
                **send_left** $(1, right\_neighbor_i)$
                    /* play the role of right_neighbor */
            **else**
                $state_i :=$ idle
            **fi**
        **else** /* $state_i$=idle */
            **send_left** $(2,j)$ ;
        **fi**
    **end**

**init**: $\forall i : max_i := right\_neighbor_i := me_i$ ;
        $\forall i : state_i :=$ unconcerned ;

**Program 7.3** Simulated bidirectional ring election algorithm.

(2,...) are sent. At the third step, we see that upon receiving the (2,...) messages, many nodes turn idle and only two nodes remain representatives of the candidates. The third and fourth steps are then carried out for these two nodes only, and the fifth step completes the election.

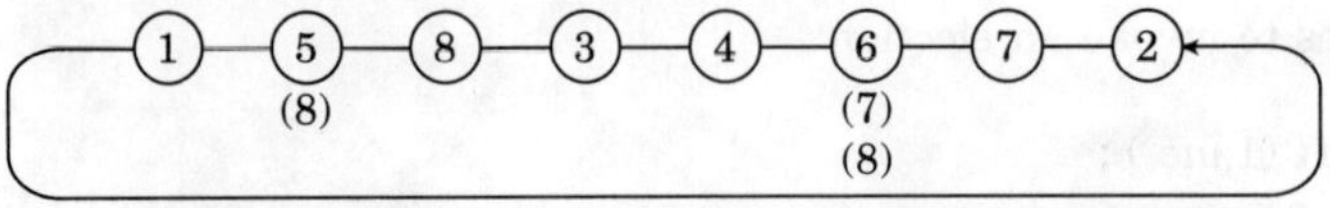

## 154 Chapter Seven

```
  ┌──(1)──(5)──(8)──(3)──(4)──(6)──(7)──(2)──┐
  │       (8)                 (7)             │
  │                           (8)             │
  └────────────────────────────────────────────┘
```

|   |       | ←1    | ←5    | ←8    | ←3    | ←4    | ←6    | ←7    | ←2    |
|---|-------|-------|-------|-------|-------|-------|-------|-------|-------|
|   |       | node  |       |       |       |       |       |       |       |
| I | max   | 1     | 5     | 8     | 3     | 4     | 6     | 7     | 2     |
|   | right | 1     | 5     | 8     | 3     | 4     | 6     | 7     | 2     |
|   | send  | (1,1) | (1,5) | (1,8) | (1,3) | (1,4) | (1,6) | (1,7) | (1,2) |
| II| max   | 1     | 5     | 8     | 3     | 4     | 6     | 7     | 2     |
|   | right | 5     | 8     | 3     | 4     | 6     | 7     | 2     | 1     |
|   | send  | (2,5) | (2,8) | (2,3) | (2,4) | (2,6) | (2,7) | (2,2) | (2,1) |
|III| max   |       | 8     |       |       |       | 7     |       |       |
|   | right |       | 8     |       |       |       | 7     |       |       |
|   | send  | ←     | (1,8) | ←     | ←     | ←     | (1,7) | ←     | ←     |
| IV| max   |       | 8     |       |       |       | 7     |       |       |
|   | right |       | 7     |       |       |       | 8     |       |       |
|   | send  | ←     | (2,7) | ←     | ←     | ←     | (2,8) | ←     | ←     |
| V | max   |       |       |       |       |       | 8     |       |       |
|   | right |       |       |       |       |       | 8     |       |       |
|   | send  | ←     | ←     | ←     | ←     | ←     | (1,8) | ←     | ←     |

**Figure 7.4** Unidirectional-ring election example.

This algorithm is an $O(n \log n)$ algorithm. For the simplified version [9] discussed here, the traffic involved in electing a node is

$$1\frac{1}{2} n \log n + O(n) \tag{7.3}$$

and in more efficient implementations it has been reduced to

$$0.15 \cdot n \log n + O(n).$$

| Class | Metrics | Effects | Remarks |
|-------|---------|---------|---------|
| S1    |         | +       | symmetry degree |
| G2    |         | ring    | topology dependence |
| N2    |         | −       | peer node failure |
| L1    |         | −       | message loss |
| L2    |         | +       | message duplication |
| L3    |         | +       | message desequencing |
| L4    |         | +       | modified message control |
| L5    |         | +       | bounded transmit time |
| T1    | $0.15 \cdot n \log n + O(n)$ | | average load |
| T2    | $0.15 \cdot n \log n + O(n)$ | | peak load |
| P1    | $O(n)$  |         | response time |

**var**
    $me_i$: integer /* my identifier */
    $in\_game_i$: boolean /* initialized to False */
    timer: time /* initialized zero */
    $winner_i$: integer
when $i$ decides to invoke an election::
    **begin**
        $in\_game_i$ := True ;
        $winner_i$ := $i$ ;
        **broadcast** ($REQ,i$) ;
        timer := timeout_constant ;
    **end**
upon REC($REQ,j$) by $i$ ::
    **begin**
        **if** $j > winner_i$ **then**
            $in\_game_i$ := False
            $winner_i$ := $j$ ;
        **else if** $\neg(in\_game)$ **then**
            **if** $j < me$ **then**
                $in\_game_i$ := True ;
                **broadcast** ($REQ,me$)
                timer := timeout_constant ;
            **else** /* ($me < j < winner_i$) $\wedge \neg(in\_game)$ */
                $winner_i$ := $j$ ;
        **fi fi**
    **end**
upon timer_expired$\wedge in\_game_i$ ::
    **begin**
        $in\_game_i$ := False ;
        $winner_i$ := $me$ ;
        Restart nominee tasks
    **end**

**Program 7.4** Broadcast election algorithm.

## 7.3 Broadcast Elections

A bully solution [112] can be used to enforce a nominee with a higher identifier. This approach requires $O(n^2)$ messages, and is effective even when the current nominee is functioning but has a lower identifier.

Program 7.4 shows an implementation of this approach with broadcasting message passing. The node which invokes the election broadcasts this intention to all other nodes. As long as no negative response has been received, it is waiting with $in\_game$ marked. It uses a timer variable to allow time for objections to its nomination to be received. If the timer expires and no objection has arrived, the node declares itself winner, and restarts the nominee tasks. In other words, $\neg in\_game \land winner_i = me_i$ means elected.

Every node which receives an election request with a candidate identifier compares it to its current winner. If the request is for a candidate with a better identifier, it is declared as the current winner. If not, the receiving node selects the following alternatives:

- If it has already initiated an election (it is $in\_game_i$ and recall that $j < me \leq winner_i$), it ignores the request since its identifier is preferable.
- If it is not $in\_game_i$ and the request is from a node with a less preferable identifier than itself, it becomes $in\_game_i$ and broadcasts its own candidacy. Here again, if the timer expires and no objection has arrived, this node will eventually declare itself the winner, and restart the nominee tasks.
- Otherwise (($me < j < winner_i$) $\land \neg(in\_game)$), it accepts the request and the requesting candidate is regarded as the nominee.

We note that the case of $in\_game_i$ and $j < me \leq winner_i$, in which the receiving node ignores the request, appears implicitly in the algorithm.

If all the nodes decide simultaneously to invoke an election, $n$ requests are broadcasted. If a broadcast is considered as $n-1$ messages, we have $O(n^2)$ messages sent. Once a broadcast has been made by a node, a second one is blocked by $in\_game_i$. Therefore, the algorithm is bounded. The bound can be decreased if instead of broadcasting to all nodes, each node sends messages to those of higher identifiers only. However, this will result in a longer sequence, as the low-identifier nodes must be told which node was eventually nominated.

The $in\_game_i$ also protects from message duplication effects, blocking broadcasts which are not required. The variable $winner_i$ also protects acceptance of multiple copies of a request from an unfavorable node. On the other hand, message losses and link failures are very critical as they can result in inconsistent decisions by nodes regarding the the winner's identity. Failed nodes are updated on election results on recovery.

| Class | Metrics | Effects | Remarks |
|-------|---------|---------|---------|
| S1    |         | +       | symmetry degree |
| L1    |         | +       | message loss |
| L2    |         | +       | message duplication |
| L3    |         | +       | message desequencing |
| L4    |         | +       | modified message control |
| L5    |         | +       | bounded transmit time |
| T1    | $O(n^2)$ |        | average load |
| T2    | $O(n^2)$ |        | peak load |
| P1    |         | +       | response time |
| P2    |         | −       | resources required |

## 7.4 Concluding Remarks

The election is a simple agreement procedure, in which the participants agree on an identifier of a chosen participant among them. This is a critical step for achieving fault tolerance in systems in which a distinguished node has unique responsibilities. Clearly a mechanism for detecting failure of a distinguished node is required, so that the election can be started. Simple time-outs may suffice for systems with periodic behavior or deterministic temporal properties.

Note that these algorithms can be used for other functions also. For example, when a token is corrupted or lost in a token ring topology, an election algorithm can be used to generate a new token.

# Chapter 8

# Deadlock and Termination

Deadlock and termination of distributed computations are global states in which the progress of the computation is denied. In deadlock situations the progress is disabled prematurely, while the termination of a computation marks the end of its progress. The deadlock situation can be either prevented or be detected after it occurs. We will discuss the two approaches. They both support resource utilization in avoiding or detecting states of blocked resources. Termination is always detected after it occurs, allowing the resource manager to reallocate resources properly. We begin the chapter with a discussion of the deadlock problem, followed by discussions of prevention techniques and detection methods. We then discuss in detail a robust algorithm for deadlock and termination detection.

## 8.1 The Deadlock Problem

The deadlock problem has received considerable attention in concurrent programming due to its critical consequences. A deadlock phenomenon occurs when the criteria governing the progress of a system are not satisfied. Hence, a system in a deadlock state is practically useless, unable to make any progress in accomplishing its goals. It is, therefore, understandable that this issue has been carefully examined, receiving both pessimistic solutions which try to prevent the occurrences of deadlock situations and optimistic solutions which deal with it when it occurs.

We distinguish two classes of deadlocks as determined by their origins: deadlock in resource allocation and communication deadlocks.

### 8.1.1 Deadlock in resource allocation

The resource allocation problem is a key issue for all properties of a system which include liveness, safety, and tolerance. In this section let us examine and properly define the deadlock situations that it can generate.

In our system model a set of $n > 1$ processes $\mathcal{V}_p = \{p_1, \ldots, p_n\}$ are related to each other through a set of logical links $\mathcal{E}_p$, to form a graph

$$\mathcal{G}_p = (\mathcal{V}_p, \mathcal{E}_p). \tag{8.1}$$

We assume that these processes need a mutually exclusive allocation for each resource that they access.

A set of $m > 1$ distinct resources $\mathcal{V}_P = \{P_1, \ldots, P_m\}$ are related to each other through a set of physical links $\mathcal{E}_P$, to form a graph

$$\mathcal{G}_P = (\mathcal{V}_P, \mathcal{E}_P). \tag{8.2}$$

We assume that a process is able to access multiple resources in a given time interval.

Allocating processes to resources is a dynamic function whose domain is the set $\mathcal{V}_p$ and whose range is $\mathcal{V}_P$. The dynamics of this function are created by *lock* and *unlock* (*release*) operations, which are initiated by the corresponding processes. During these operations, processes which acquire resources are assumed not to release them as long as they wait to complete, continuing to acquire other resources as needed. Furthermore, no external preemption is allowed.

> DEFINITION 17. The *wait-for-resource* graph, is a directed graph $\mathcal{W}_p^R$ defined as follows:
>
> 1. A set of vertices $\mathcal{V}_p$, as defined in Equation 8.1.
> 2. A set of directed arcs $\mathcal{A}_p$, such that $p_i \longrightarrow p_j$ if and only if $p_i$ is waiting for a resource locked by $p_j$.
> 3. This is an *AND* graph. A process $p_i$, such that $p_j \longleftarrow p_i \longrightarrow p_k$, will be "unblocked" when receiving the resources from both $p_j$ and $p_k$. □

We note that the directed arcs represent an *AND graph* relationship. When several arcs leave a particular vertex, this vertex waits for all of them. From this definition we can conclude that if there exists a cycle in $\mathcal{W}_p^R$ we are facing a deadlock situation.

We can also define a bipartite graph $\mathcal{B}_{p,P}$ as follows:

1. A set of vertices $\mathcal{V}_p$, as defined above.
2. Another set of vertices $\mathcal{V}_P$, as defined above.
3. A set of directed arcs $\mathcal{A}_p$, such that $p_i \longrightarrow P_j$ if and only if $p_i$ is waiting for the resource $P_j$.

4. A set of directed arcs $\mathcal{A}_P$, such that $P_i \longrightarrow p_j$ if and only if $P_i$ is locked by $p_j$.

Again, if there exists a cycle in $\mathcal{B}_{p,P}$ we are facing a deadlock situation. The reader may notice that the latter representation of the problem does not provide any additional information with respect to the deadlock situation.

The resources deadlock condition can therefore be summarized as follows:

> CONDITION 2. Let the system computations' resource-requirements be modeled by the $\mathcal{W}_p^R$ or $\mathcal{B}_{p,P}$, respectively. If the following are true, the system is *deadlocked*:
>
> 1. *Mutual exclusion* is required by computations when using the resource.
> 2. *No preemption* is permitted once a resource is locked until it is voluntarily released by the node.
> 3. *Wait-for* relation is maintained for all resources required by the node (in $\mathcal{B}_{p,P}$ or to the nodes locking these resources in $\mathcal{W}_p^R$) until they are locked by this node.
> 4. A *cycle exists* in the $\mathcal{W}_p^R$ or $\mathcal{B}_{p,P}$ graphs. □

Figure 8.1 describes resource deadlock in both approaches: (*a*) wait-for-resource graph, and (*b*) bipartite graph. In the wait-for-resource graph only $p_i \in \mathcal{V}_p$ are represented as nodes, and the arcs $\mathcal{A}_p$ show that:

- $p_1$ awaits a resource possessed by $p_2$
- $p_2$ awaits a resource possessed by $p_3$
- $p_3$ awaits a resource possessed by $p_1$

In the bipartite graph $p_i \in \mathcal{V}_p$ as well as $P_i \in \mathcal{V}_P$ are represented as nodes, and the arcs $\mathcal{A}_p$ and $\mathcal{A}_P$ present the same case as follows:

- $p_1$ possesses resource $P_1$
- $p_2$ possesses resource $P_2$
- $p_3$ possesses resource $P_3$

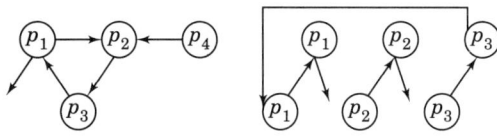

(a) Wait-for-resource graph    (b) Bipartite graph

**Figure 8.1** Deadlock in resource allocation.

- $p_1$ awaits resource $P_2$
- $p_2$ awaits resource $P_3$
- $p_3$ awaits resource $P_1$

The cycle in $\mathcal{B}_{p,P}$ shows the deadlock exists.

### 8.1.2 Deadlock in communication

Communicating processes are also deadlock sensitive. Processes which wait for receiving messages (e.g., rendezvous based communication) are imminent candidates for communication deadlocks.

> DEFINITION 18. The *wait-for-communication* graph, is a directed graph $\mathcal{W}_p^C$ defined as follows:
>
> 1. A set of vertices $\mathcal{V}_p$, as defined in Equation 8.1.
>
> 2. A set of directed arcs $\mathcal{A}_p$, such that $P_i \longrightarrow P_j$ if and only if $P_i$ is waiting for a message from $P_j$.
>
> 3. This is an *OR* graph. A process $p_i$, such that $p_j \longleftarrow p_i \longrightarrow p_k$, will be "unblocked" when receiving a message either from $p_j$ or from $p_k$. □

We note that unlike in the resource allocation case, the directed arcs represent an *OR graph* relationship. When several arcs leave a particular vertex, this vertex waits for one of them. Therefore, a cycle in the communication wait-for graph is not a sufficient condition for deadlock.

Figure 8.2 illustrates an example where the wait-for graph contains a cycle and deadlock is not unavoidable. In this example, $p_1$ waits for $p_2$, which waits for $p_3$, which waits for $p_1$. However, $p_1$ can also be released from that deadlock by $p_5$, which is another alternative to the wait-for relation $p_1$ has with $p_2$.

In OR-graphs the deadlock condition is much stronger. Only a strongly connected graph[1] is a sufficient indication for communication deadlock. This type of deadlock is extensively examined in Section 8.3.2.

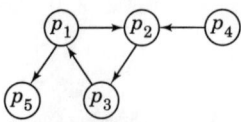

**Figure 8.2** Release blocking in an OR graph.

---

[1]In a strongly connected graph, every two nodes are mutually reachable.

### 8.1.3 Axiomatic definition of deadlock

We introduced earlier an axiomatic approach [109] with guarded variables gathered in resources ($r$) and used in the parallel execution of statements $S_i$. Let us recall that the invariant $I(r)$ is true when parallel execution begins and remains true whenever it is observed outside the critical sections of $r$. In addition, recall that a statement $S_{cr}$ in a critical section is protected by a condition $C$ in **with** statement of the form

**with** $r$ **when** $C$ **do** $S_{cr}$.

Logically, we use the notation $\{P\}\ S\ \{Q\}$, which means: if $P$ is true before executing $S$, and $S$ halts, then $Q$ is true after executing $S$. In this notation, $S$ is a statement, $P$ a precondition, and $Q$ a postcondition. With this model, a program $S$ with concurrent processes $S_k$, does not include a deadlock if the following condition holds.

CONDITION 3. If

1. The critical section statements of processes $S_k \in S$ are

    **with** $r$ **when** $C_k^j$ **do** $S_k^j$

    $1 \leq j \leq n_k$, and

2. $I(r)$ and $Pre(S_k^j)$ are assertions derived from the proof of $\{P\}S\{Q\}$, and

3. $P$ is true when execution of $S$ begins, and

4. Define $D_2$ as

$$D_2 \equiv \bigvee_k [\bigvee_j (\neg C_k^j \wedge Pre(S_k^j)\,)] \tag{8.3}$$

    and

5. Define $D_1$ as

$$D_1 \equiv \bigwedge_k [Post(S_k) \vee \bigvee_j (\neg C_k^j \wedge Pre(S_k^j)\,)] \tag{8.4}$$

    and

6. The assertion

$$D_1 \wedge D_2 \wedge I(r) \implies false \tag{8.5}$$

    holds,

    then

    $S$ cannot be deadlocked.

□

### 8.1.4 Optimistic and pessimistic solutions

In both of these approaches, we base our identification of the deadlock situation on the cyclic behavior of a graph representation of the system state. We find both optimistic and pessimistic solutions to this undesirable situation. The pessimistic approach is the prevention of deadlocks through rejection of jobs that can create a cycle in the system graph even before they are accepted into the job stream. The optimistic approach defers no job, and tests periodically for the existence of such a cycle, detecting it only after it has already occurred.

A widespread prevention technique is based on ordering the resources. This approach necessitates an *a priori* declaration of all the resource requirements for the following transaction by the requesting process. All the requests acquire resources in the same order as the predefined ordered set of resources. This order adds an acyclic nature to the system graph by imposing priority through the order of request. This technique is quite high in overhead and low in parallelism and needs additional rules (e.g., FIFO queue) in order to prevent starvation.

Graph analysis techniques support both pessimistic and optimistic methods. In the prevention techniques a predeclaration of all the requirements is set down, and the analysis is carried out prior to job acceptance. If the analysis fails, the requesting process is deferred. On the other hand, in the detection techniques, the analysis detects the deadlock only after it has already occurred.

## 8.2 Deadlock Prevention in Multiple Resource Allocation

### 8.2.1 Basic principles

1. Each node maintains a copy of a *wait for* graph.
2. All resource needs are *predeclared* via broadcast upon entering.
3. Each resource is managed by a mutual exclusion algorithm, say the timestamp algorithm in Section 6.3.1, providing the knowledge of who is requesting, who locks, etc.
4. A *transaction* layout starts with declaring requirements up front, followed by sequences of acquiring resources, using them and then releasing them.

$$\textbf{pre-declare}(r_1, r_2, \ldots, r_k)$$
$$\vdots$$

```
begin transaction
        ⋮
    request(r₁) ;
    request(r₂) ;
        ... use r₁ and r₂
    release(r₁) ;
        ⋮
    request(r₃) ;
    request(r₄) ;
        ... use r₂, r₃ and r₄
    release(r₂, r₃, r₄) ;
end transaction
```

### 8.2.2 The Global Graph solution

The *global graph* solution [11] assumes that each site maintains the global knowledge of all transactions and all resources.

- **Knowledge required for each site: all transactions, all resources**

The deadlock prevention is based on

```
upon entrance to a new transaction::
begin
    if ∃ an eventual cycle
        then defer new
        else accept new
    fi
end
```

We note here that we must take into account all possible sequences of acquiring the declared resources. However, that is not the actual case, since not all the resources are either locked or waited for although some may be in neither lock nor wait-for states at a particular point of time.

| Class | Metrics | Effects | Remarks |
|-------|---------|---------|---------|
| S1    |         | +       | symmetry degree |
| G1    |         | −!!!    | global info necessity |
| G2    |         | +       | topology dependence |
| L1    |         | +       | message loss |
| L2    |         | +       | message duplication |
| L3    |         | +       | message desequencing |
| L4    |         | +       | modified message control |
| L5    |         | +       | bounded transmit time |
| T1    | 0       |         | average load |

### 8.2.3 The Local Graph solution

The global graph can become extremely large, and if one can achieve deadlock prevention based on less demanding requirements, the overhead may be substantially reduced and so does the local graph solution [12].

- Knowledge required for each site: some transactions, local resources, timestamps

The basic idea is to use only a partial wait-for graph, which contains information only on the local resources. However, by doing that we lose the ability to locally detect a cycle in the graph since we do not have knowledge of the whole graph. The solution is to add additional ordering to the graph by ordering the transactions with timestamps.

Figure 8.3 demonstrates the cycle generated in this ordering approach when a task ($T_1$) waits for a resource ($R$) locked by another task ($T_2$) with a "younger" timestamp.

**upon** entrance to a *new* transaction::
**begin**
    **if** ∃ an eventual cycle
        **then** reject *new*
        **else** accept *new*
    **fi**
**end**

The timestamp is broadcast to all sites with pending resource requirements. The time service imposes an ordering which compensates for the absence of information of the global wait-for graph. Since all resources get the same timestamp with a specific request, the case of an undetected cycle in the global graph does not exist.

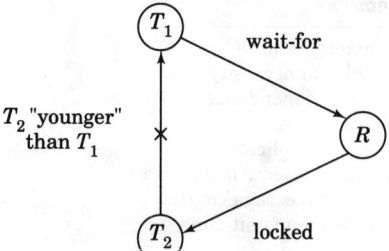

**Figure 8.3** Deadlock prevention in local graph with timestamps.

**Deadlock and Termination**   167

| Class | Metrics | Effects | Remarks |
|---|---|---|---|
| S1 |  | + | symmetry degree |
| G1 |  | −! | global info necessity |
| G2 |  | + | topology dependence |
| T1 | 0 |  | average load |

## 8.2.4 "On Each Resource" solution

A better solution allows the resources to manage the timestamp ordering of the transactions [19], assuming a timestamp ordering as the classical clock synchronization algorithm [79].

- Knowledge required for each site: two transactions, one resource, timestamps

The following properties are exhibited by this approach:

- Each transaction $T$ is associated with a timestamp $e(T)$
- Each resource is allowed to be in one of two states: (idle, locked)
- When a resource becomes locked, it receives the locking transaction's timestamp $e(r_k)$

**Wait and Die scheme.**

```
when T wants to lock r_k::
begin
    if state(r_k)=idle
        then
            state(r_k):=locked
            e(r_k):=e(T)
        else                 /* state(r_k)=locked */
            if e(T) < e(r_k)
                then                 /* r_k is younger */
                    T waits
                else                 /* r_k is older */
                    T dies
        fi    fi
end
```

When a request arrives for an idle resource, the resource is allocated to it. When a request arrives for a locked resource, and the request is younger than that of the locking transaction, it "dies" and must be restarted over again with the same timestamp. Only "older" transactions are allowed to wait; a young transaction will eventually become old enough to be able to either wait or acquire the resource.

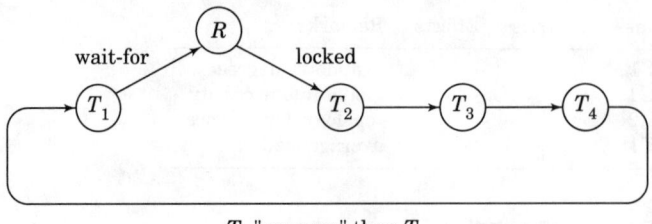

$T_4$ "younger" than $T_1$

**Figure 8.4** "Wait & die" prevention with timestamps.

Figure 8.4 shows a case of a timestamp cycle where $T_1$ awaits a resource locked by $T_2$, where $T_2$ is younger than $T_3$ which is younger than $T_4$ which is younger than $T_1$. In this case, $T_2$, $T_3$ and $T_4$ will die upon requesting $R$ which is locked by $T_1$.

| Class | Metrics | Effects | Remarks |
|-------|---------|---------|---------|
| S1 | | + | symmetry degree |
| G1 | | + | global info necessity |
| G2 | | + | topology dependence |
| P1 | | see WD vs WW | response time |

**Wound and Wait scheme.** Another approach allows preempting the execution of the locking transaction when an incoming transaction is older. In this case a deferred transaction does not have to die, and ordering is maintained at the waiting queue. The preempted transaction is called "wounded" compared to "dead" in the previous scheme, as preemption may save all which has been done. The wounded transaction is restarted with the same timestamp when it is the oldest in the waiting list and the resource becomes idle.

**when** $T$ wants to lock $r_k$::
**begin**
    **if** state($r_k$)=idle
        **then**
            state($r_k$):=locked
            $e(r_k):=e(T)$
        **else**    /* state($r_k$)=locked */
            **if** $e(T) < e(r_k)$
                **then**              /* $r_k$ is younger */
                    locking transaction (current) is wounded
                **else**              /* $r_k$ is older */
                    $T$ waits
    **fi**    **fi**
**end**

| Class | Metrics | Effects | Remarks |
|---|---|---|---|
| S1 | + | | symmetry degree |
| G1 | + | | global info necessity |
| G2 | + | | topology dependence |
| P1 | | see WD vs WW | response time |

**Wait and Die versus Wound and Wait.** The following table compares the wait and die versus the wound and wait schemes. We note that the wound and die is closer to the first-in first-out logic.

| Situation | Wait & Die | Wound & Wait |
|---|---|---|
| New Older than Current | New Waits | New locks Current Wounded |
| New Younger than Current | New Dies | New Waits |

## 8.3 Deadlock and Termination Detection

Opposite to the pessimistic prevention approach, optimistic detection techniques carry out analyses which detect deadlocks only after they have already occurred. Termination is also detected in a post mortem manner. Let us first describe the properties of both deadlock and termination situations. Then, in the following section, we introduce distributed algorithms, their concepts, their structures, and examine the properties of these algorithm. These properties include their robustness to network failures, their deadlock and termination detection capabilities, and some performance aspects.

### 8.3.1 Computation deadlock

Let us continue our discussion from Section 8.1, with the system model defined in Equations 8.1 and 8.2 of processes graph $\mathcal{G}_p$ and resources graph $\mathcal{G}_P$. The directed wait-for-resource graph, $\mathcal{W}_p^R$ of Definition 17, is an AND graph with edges pointing from processes which await resources to the processes which hold them. Therefore, if there exists a cycle in $\mathcal{W}_p^R$, we are facing a deadlock situation.

For each vertex (process) in $\mathcal{W}_p^R$ we can define a *dependency set* as follows.

DEFINITION 19. A *dependency set* of a process $p_i$, denoted $DS(p_i)$ is a set of process identifiers $p_j \in W_p^C$, such that

$$DS(p_i) \equiv \{\forall p_j : p_i \longrightarrow p_j\}$$

□

This set of process identifiers includes all resource holders on which the process waits for resources so it can execute its tasks. The resources deadlock condition (Condition 2) can be rephrased as follows:

CONDITION 4. A deadlock set $S$ is a set such that:

1. Each $p_j \in S$ is in a waiting state, and
2. $\exists p_k \in S : DS(p_k) \neq empty$, and
3. $\forall p_j \in S : DS(p_j) \subseteq S$. □

If there exists a deadlock set for a subgraph of $\mathcal{W}_p^R$, this subgraph is blocked. A structure of the this type is such that each of its connected subgraphs is a *terminal strongly connected component* in the global system wait-for-resource graph.

### 8.3.2 Communication deadlock

In a distributed system, processes communicate with each other in order to share information, to synchronize, or to verify their status. Each process which needs external data, enters a "wait–for–communication" state, and therefore, its forthcoming activity depends on communication initiated by another process. Such dependency can lead to a deadlock situation: a process $p_i$ can wait for communication from process $p_j$, while at the same time $p_j$ can be waiting for communication from $p_i$. $p_j$ can also be waiting for $p_k$, which waits for $p_i$ (or even may be already terminated). Situations as described above will be called a *communication deadlock situation*. This situation requires as prerequisite that:

- There exists a set $S = \{p_i\}$ of processes which send messages to each other, and
- Processes can enter *waiting states* where they cannot proceed with their computation, but have to wait for messages.

We can model such systems with a graph. Vertices represent the processes, with a vertex for each individual process. Directed edges connect the communicating vertices.

For each process in the wait-for-communication graph (see Definition 18) we define a *dynamic dependency set* as follows, analogously to our previous definition of a dependency set.

DEFINITION 20. A *dynamic dependency set* of a process $p_i$, denoted $DDS(p_i)$ is a set of process identifiers, $p_j \in W_p^C$, such that

$$DDS(p_i) \equiv \{\forall p_j : p_i \longrightarrow p_j\}$$

□

In other words the dependency set of a process is the set of all processes that might awake it from a waiting state. This set is updated during the task execution. Hence, we can formulate the *communication deadlock condition*.

CONDITION 5. A deadlock set $S$ is a set such that:

1. Each $p_j \in S$ is in a waiting state, and
2. $\exists p_k \in S : DDS(p_k) \neq empty$, and
3. $\forall p_j \in S : DDS(p_j) \subseteq S$, and
4. There is no message that was sent and not yet received within processes in $S$. □

Here again, we have a structure such that each of its connected subgraphs is a *terminal strongly connected component* in the global system wait-for-communication graph.

### 8.3.3 Termination detection

Let us resume the description of our model and expand it to include the termination property. As mentioned before, there is a difference between deadlock detection and termination detection. After termination, the dynamic dependency set of a process [5] is empty. Therefore, dealing with termination-deadlock detection, we have to distinguish between two different classes of processes that can change the state of a process.

- A process $p_{me}$ which has not finished its current computation task yet is a *black* process. It can be awakened by one of the processes it is currently waiting for. These process ID's are the elements of the *dynamic dependency set*, denoted $DDS(p_{me})$.

- A process $p_{me}$ which has finished its current computation task is a *white* process. It can be awakened only by *reactivation* of its current task, triggered from other tasks, which included the ID *me* in their *declaration* phase. Process $p_{me}$ knows this set of process IDs, which is called its *static dependency set* and denoted $SDS(p_{me})$.

In our model an idle process with an empty DDS is a potentially terminated process. So, we use an *extension* of the wait-for-graph defined earlier: the nodes of this graph are all the processes involved in the underlying computation, and the vertices are defined as before by the DDS when the node is black, otherwise by the SDS. To assure us that the underlying computation is an actual distributed computation (i.e., that the communication pattern is not partitioned), we require that this graph be connected when all processes are white. Under normal computation conditions, all those structures are dynamic, and detecting deadlock or termination is identical to detecting the static parts of these structures. Since we have defined the notion of a deadlock set in the reduced model, we want to define the notion of terminated set in our new extended model.

CONDITION 6.  A terminated set $S$ is a set such that:

1. Each $p_i \in S$ has locally terminated its computation, and
2. $\forall p_i \in S : SDS(p_i) \subseteq S$, and
3. There is no message that was sent and not yet received within members of $S$. □

The global termination of the underlying computation is guaranteed when the set of all processes has this property. Nevertheless, we may want to define a partial termination. In this model, a white process can participate in a deadlock situation. So, we have this new definition of a deadlock set.

CONDITION 7.  A deadlock set $S$ is a set such that:

1. Each $p_j \in S$ is in the waiting state, and
2. $\exists p_k \in S : DDS(p_k) \neq empty$, and
3. $\forall p_j \in S : DDS(p_j) \subseteq S$, and if $DDS(p_j) = empty$ we require that $SDS(p_j) \subseteq S$, and
4. There is no message that was sent and not yet received within processes in $S$. □

So, a process is said to be locally deadlocked (or locally terminated) if it belongs to a deadlock (or terminated) set.

For each process $p_i$, we define its *reachability set* relative to our extended definition of its wait-for-graph, and denote it by $R(p_i)$.

DEFINITION 21.  A *reachability set* of a process $p_i$ is the set of all processes $p_k$, such that there exists a finite path from $p_i$ to $p_k$ in the extended wait-for-graph of the underlying computation. □

In order to distinguish this property from the global deadlock property, a different approach should be taken. For deadlock detection the question is: "Is there *any* set $S$ of processes with the deadlock property?" For termination detection of the global computation, the *set $S$ is known a priori*, and the question is: "Does this set have the termination property?" The knowledge of the members of this set is obtained from the *declaration* part of the programs involved in the underlying computation. But, one can be interested in the termination of a dedicated process $p_{me}$, which is equivalent to asking the question: "Does a terminated set $S$ exist that contains $p_{me}$?" Although these properties differ, the same method of a two–fold wave of messages can be used for both (global and local) termination and deadlock detection, and that is what we propose here. We refer to the messages of the querying and responding waves as a *detection session*. Basically, the algorithm initiated by a process $p_{me}$ will answer the question: "Am I terminated or deadlocked?" When this algorithm is executed by certain processes of the underlying computation, they can conclude that their local termination is also a global termination.

We can formulate the termination property based on the axiomatic approach [109] introduced earlier. Recalling that the parallel program $S$ is of the form

**resource** $r_1, r_2$; **cobegin** $S_1 \| S_2 \| \ldots \| S_n$ **coend**

where $r_1$ and $r_2$ are resources of guarded variables and $S_1 \ldots S_n$ are the parallel statements executed between the cobegin and coend. The following defines $S$ conditional termination property.

CONDITION 8. If

1. $S$ satisfies Condition 3 for not having a deadlock, and
2. $\forall S_i \in S$ terminates conditionally, that is, it is proved to terminate under the assumption of not being deadlocked,

then

$S$ can terminate successfully.

□

### 8.3.4 Detection algorithms

Several solutions for dealing with the global termination detection problem have appeared in the literature. We have chosen to discuss a few of them: a diffusion approach [7], a ring structured [8, 1] solution, a spanning tree solution [22], a circuit [15] solution, a timestamp [18] solution, and a quiesent detection solution [20] in which we want to detect the so called "monotonic" properties of an underlying computation.

Then, we present a solution based on an algorithm by Chandy, Misra, and Haas [5] free of limitations demonstrated by the previous solutions. In this solution our model of computation takes into account the reactivation of a process, and takes advantage of the monotonicity property to decrease the number of messages in the detection.

**A communication deadlock detection algorithm.** The goal of a deadlock detection algorithm is to pin-point a terminal strongly connected component. The algorithm [5] uses only local information for each process: its dynamic dependency set (DDS). A process which is "tired" of waiting for communication, can initiate a query sequence concerning its deadlock status. It is done by transmitting a *(Question,...)* message to its dependency set. If a potential deadlock situation is detected by any of its successors, an *(Answer,...)* message is sent back to the querying process. Otherwise, no answer is sent back. Therefore, a querying process which has not yet received any answer is assumed to be eventually unblocked. On the other hand, if a querying process receives any message from the underlying computation (sent by any member of its dynamic dependency set), the deadlock condition is not satisfied. Distinguishing activity status is possible through the use of a *state* description. In their model processes can be either *idle* or *active*. The notion of a terminated process does not exist. So, the case of an idle process with an empty DDS is not considered as a potential deadlock source in their model.

In order to ensure deadlock detection, two *waves* of messages are used: a forward *query* diffusion, and a backward *response* feedback. The number of queries a process can send before receiving an answer is not limited. The two-fold wave, that scans all possible communication links of the underlying computation, also ensures there are no *outstanding* messages in the network. With the given model, this algorithm addresses only the deadlock detection problem. In Section 8.3.6, we present an extension of this algorithm to provide solution to the termination problem.

| Class | Metrics | Effects | Remarks |
|---|---|---|---|
| S1 |  | + | symmetry degree |
| G1 |  | + | global info necessity |
| G2 |  | + | topology dependence |
| N1 |  | + | coordinator node failure |
| N2 |  | + | peer node failure |
| L1 |  | − | message loss |
| L2 |  | + | message duplication |
| L3 |  | + | message desequencing |
| P1 | $O(d)$ |  | response time |

**Termination of diffusing computation.** The diffusion approach [7] is limited to a tree structured computation. Termination is detected locally according to the following principle of diffusion: invocation can be followed by invocations of others and is answered by an acknowledgement upon computation termination. Therefore, in a tree structured computation, a parent process invokes its children, and acknowledges its invokee only after all its children have acknowledged.

**Message types used.**

- $(Msg, j)$: a message from the underlying computation, with $j$ the sender process ID
- $(Ack, j)$: acknowledgement from the control procedure, associated with a $Msg$ message

**Local variables of process $p_{me}$.**

**def_in, def_out :** integer (0...nmax), represent the "deficit" in communication, counting $Msg$s out and $Ack$s in, initialized to 0.

**me :** integer (1...nmax) the process ID.

**parent :** integer (1...nmax), the ID of the awakening process of $p_{me}$, that is, the first to have its $Msg$ received by $p_{me}$.

**SDS :** multiset of (1...nmax), initialized empty, which keeps track of all processes which have sent $Msg$s to $p_{me}$.

**Principle.** If def_in= 0 (all answers were sent upward), we want def_out= 0 (all children have answered). We note that the computation itself decides if it is willing to send a $(Msg, ...)$ or an $(Ack, ...)$ message. However, the computation itself can only be willing to send a $(Msg, ...)$ message if it has been invoked[2] (def_in$\neq$ 0). Furthermore, the computation is willing to send an $(Ack, ...)$ message only if def_in> 1 or if def_in= 1 and def_out= 0. The latter represents the last response which is sent to the parent. On the other hand, if def_in> 1 one can generate an immediate $(Ack, ...)$ response, since answering the parent will await all the children answers.

**The Algorithm.**

1. *Willing to send $(Msg, me)$ to $j$*::

    **do**
       def_out := def_out+1
       **send**$(Msg, me)$ to $j$
    **endo**

---

[2]Unless it is the root process.

2. *Upon receiving* $(Msg, j)$::

   **do**
       **if** def_in= 0 **then**
           parent := $j$
       **else**
           SDS := SDS $\oplus \{j\}$
       **fi**
       def_in := def_in+1
   **endo**

3. *Willing to send* $(Ack, me)$::

   **do**
       **if** def_in= 1 **then**
           **send**$(Ack, me)$ to parent
       **else**
           aux := one element of SDS ;
           SDS := SDS $\ominus \{\text{aux}\}$ ;
           **send**$(Ack, me)$ to aux
       **fi**
       def_in := def_in$-1$
   **endo**

4. *Upon receiving* $(Ack, j)$::

   **do**
       def_out := def_out$-1$
   **endo**

**Pros and Cons.**

| Class | Metrics | Effects | Remarks |
|---|---|---|---|
| S1 |   | + | symmetry degree |
| G1 |   | + | global info necessity |
| G2 | − |   | topology dependence |
| L1 | − | See 2 below | message loss |
| L2 | − | See 1 below | message duplication |
| L3 |   | + | message desequencing |

**Effects summary.**

1. Counters will function improperly due to message duplications. For example, duplicated $(Ack, j)$ results in an undesirable decrease of def_out.

2. Message losses can result in having no $(Ack, me)$ sent to the parent since def_out will never reach zero.

**Termination in ring topologies.** The ring structured token control [8, 1] is proposed for the termination detection of a computation of an arbitrary graph structure. Processes which have terminated are colored white, and processes which have not are colored black. The process which receives a token waits to become idle, and then colors the token as follows: if it is white it passes the token color, if it is black it sends a black token. An idle black process which has sent a black token then paints itself white, since the black token retains the proper global state, while the process's relevant color is white. This property is called "sin forgiveness." The root detects termination if it is white and has received a white token.

**Message types used.**

- $(Msg, j)$: a message from the underlying computation, with $j$ the sender process ID
- $(Token, color)$: a control message

**Local variables of process p$_{me}$.**

**me :** integer (1...nmax) the process ID.

**token_present :** boolean, initialized to false for all processes which have not initiated the detection

**state :** (idle,active) initialized to active

**proc_color :** (black,white) initialized to white

**token_color :** (black,white)

**Assumptions.**

- The token control uses a ring with ID's whose value decreases as the token advances. Therefore, only $p_0$ is allowed to initiate a token cycle, otherwise the above assumption is disclaimed.
- The transmission of a $Msg$ (the underlying computation message) takes zero time. This is required as a consequence of having a different topology for the underlying computation and the control ring. Figure 8.5 shows the reason for this requirement: a token can visit $p_i$ after it sent a $Msg$ to $p_j$ and can arrive at $p_j$ before $Msg$, since the two use different routings.

When a node sends an underlying computation message ($Msg$) to another node, it must allow for of the case of having the token initiator ($p_o$) between the communicating processes, as depicted by Figure 8.5. Process $p_i$ must notify the token that there is a message on its way to $p_j$, because the token could arrive at an idle $p_i$ and reach $p_j$ before $Msg$ does.

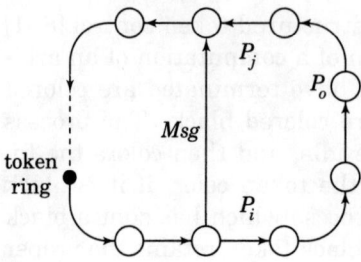

Figure 8.5 Token race with underlying computation message.

### The Algorithm.

1. *Upon receiving* $(Msg, j)$::

    **do**
        state := active
    **endo**

2. *Willing to send* $(Msg, me)$ *to* $j$::

    **do**
        **if** $me < j$ **then**
            proc_color := black
        **fi**      /* marks child invocation for token
                      which arrive after $Msg$ was sent
                      and $me$ becomes idle */
        **send**$(Msg, me)$ to $j$
    **endo**

3. *Upon receiving* $(Token, token\_color)$ *from* $me+1$::

    **do**
        token_present := true ;
        **if** $me = 0$ **then**
            **if** (proc_color=white)∧(token_color=white) **then**
                Termination Detected
            **else**
                Start New Detection
        **fi  fi**
    **endo**

4. *When* token_present∧(state=idle)::

    **do**
        **if** proc_color=black **then**
            token_color := black
        **fi**

            token_present := false ;
            **send**($Token$,token_color) to $me - 1$ ;
            proc_color := white
    **endo**

5. *When* wait($Msg, j$)∨(locally terminated)::

    **do**
        state := idle
    **endo**

### Pros and cons.

| Class | Metrics | Effects | Remarks |
|---|---|---|---|
| S1 | – | See 1 below | symmetry degree |
| G1 |   | + | global info necessity |
| G2 |   | – | topology dependence |
| N1 | – | See 1 below | coordinator node failure |
| L1 |   | – | message loss |
| L2 |   | + | message duplication |
| L4 | – | See 3 below | modified message control |
| L5 | – | See 2 below | bounded transmit time |
| P4 | – | See 4 below | resource efficient use |

### Effects summary.

1. The process $p_o$ is the only node which can initiate detection cycle, since this is the only way to maintain the token progress through decreasing identifiers. This is a strong asymmetry, and a single-node failure point.

2. The algorithm requires a zero transmission time for $Msg$. This can be solved by modifying the algorithm to paint the process black for any $Msg$ transmission (not only $me < j$). Such a modification can increase communication traffic but it can be decreased by the following note.

3. If $Msg$s routing is totally disjoint of that of $Tokens$, instantaneous communication is mandatory to assure Conditions 5 and 6 requirement that there be no outstanding messages in the network. Two white token rotations through $p_o$ can solve this race, in addition to painting the process black for any $Msg$ transmission.

4. The ring is not efficiently used. There is no need to continue passing a black token, instead a local time-out can produce the same results.

**Termination in spanning tree topology.** Instead of using the ring topology for detection, one can use a spanning tree structure, independent of the underlying computation topology. In the spanning tree solution [22], the root which decides on detection sends a *signal* to its children and waits for their returned tokens. An idle process paints itself white, and while waiting, every message from the underlying computation turns it back to an active state painted black. The signal wave advances through the tree until it reaches leaf processes, and idle leafs return white tokens. A white token which travels back through white idle processes arrives white. Only a path through idle processes allows the backward wave to advance.

**Message types used.**

- $(Msg, j)$: a message from the underlying computation, with $j$ the sender process ID
- $(Signal, j)$: a querying message
- $(Token, color, j)$: a feedback message

**Local variables of process $p_{me}$.**

**me** : integer (1...nmax) the process ID

**nb_child, my_parent, children** : network structure characteristics

**nb_token** : integer (0...nb_child), initialized to 0, the number of answers *me* is waiting for

**state** : (idle,active) initialized to active

**proc_color** : (black,white) initialized to white

**token_color** : (black,white)

**Assumptions.**

- Only the tree's root ($p_o$) is allowed to start a detection cycle.
- The transmission time of a $Msg$ (the underlying computation message) takes zero time. This is required as a consequence of having a different topology for the underlying computation and the control tree.

**The Algorithm.**

1. *When $p_o$ is willing to start detection*::

    **do**
        $\forall j \in$ children **send**($Signal, j$) ;
        proc_color := white ;
        nb_token := nb_child
    **endo**

**2.** *Upon receiving* $(Signal, me)$, $p_{me} \neq p_o$::

   **do**
        proc_color := white ;
        **if** children$\neq \emptyset$ **then**
            $\forall j \in$children **send**$(Signal, j)$ ;
            nb_token := nb_child
        **else**    /* leaf node */
            **wait until** state=idle ;
            **send**$(Token, white, my\_parent)$
        **fi**
   **endo**

**3.** *Upon receiving* $(Msg, me)$::

   **do**
        state := active
   **endo**

**4.** *Willing to send* $(Msg, j)$::

   **do**
        proc_color := black
   **endo**

**5.** *When* wait$(Msg, j) \vee$(locally terminated)::

   **do**
        state := idle
   **endo**

**6.** *Upon receiving* $(Token, color, me)$::

   **do**
        nb_token := nb_token-1 ;
        **if** color=black **then**
            token_color := black
        **fi**
        **if** nb_token= 0 **then**
            **if** $(me \neq 0) \wedge$(state=idle) **then**
                **if** (proc_color=black)$\wedge$(token_color=black) **then**
                      **send**$(Token, black, my\_parent)$
                **else**
                      **send**$(Token, white, my\_parent)$
            **fi fi**
            **if** $(me = 0) \wedge$(state=idle) **then**

                if (proc_color=white)∧(token_color=white) then
                    Termination Detected
            fi  fi
        fi
    endo

**Pros and cons.**

| Class | Metrics | Effects | Remarks |
|---|---|---|---|
| S1 | − | See 1 below | symmetry degree |
| G1 |   | + | global info necessity |
| G2 |   | + | topology dependence |
| N1 | − | See 1 below | coordinator node failure |
| L1 | − | See 2 below | message loss |
| L2 | − | See 3 below | message duplication |
| L5 | − | See 4 below | bounded transmit time |

**Effects summary.**

1. The process $p_o$ is the only node which can initiate detection cycle, since it is the only node with a reachability set which covers the whole system.

2. Message loss affects the detection drasticly. The loss of both $(Signal, j)$ and $(Token, color, j)$ will prevent detection because nb_token will not reach zero.

3. Message duplication, on the other hand, can cause a premature detection, causing nb_token to reach zero even though not all children have responded. In order to overcome this problem, or to add the possibility of working with multiple tokens, we need to tag the signals and the corresponding tokens. Furthermore, we must avoid duplicated counter decrements per tag and manage a different counter per tag or a single counter for the latest tag only.

4. The algorithm requires a zero transmission time for underlying computation messages, since the tree structured communication of detection does not cover the whole computation network. Therefore, there is no other way to assure that there is no outstanding message in the network, as required by Conditions 5 and 6.

**Termination in Eulerian circuit topology.** An Eulerian circuit topology $C$ [15] means that every link in the computation network is included at least once in the control communication. The network size ($\|C\|$) and node successor in the circuit ($succ(i)$) are known at each node $i$. A token advances through the circuit, counting the consecutive white nodes it has visited. Termination is ascertained when the count reaches $\|C\|$.

## Message types used.

- $(Msg, j)$: a message from the underlying computation, with $j$ the sender process ID
- $(Token, nb\_visit, j)$: a feedback message, with nb_visit containing the number of consecutively visited white processes

## Local variables of process $p_{me}$.

**me :** integer (1...nmax) the process ID

$\|C\|$, $succ(me)$ **:** network structure characteristics

**nb_token :** integer (0... $\|C\|$), initialized to 0, the number of consecutive white processes the token has passed

**state :** (idle,active) initialized to active

**proc_color :** (black,white) initialized to black, being black[3] indicates that $me$ became active between two token visits

**token_present :** boolean initialized to false

## Assumptions.

- In addition to local knowledge ($succ(me)$), we assume processes have global knowledge ($\|C\|$) about the network.

## The Algorithm.

1. *When willing to start detection::*

    **do**
        **send**($Token$, 0, succ(me))
    **endo**

2. *Upon receiving $(Msg, me)$::*

    **do**
        state := active ;
        proc_color := black
    **endo**

---

[3]Unlike the previous algorithm, where black was painted after sending $Msg$.

**3.** *When* wait($Msg, j$)∨(locally terminated)::

   **do**
       state := idle
   **endo**

**4.** *Upon receiving* ($Token,n,me$)::

   **do**
       nb_token := $n$ ;
       token_present := true ;
       **if** nb_token= $\|C\|$∧(proc_color=white) **then**
           Termination Detected
       **else**
           **when** (state=idle) **do**
               **if** proc_color=black **then**
                   nb_token := 0
               **else**
                   nb_token := nb_token+1
               **fi**
           **endo**
           **send**($Token$,nb_token, $succ(me)$)
       **fi**
       proc_color := white;
       token_present := false
   **endo**

**Pros and cons.**

| Class | Metrics | Effects | Remarks |
|---|---|---|---|
| S1 | | + | symmetry degree |
| G1 | | + | global info necessity |
| G2 | | − | topology dependence |
| N2 | | − | peer node failure |
| L1 | | − | message loss |
| L2 | − | See 1 below | message duplication |
| L3 | − | See 2 below | message desequencing |
| L5 | + | See 3 below | bounded transmit time |

**Effects summary.**

1. Message duplication can result in a premature detection. The first token paints the process white, and the second counts the process inappropriately. Furthermore, as every node can initiate a detection session, the above scenario may happen for these multiple tokens as well.

2. In a circuit, $succ(i)$ can be predecessor dependent. Figure 8.6 demonstrates this case. For a token that arrives from below, the

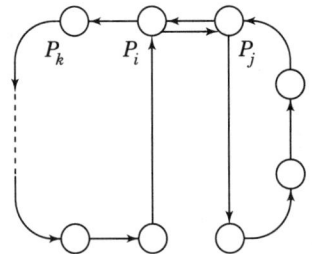

Figure 8.6 Successor dependence on predecessor in a circuit.

predecessor of $p_i$ is $p_j$, and for a token which comes from $p_j$, the predecessor of $p_i$ is $p_k$.

3. The fact that every link in the computation network is included at least once in $C$, assures that there is no outstanding message in the network, as required by Conditions 5 and 6. The zero transmission time requirement can be removed.

**Timestamped termination detection.** The timestamped detection [18] solution also imposes a circuit $C$ topology on the control communication, such that every link in the computation network is included at least once. The network size ($\|C\|$) and node successor in the circuit ($succ(i)$) are known at each node $i$. The algorithm employs Lamport's logical clocks conditions [79] (see Section 6.3) to synchronize logical clocks through communication atomicly, in its *clock-set* function.

- There are certain local events (ticks) for which $i$ advances its clock

$$C_i := C_i + 1. \tag{8.6}$$

- Upon each REC(msg,$C_j$,$j$) at $i$, it advances its clock as follows

$$C_i := \max(C_i, C_j) + 1. \tag{8.7}$$

**Message types used.**

- ($Msg, j$): a message from the underlying computation, with $j$ the sender process ID
- ($Token, clk, nb\_token, j$): a feedback message, with $clk$ being the local time of detection initiation and nb_visit containing the number of processes visited so far
- ($Finished, j$): termination message

**Local variables of process p$_{\text{me}}$.**

**me :** integer (1...nmax) the process ID

$\|C\|$, $succ(me)$ : network structure characteristics

**proc_color** : (black,white) initialized to black, being white indicates $me$ has either initiated the detection or locally terminated

**clk$_{me}$** : timestamp of local clock

**last_date** : timestamp of proc_color becoming white the last time

**finished** : boolean initialized to false

### Assumptions.

- In addition to local knowledge (of $succ(me)$), we assume processes have global knowledge (of $\|C\|$) about the network.
- Atomic nature of clock synchronization is mandatory to assure that no messages will sneak in during the exchange among clocks.

### The Algorithm.

1. *When willing to start detection*::

    **do**
       last_date := $clk_{me}$ ;
       proc_color := white ;
       **send**($Token$,last_date, 1, $succ(me)$)
    **endo**

2. *Upon receiving* $(Msg, j)$::

    **do**
       clock-set($clk_{me}, clk_j$) ;
       proc_color := black
    **endo**

3. *When locally terminated*::

    **do**
       proc_color := white
    **endo**

4. *Upon receiving* $(Token, clk_j, nb\_token, j)$::

    **do**
       clock-set($clk_{me}, clk_j$) ;
       **if** nb_token= $\|C\| \wedge$(proc_color=white) **then**
          finished := true ;
          **send**($Finished, me$) to $succ(me)$
       **else if** nb_token$\neq \|C\| \wedge$(proc_color=white) **then**

           **if** $clk_j >$last_date **then**
               **send**($Token, clk_j, nb\_token + 1, me$) to $succ(me)$
      **fi**  **fi**
**endo**

5. *Upon receiving* $(Finished, j)$::

    **do**
        **if** ¬finished **then**
           finished := true ;
           **send**($Finished, me$) to $succ(me)$
        **fi**
    **endo**

**Pros and cons.**

| Class | Metrics | Effects | Remarks |
|---|---|---|---|
| S1 |  | + | symmetry degree |
| G1 |  | + | global info necessity |
| G2 |  | − | topology dependence |
| L5 | − | See 1 below | bounded transmit time |

**Effects summary.**

1. The algorithm requires the atomic clock setting to take place at two different processes: a requirement equivalent to the instantaneous communication.

### 8.3.5 Quiescence

In this approach [20], we want to detect the monotonic properties of an underlying computation. Both deadlock and termination are such properties, since once they hold, they continue in force as long as there is no external intervention. Detection of quiescence, in which processes do not initiate communication with each other, serves our goal appropriately.

**Message types used.**

- $(Msg)$: a message from the underlying computation
- $(Msg\_reply)$: a reply message for the underlying computation message by the control scheme
- $(Test, j)$: a test message of the diffusing control, with $j$ the test initiator ID
- $(Test\_reply, j)$: a reply message for the diffusing control, with $j$ the test initiator ID

**Local variables of process $p_{me}$.**

1. Underlying computation variables

    **me :** integer (1...nmax), the process ID.
    **I :** boolean (initiator=true, noninitiator=false), initialized to true
    **proc_color :** (white=stable, black=unstable) local stability indicator
    **Msg_def_in, Msg_def_out :** integer (0...nmax), represent the "deficit" in underlying computation communication, counting incoming and outgoing $Msg$s minus $Msg\_reply$s, initialized to 0.
    **Msg_parent :** integer (1...nmax), the ID of the awakening process of $p_{me}$, that is, the first to have its $Msg$ received by $p_{me}$.
    **SDS :** multiset of (1...nmax), initialized empty, which keeps track of all processes which have sent $Msg$s to $p_{me}$.

2. Detection computation variables

    **test_color :** (white=stable, black=unstable) local stability indicator
    **Test_def_in, Test_def_out :** set of integers (0...nmax), each representing the "deficit" in detection communication, counting incoming and outgoing $Test_j$s minus $Test_j\_reply$s, initialized to empty.
    **test_parents :** set of integers (1...nmax), the ID of the processes which have awakened test sessions
    **TIDS :** set of multisets of (1...nmax), initialized empty, which keeps track of all processes which have sent $Test_j$s to $p_{me}$.
    **TODS :** set of multisets of (1...nmax), initialized empty, which keeps track of all processes to which $p_{me}$ has sent $Test_j$s.

**Principle and assumptions.** The approach here follows the diffusing computation model of Section 8.3.4 and the principle of deficits of incoming and outgoing messages. As above, verification of having no outstanding message in the system is performed by replying the sender. The process is painted to present local stability indications (not necessarily monotonic), and under the process invariants, an initiator process can deduce the global stability of the system.

However, this algorithm is symmetric and does not distinguish the initiator of the computation from its assisting processes. Furthermore, it lets the initial state include a few initiating processes, and lets each of them be contacted by its assistants. In order to have local stability indicators which will support the above without deadlocks, the guard on generating the reply is changed along with the establishment of monotonic stability indicators. The algorithm includes a tracing phase which is intertwined with the basic computation and a testing phase in which each process can initiate termination detection query.

- It is assumed that the processes communicate faultlessly.
- It is assumed that the message communication is time bounded, but that there is no need for zero transmission time.

**The Algorithm.**

1. The Underlying Computation ::
   a) *Init*::

       **do**
           **if** $me \in \{noninit\}$ **then**
               proc_color := white ;
               $I$ := false      /* noninitiator */
           **else**
               proc_color := black      /* initiator */
           **fi**
       **endo**

   b) *Willing to send* $(Msg)$ *to* $j$::

       **do**
           $Msg\_def\_out := Msg\_def\_out + 1$
           **send**($Msg$) to $j$
       **endo**

   c) *Upon receiving* $(Msg)$ *from* $j$::

       **do**
           proc_color := black ;
           **if** $I$ **then**
               **send**($Msg\_reply$) to $j$
           **else**      /* $\neg I$ */
               **if** $(Msg\_def\_in = 0)$ **then**
                   Msg_parent := $j$ ;
                   $Msg\_def\_in$ := 1
               **else**
                   **send**($Msg\_reply$) to $j$ ;
                   SDS := SDS $\oplus \{j\}$
               **fi**
           **fi**
       **endo**

   d) *Upon local termination*::

       **do**
           proc_color := white
       **endo**

e) *Upon* $I \wedge$ (proc_color=white)$\wedge(Msg\_def\_in = 0) \wedge (Msg\_def\_out = 0)$::

   **do**
       $I$ := false      /* become noninitiator */
   **endo**

f) *Upon* $\neg I \wedge$(proc_color=white)$\wedge(Msg\_def\_in = 1)\wedge(Msg\_def\_out = 0)$::

   **do**
       **send**($Msg\_reply$) to Msg_parent ;
       $Msg\_def\_in$ := 0
   **endo**

g) *Upon receiving* ($Msg\_reply$) *from* $j$::

   **do**
       $Msg\_def\_out := Msg\_def\_out - 1$
   **endo**

2. The Detection Computation::
   a) *Willing to initiate detection*::

   **do**
       **wait** $\neg I$ ;
       **allocate** TIDS.$me$ := $\emptyset$ ;
       **allocate** TODS.$me$ := SDS($me$) ;
       **allocate** $test\_color.me$ := white ;
       **allocate** $Test\_def\_in.me$ := 0 ;
       **allocate** $Test\_def\_out.me$ := $\|$TODS.$me\|$ ;
       **for** $\forall j \in$TODS.$me$ **do**
           **send**($Test, me$) to $j$ ;
       **endo**
       test_color.$me$ := white
   **endo**

   b) *Upon receiving* ($Test, j$) *from* $k$::

   **do**
       **if** (TIDS.$j$)$\notin$ TIDS **then**
           **allocate** TIDS.$j$ := $\{k\}$ ;
           **allocate** TODS.$j$ := SDS($me$) $\ominus \{k\}$ ;
           **allocate** test_parent.$j$ := $k$ ;
           **allocate** $test\_color.j$ := black ;
           **allocate** $Test\_def\_in.j$ := 1 ;

```
            allocate Test_def_out.j := ||TODS.me|| ;
            for ∀n ∈TODS.j do
                send(Test, j) to n
            endo
        else
            send(Test_reply, j) to k ;
            TIDS.j := TIDS.j ⊕{k}
        fi
    endo
```

c) *Upon* $\neg I \wedge (\text{TIDS}.j \in \text{TIDS}) \wedge (\|\text{TIDS}.j\| = \|\text{SDS}\|)$::

   **do**
         test_color.$j$ := white
   **endo**

d) *Upon* (test_color.$j$=white) $\wedge (Msg\_def\_in.j = 1) \wedge (Msg\_def\_out.j = 0)$::

   **do**
         **send**($Test\_reply, j$) to test_parent.$j$ ;
         $Test\_def\_in.j := 0$
   **endo**

e) *Upon receiving* ($Test\_reply, j$) *from* $k$::

   **do**
         $Test\_def\_out.j := Test\_def\_out.j - 1$
   **endo**

**Pros and cons.**

| Class | Metrics    | Effects     | Remarks              |
|-------|------------|-------------|----------------------|
| S1    | +          |             | symmetry degree      |
| G1    | +          |             | global info necessity |
| G2    | +          |             | topology dependence  |
| L1    | −          | See 2 below | message loss         |
| L2    | −          | See 1 below | message duplication  |
| T1    | $o(M+2NE)$ | See 3 below | average load         |
| T4    | −          | See 4 below | peak correlation     |

**Effects summary.**

1. Counters will function improperly due to message duplications. For example, duplicated ($Msg\_reply$) results in an undesirable decrease of Msg_def_out.

2. Message losses can result in having no ($Msg\_reply$) sent to the parent since Msg_def_out will never reach zero.

3. For $M$ underlying computation messages we need $M$ ($Msg\_reply$) messages. There is at most one ($Test, j$) message sent by each process (out of $N$) in each incident edge (out of $E$). Hence, $o(NE)$ of ($Test, \ldots$) messages and $o(NE)$ of ($Test\_reply, \ldots$) messages.

4. Each underlying computation $Msg$ has an overhead $Msg\_reply$. This overhead is performed not when quiescence has been achieved, but rather during the performance of the basic computation.

### 8.3.6 Algorithm for termination and communication deadlock detection

This section outlines a solution to both deadlock and termination detection situations, both of which are monotonic in nature. It is done by extending [5], which is suited to deadlock only. The incentive to extend the deadlock detection algorithm, and the way it is extended, originate from the following scenario. Let a set of processors participate in a distributed session of a computation, a session we call the "underlying computation." Each of the processors is assigned with a task, and when it finishes its task, a new task may be assigned to it.

The local information in each processor does not contain any definite view of the global state of the distributed environment. Hence, a processor cannot decide locally if it can accept a new task, or if one of the other participants is going to reactivate it eventually for the same task. Furthermore, in the case that it has not yet terminated locally, but is waiting for messages from one or more other participants in the session, if a deadlock exists then the waiting is in vain.

Subsequently, in either of the above cases, a distributed dynamic allocation of processors can be more efficient if it is able to detect termination and deadlock situations, to distinguish them, and to inform the process which is in charge of the dynamic allocation.

**Assumptions and features.**

- Only two network properties are necessary:
  1. The transmission time of a message is finite, but not necessarily zero.
  2. Messages are delivered in the order they were sent.
- Each process can initiate as many queries as it "wants" regarding deadlock or termination detection, as long as it is idle and different queries are distinguishable.
- A two-fold wave is used for this detection, to ensure that the network is clean from underlying computation messages. The Answer to a query will distinguish deadlock detection from termination detection, and a no-answer is interpreted as no detection.

- Each process contains its own status knowledge (its state can be idle or active, the execution of its local computation be completed or not, etc.). In addition, each process knows the processes that can awake it from an idle state, that is, its dependency sets, so that only local information is required. We assume that a process executes only one task at a time. After completion, another task can be assigned to it.

**Principles.**

1. Considering the Wait–for–Graph defined in Section 8.3.2, we call a *leaf node*, a node whose static dependency set is empty. For example, in a tree-structured hierarchy of tasks, the "root" of the tree from the computation point of view (the process that activates the underlying computation), is a leaf node from the detection point of view. As we will see later, a white leaf node plays a special role in this termination-deadlock detection algorithm.

2. The color of a process is a part of its status information, and can be black or white. The *state* of a process can be *terminated, blocked, idle,* or *active*. A process is *active* when it is actively performing a task of the underlying computation, and *idle* otherwise, as long as it has detected no deadlock or termination. When a process starts its execution it becomes *black* and *active*. When it waits for communication it becomes *idle*, but it remains *black* as long as the task is not completed. Only upon completing the task does the process changes its color to *white*. In other words, a process has five feasible status: (active,black), (idle,black), (idle,white), (blocked,black), (terminated,white). Even when a process is either blocked or terminated or idle, it is assumed to be capable of responding to the termination-deadlock detection protocol messages.

3. When a process $P_{me}$ becomes *idle* it is allowed to initiate a detection session. After sending a query message $-(Question,...)-$ to all members of its DDS, and if it is empty, to all members of its SDS, process *me* waits for the eventual answers. In the following, to send a message to its dependency set will mean, to its DDS if not empty, and to its SDS otherwise. If both are empty, a special action should be taken. There is no restriction on the number of detection sessions, initiated by different nodes that are concurrently executed. The query messages carry the identity of the initiator. Each process can initiate more than one detection session and therefore each query of a particular process is distinguished by a *tag*. Only the last query of each initiator is relevant. In the following paragraphs we examine one of these query waves.

4. An idle process that receives a *(Question,...)* message that is strictly "younger" than the previous message related to the same initiator (younger refers to a greater tag number), forwards it to its dependency set. This *forward wave* proceeds until a query message reaches a process which already knows about this query, or a leaf process. We show later that another reason for the wave to stop is the case where a monotonic property has already been detected. However, since an idle process can receive the same query from several origins, we name as its *parent*, the process that is the sender of the first query it has received.

5. Before telling how the back wave of *Answers* is generated, let's define how it propagates. Each time process $P_{me}$ sends a query, it expects an answer message as an acknowledgement, if that detection wave is still up-to-date. So, if process *me* has been permanently in the idle state since it has received the query from its parent, and if it has received *all* the answers it expected, then it sends back the answer message to its parent. Thus, each node performs a boolean AND of all the positive answers. We notice that an active process does not transmit any *(Answer,...)* messages. An active process prevents the algorithm from any conclusion. No deadlock or termination can be detected in a set where there is an executing process.

6. We now examine how the answer messages are generated. If an idle process $P_{me}$ receives a query which $P_{me}$ already knows, then two cases are possible:

   - If it is a query message that process $P_{me}$ has sent itself, then there exists a *cycle* in the path followed by that message. Therefore, the query wave is stopped locally, and $P_{me}$ generates an *Immediate Answer* message as an acknowledgement. This answer message will possibly come back to process $P_{me}$, reversing the same cycle the query passed in the forward wave.
   - If it is a query message that comes to process $P_{me}$ through a different route than the first one, $P_{me}$ also sends an *Immediate Answer* message as an acknowledgement. Indeed, there is no point in postponing this *(Answer,...)* message: the two query paths have at least a node in common before $P_{me}$, and thus the answer paths will have this node in common. The AND function, applied to the *(Answer,...)* messages, will be performed by this common node.

   There are three types of nodes that also send back an *(Answer,...)* upon receiving a *(Question,...)*: an *idle white leaf node*, which is a node that has already terminated its computation, and cannot be reactivated during this specific computation session, a *terminated* node (since termination is a monotonic property, there is no reason to verify it again and again), a *blocked* node for the same reason of monotonicity.

7. The *(Answer,...)*, when generated, is colored by the process color (i.e. is white if this process has already completed its task, black otherwise). As long as the *(Answer,...)* passes through white processes, it retains its color. The moment it meets an incomplete task, it changes its color to black and travels black thereafter.

8. As in the deadlock detection algorithm [5], a very important role is played by what the processes do not do. Since no answer means no detection, discarding messages when no detection should be announced is an "active" part in this algorithm. On the other hand, if an answer message comes back to the initiator with a *black* color or if the initiator process is *black* itself, it means that there are still incomplete tasks, therefore only a *deadlock* is detected. The process consequently takes the *blocked* state. If the answer message is returned with a *white* color, and the process is *white* itself, it means that no incomplete task was met by the detection session, hence a *termination* is detected, and the process switches to a *terminated* state.

**Message types used.**

- *(Question,$k,m,j,n$)* is the query message:

  k - query's initiator ID.
  m - tag number of this detection session from process k.
  j - current sender process ID.
  n - current destination process ID.

- *(Answer,color,$k,m,j,n$)* is the feedback message:

  color - message color: it can be white or black.
  k - query's initiator ID.
  m - the tag number referencing the detection session from process k.
  j - current sender process ID.
  n - current destination process ID.

- *(Msg,.....)*. Any message sent by the underlying computation, not related to the detection protocol.

**196    Chapter Eight**

**Local variables to process $P_{me}$.**

**nmax :** constant that counts the total number of processes in the distributed system[4].

**me :** integer (1...nmax). This is the process ID.

**last :** array [1...nmax] of integer. last[j] contains the tag number of the last *(Question,j,...)* received by process *me*. This variable is initialized to 0.

**number :** array [1...nmax,1...nmax] of boolean. number[j] is a safer implementation of a counter. It is a vector of indicators marking which *Answers* have been received so far for *(Question,last[j],...)*, a query which was forwarded by process *me* . This variable is initialized to False.

**wait :** array [1...nmax] of boolean. wait[j] is set to true after receiving *(Question,last[j],...)*, and as long as process *me* is idle and waiting for answers to that query. This variable is initialized to False.

**parent :** array [1...nmax] of integer (1...nmax). parent[j] is the ID L of the process from which the first *(Question,j,last[j],L,k)* has been received.

**state :** (idle,active,blocked,terminated).

**proc-col:** (black,white).

**token-col:** (black,white). Determines the color of the *(Answer,...)* message to be forwarded according to the process status and the *Answers* it received from its successors.

**DDS(me) :** a set of integers in 1...nmax. The dynamic dependency set of this process.

**SDS(me) :** a set of integers in 1...nmax. The static dependency set of this process.

**The Algorithm.** The following algorithm is implemented in all processes $P_i$ that share the distributed environment . Here $i = me$.

1. *Upon receiving (Msg,...,j,me)::*

    *do*
        *state ← active* ;

---

[4]The knowledge of *nmax* is not necessary. Dynamic arrays may be implemented, and thereby reduce this requirement. It is left as such to improve the readability of the algorithm.

$DDS(me) \leftarrow DDS(me) - \{$ the set of processes from which
                               Msg was expected$\}$ ;
$\forall k \in (1\boxed{\ldots}nmax)$ do $wait[k] \leftarrow False$ ;
$proc - col \leftarrow black$
endo

**2. Upon waiting for communication (computation not completed)::**

do
  $state \leftarrow idle$ ;
  $DDS(me) \leftarrow DDS(me) \cup \{$ the set of processes which may
                                send the expected Msg $\}$ ;
endo

**3. Upon completing local computation::**

do
  $state \leftarrow idle$ ;
  IF DDS(me)=empty THEN
    /* Completing locally with non-empty DDS is an ERROR.*/
    $proc - col \leftarrow white$
endo

**4. When (deciding on Detection) AND (state=idle)::**

do
  $last[me] \leftarrow last[me] + 1$ ;
  $wait[me] \leftarrow True$ ;
  IF $DDS(me) \neq empty$ THEN
    $\forall j \in DDS(me)$ do begin
             Send *(Question,me,last[me],me,j)* ;
             $number[me, j] \leftarrow True$
           end
  ELSE /* DDS(me) = empty */
    $\forall j \in SDS(me)$ do begin
             Send *(Question,me,last[me],me,j)* ;
             $number[me, j] \leftarrow True$
           end
endo

**5. Upon receiving (Question,k,m,j,me) AND (state≠active)::**

do
  IF $m > last[k]$ THEN begin
    $last[k] \leftarrow m$ ;
    $parent[k] \leftarrow j$ ;

$wait[k] \leftarrow True$ ;
IF $state = terminated$ OR $state = blocked$ THEN
   Send *(Answer,token-col,k,m,me,j)*
ELSE /* $state = idle$ */ begin
   $\forall r \in DDS(me)$ do begin
        Send *(Question,k,m,me,r)* ;
        $number[k,r] \leftarrow True$
        end;
   $token - col \leftarrow white$ ;
   IF SDS(me)=empty AND proc-col=white THEN
        /* Leaf Node */
        Send *(Answer,token-col,k,m,me,j)*
   end
end
ELSE /* $m \leq last[k]$ */
   IF wait[k] AND $(m = last[k])$ AND $(j \neq parent[k])$ THEN
   /* Immediate Answer */
   Send *(Answer,proc-col,k,m,me,j)*


endo [5]

**6.** *Upon receiving (Answer,color,k,m,r,me) AND (state=idle)::*

do
   IF $(m = last[k])$ AND $(wait[k])$ THEN begin
   $number[k,r] \leftarrow False$ ;
   IF $color \neq white$ OR $proc - col \neq white$ OR $token - col \neq white$
      THEN $token - col \leftarrow black$ ;
   IF $proc - col = black$
      AND$(\forall j \in DDS(me) : number[k,j] = False)$
      THEN $state \leftarrow blocked$ ;
   IF $token - col = white$
      AND $(\forall j \in SDS(me) : number[k,j] = False)$
      THEN $state \leftarrow terminated$ ;
   IF $(state = terminated)$ OR $(state = blocked)$ THEN begin
      $\forall j \neq me$ AND $wait[j]$ : begin
         /* Locally detected – Inform all waiting parents */
         Send*(Answer,token-col,j,last[j],me,parent[j])*;
         $wait[j] \leftarrow False$
        end ;
      IF $(k = me)$ THEN begin
         DETECTION OF *state* FOR *me*: QUERY SESSION IS OVER ;

---

[5]*Leaf* Node and *Immediate* Answer: see section 8.3.6 No 6.

                    $wait[me] \leftarrow False$
                    end
                end
            end
    endo

**7.** *Upon Reactivation or Initialization*::

   do
       Set $SDS(me)$ according to *Declaration*;
       $DDS(me) \leftarrow empty$ ;
       $\forall k \in (1\boxed{\ldots}nmax)$ do $wait[k] \leftarrow False$ ;
       $proc - col \leftarrow black$
   endo

*end algorithm*

**Important remarks.**

1. The model of computation we use in this section is different from the model used for other deadlock [5] or termination [20] detection. In our model of a deadlock there can exist terminated processes, and they can interfere in deadlock situations. Only for the particular instance, in which *no* processes in the underlying computation are terminated, the other models are equivalent.

   Furthermore, in our model a process $P_i$ can be waiting for a (white,idle) process $P_j$. In this case we assume $P_i$ is waiting for the next instance of process $P_j$ task. Therefore, $P_i$ is potentially in a deadlock set, only if this (white,idle) process $P_j$ will not be eventually reactivated for this particular underlying computation. So, our model is broader than the above, as well as useful and practical.

2. In order to allow message duplications two precautions are taken:
   - A boolean array *number[k,r]* is implemented, instead of a simple counter which may count duplicated replies.
   - A condition $j \neq parent[me]$ is ANDed with the condition of the *immediate* response, in order to prevent a false response which may lead to a false detection (Rule 5 of the algorithm).

3. The variable *proc-col* is not really needed. Its white status can be expressed as "$state = idle$ and $DDS(me) = empty$", and its black status as the complement of this expression. It is left as such in the algorithm for better readability. One may benefit by using it to detect an erroneous implementation, in which a locally terminated process (white) has a nonempty DDS.

4. The algorithm initiated by $P_{me}$ eventually gives the general answer: $P_{me}$ is in a deadlock set or in a terminated set. When the reachability set[6] of $P_{me}$ is the whole set of processors (for example when $P_{me}$ is the root of hierarchical underlying computation and that termination is to be detected), then those properties are global.

5. The monotonicity of termination and deadlock is effective in reducing the number of messages in a query session. We take advantage of this property in using the *terminated* and *blocked* states as "barriers" for further query waves.

| Class | Metrics | Effects | Remarks |
| --- | --- | --- | --- |
| S1 |  | + | symmetry degree |
| G1 |  | + | global info necessity |
| G2 |  | + | topology dependence |
| L1 |  | − | message loss |
| L2 |  | + | message duplication |
| L3 |  | + | message desequencing |
| L4 |  | + | modified message control |
| L5 |  | + | bounded transmit time |
| P1 | $O(d)$ |  | response time |

## 8.4 Concluding Remarks

The existence of a deadlock condition is a serious problem for any distributed computation. The detection of termination is its converse problem. In this chapter we have studued several algorithms for deadlock and termination detection. Each algorithm requires the implementation of a protocol to be followed by each participant in exchanging messages and responding to incoming messages. The complexity of the algorithms and protocols depends on the complexity of the situations they are addressing. The algorithm for termination and communication deadlock detection is rather complex, for example.

In designing a system, we have to not only select the algorithms we may want to implement, but also have to decide under what conditions the algorithm will be activated.

---

[6] See Definition 21 in Section 8.3.3.

Chapter 9

# Agreement Protocols

Decentralization of computing power in systems has strengthened the trend of increasing system robustness by the use of the redundancy of resources with respect to needs. For example, in database applications, there are applications in which it is efficient to distribute copies of the data. There, users are allowed to use more than one resource in order to assure robustness of the data used. In other cases, controllers use multiple resources to sense environmental parameters. Aircraft avionics, for example, includes an altimeter which senses the altitude, and an inertial navigation unit which also provides position parameters. Replication and multiplication increase robustness only if the computation can rationalize the various sources of data. In robust object-based architectures, the invocation of an object depends on answers collected from various resources. This type of architecture calls for agreement between the various answers to properly trigger an invocation. The use of distributed databases and the maintenance of their consistency call for a special type of agreements between objects, each of which performs a partial modification of the database. If there is no agreement on entrance to and exit from such modifications, one can find the database partially modified with contradictory facts.

There are various algorithms and protocols which attempt to work out some of the requirements of a protocol agreement. We present here a collection of such algorithms, each suited for a specific environment. We start with a nonblocking commit protocol, and then describe a voting protocol which we later extend to consensus protocols. We then discuss the issue of approximate agreements and conclude with the model of Byzantine agreements.

## 9.1 Commit

Distributed processing of data requires agreement on the commitment for carrying out the processing by all the nodes involved in the processing. We find two aspects in this commitments:

1. In case of a failure during the committing process, one must be able to recover to an overall consistent system state.
2. In case some resources cannot commit, the execution of local actions can be deferred until all participants in the distributed processing are ready.

### 9.1.1 Crash recovery orientation

Let us start our discussion on committing processing-nodes through an examination of the recovery from a crash, as analyzed in [84]. Let us assume each processing node works in basic steps of actions, in which after executing the action, the local state is defined and communication to other processing nodes takes place. The action can be described by an *intention list* which is kept in a stable storage.

Figure 9.1 describes the basic step of executing a local action and communicating with external processors in order to synchronize the distributed activities. This step can conclude in one of three possible ways: continue to the next step upon an expected response, abort due to expiration of time-outs of write-locks or other resource controllers, or abort due to a missing processing node. We note that each processing node can also fail due to local causes, and thus the recovery process must "awake" the system after a crash to a point from which the basic steps can continue.

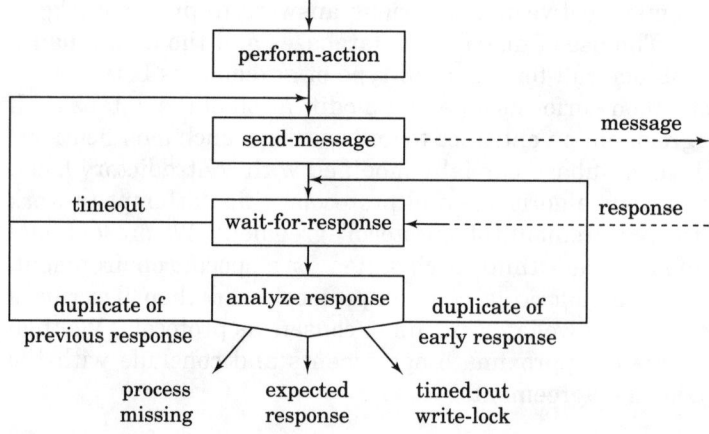

**Figure 9.1** A basic step for commit communication.

The intention lists are kept in the stable storage in order to satisfy this recovery requirement. However, they must demonstrate an idempotent nature if the recovery repeats parts of the list.

> DEFINITION 22.  A list of actions is *idempotent* if the execution of a sequence of actions, formed by a concatenation of any initial segment of the list with the entire list, produces the same results as if the list was executed only once. □

Figure 9.2 describes the committing communications, and the coordinator and cohorts states in a time-diagram. The coordinator starts in an *idle* state when it receives an "end transaction" request from the user application. It sends a "get prepared" message to its cohorts, which after cleaning up all the outstanding commands, acknowledge their state of preparedness. The coordinator then sends a "get ready" message to the cohort, which in turn changes its state to a *ready-to-finish* state after depositing its intention list. The cohort then acknowl-

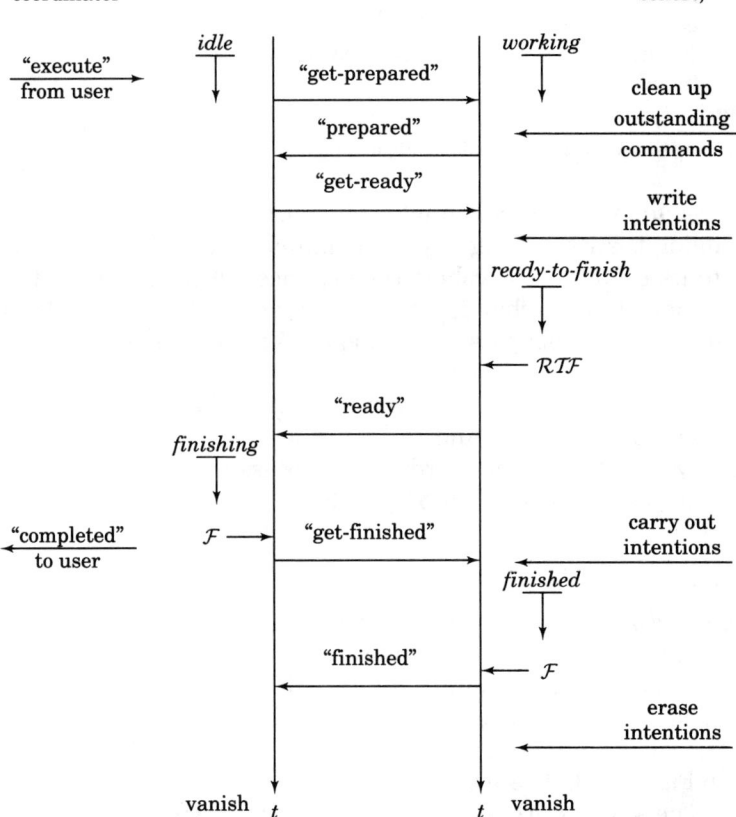

**Figure 9.2** Sequence of crash-recoverable committing.

```
For each coordinator ::
    switch on coordinator_state
        begin
            case idle:
                erase state information
                break
            case finishing:
                initialize coordinator process execution at $\mathcal{F}$
                break
        end
For each cohort ::
    switch on cohort_state
        begin
            case working:
                erase state information and new data blocks
                break
            case ready-to-finish:
                initialize cohort process execution at $\mathcal{RTF}$
                break
            case finished:
                initialize cohort process execution at $\mathcal{F}$
                break
        end
```
**Program 9.1** Crash-recoverable commiting algorithm.

edges its readiness. The coordinator changes its state to *finishing*, acknowledging the requesting application, and sends a "get finished" message to its cohorts. The cohort then carries out its intention list, and upon completion, it changes its state to *finished*, acknowledging the coordinator and erasing its intention list. At that point coordinator and cohort processes can vanish.

Program 9.1 describes the algorithm of recovery based on the state at which the crash occurred and the intention list to be carried out. The recovery points $\mathcal{F}$ in the coordinator process and $\mathcal{RTF}$ and $\mathcal{F}$ in the cohort process are marked in Figure 9.2.

### 9.1.2 Two-phase commit

Let us continue our discussion on committing processing-nodes through the second aspect of resources which are unable to commit. Deferring the execution of local actions until all participants in the distributed process are ready saves resources and avoids blocking.

The classical two-phase commit [125, 126] is described as a state machine in Figure 9.3. There are two types of nodes: a coordinator and $n$ cohorts. The states of these nodes are marked in circles, while the transitions are marked by edges. Near each edge, the communications

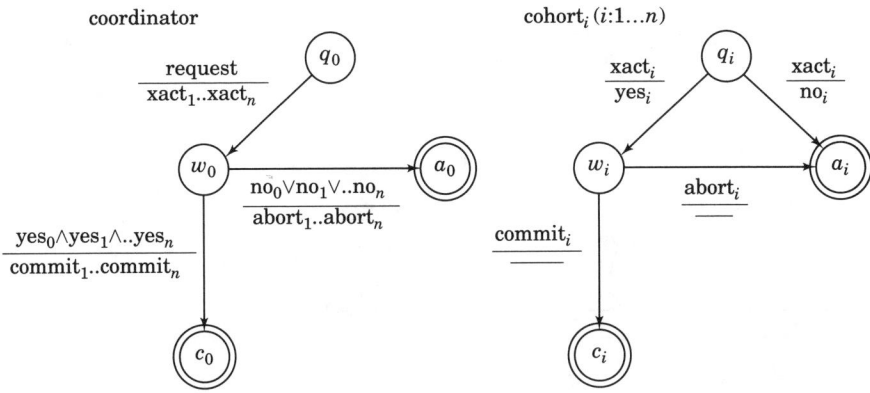

**Figure 9.3** Two-phase commit protocol.

related to this edge are presented. Above the line, we present the communication received by the node which triggered the transition. Below the line, we present the communication sent by the node in association with this transition.

The coordinator starts at state $q_0$, and is invoked by a request from the user. It issues execute messages ($\text{xact}_1$ to $\text{xact}_n$) and turns to a wait state $w_0$. Cohort $i$ starts at $q_i$ and is invoked by the $\text{xact}_i$ message. If it is capable of committing, it answers $\text{yes}_i$ and turns to the wait state $w_i$. If it cannot, it answers $\text{no}_i$, and turns to an abort state $a_i$.

If the coordinator, now at state $w_0$, is capable of committing and all cohorts agree to commit, it issues $\text{commit}_i$ messages to all the cohorts and turns to its commit state $c_0$. If, on the other hand, the coordinator or any cohort does not agree to commit, the coordinator issues $\text{abort}_i$ messages to all the cohorts which agreed to commit and turns to its abort state $a_0$.

Each cohort which agreed to commit waits in its $w_i$ state. If it receives a $\text{commit}_i$ message, it turns to its commit state $c_i$. If it receives an $\text{abort}_i$ message, it aborts into its $a_i$ state.

DEFINITION 23. A local state $\sigma$ is a *committable* state if occupancy of the state by any node implies that all nodes have voted *yes* on committing. □

Clearly, noncommittable states do not include any knowledge of whether all the nodes have voted "yes."

Figure 9.4 describes the reachable states of a simple two-phase commit protocol. For simplicity, we describe a system with a single cohort avoiding cohort concurrency description. Each state consists of the states of the nodes (written above the line) and the outstanding messages in the network (written below the line). Let us define some important sets of states in this reachability graph.

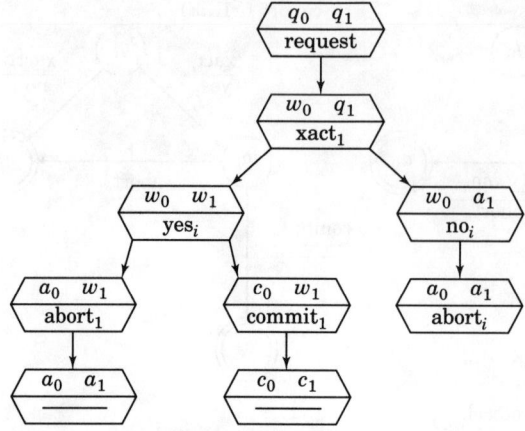

**Figure 9.4** Reachable states of two-phase commit protocol.

DEFINITION 24. The *concurrency set* of a local state $\sigma$ (denoted $C(\sigma)$) is the set of all local states which are potentially concurrent with it. □

From this definition, we note that in Figure 9.4

$$C(w_1) = \{ w_0, a_0, c_0 \}$$

DEFINITION 25. The *sender set* of a local state $\sigma$ (denoted $S(\sigma)$) is the set of all local states which potentially send messages to this state:

$$S(\sigma) \equiv \{t|\ t\ sends\ m\ \wedge\ m \in M \}$$

where $M$ is the set of all messages received by $\sigma$. □

From this definition, we note that in Figure 9.4

$$S(w_1) = \{ w_0 \}$$

In a fully decentralized approach, the asymmetry of the coordinator and the cohorts is inadequate. Figure 9.5 describes a symmetrical implementation of the above two-phase commit protocol. The performance of the symmetrical algorithm is of $O(n^2)$.

- $n$ xact messages initiate the process.
- Every node $i$ sends $yes_{i,1} \ldots yes_{i,n}$.
- Every node $i$ sends $prep_{i,1} \ldots prep_{i,n}$.

Therefore, $n + 2n(n-1) = 2n^2 - n$ messages are required for a commit.

An important condition on the blocking properties of the protocol is given in [125]:

CONDITION 9. A protocol is *nonblocking* if and only if it satisfies both following requirements:

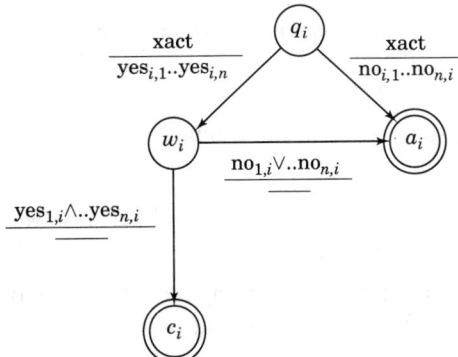

Figure 9.5 Symmetric two-phase commit protocol.

1. $\not\exists \sigma$: abort$\in C(\sigma) \bigwedge$ commit$\in C(\sigma)$.
2. $\not\exists \sigma$: $\sigma$=noncommittable $\bigwedge$ commit$\in C(\sigma)$. □

### 9.1.3 Nonblocking three-phase commit

The two-phase commit is therefore a blocking protocol, having both abort and commit in $C(w)$. Therefore, an additional buffer state must be added as a subsequent state of $w$, to allow the removal of the commit state from its concurrency state.

Figure 9.6 describes a symmetric nonblocking three-phase commit protocol, for which an additional buffer state ($p_i$) assures the nonblocking property. However, failures in the system must be considered in order to assure recovery properties. The only nonblocking recovery scheme is the independent recovery.

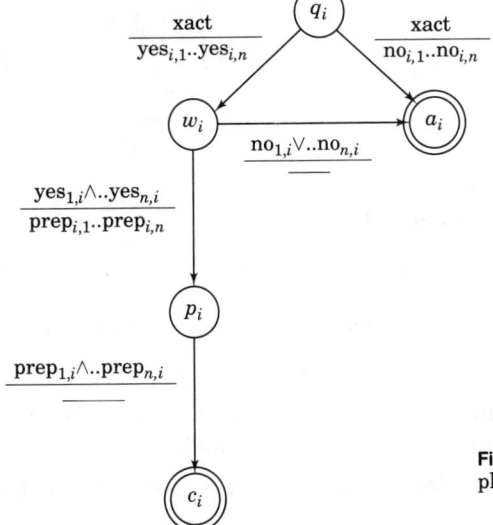

Figure 9.6 Symmetrical three-phase commit protocol.

DEFINITION 26. An *independent recovery* is a scheme where a recovering processing node makes a transition directly to the final state, without communicating with other nodes. □

Failure transitions can be assigned to accommodate faults detected in the concurrency set and in the sender set.

- Node failure: detected by the local built-in test mechanisms, result in a local incomplete processing of a basic step.
- Timeout: detected by a local timer expiration, as emphasized in Figure 9.1, mostly due to link failures or missing processes.

Analysis of single node failures yields two design rules which are sufficient to assure protocol resiliency for a single node failure.

CONDITION 10. For every intermediate state $\sigma$ in the protocol: if $C(\sigma)$ contains a commit, assign a *failure transition* from $\sigma$ to *commit*. Otherwise, assign a failure transition from $\sigma$ to *abort*. □

CONDITION 11. For every intermediate state $\sigma_i$ in the protocol: if there exists a state $\tau_i$ in $S(\sigma_i)$ which has a failure transition to commit (/abort), assign a *timeout transition* from $\sigma_i$ to commit (/abort). □

Hence, we can modify the protocol described in Figure 9.6 as follows, to support the above types of failures:

- According to Condition 10, we add node failure transitions
    1. from $q_i$ to $a_i$, as $a_i \in C(q_i)$
    2. from $w_i$ to $a_i$, as $a_i \in C(w_i)$
    3. from $p_i$ to $c_i$, as $c_i \in C(p_i)$
    
    The set of failure transitions, $F_t$ is therefore

$$F_t = \{ (q_i, a_i) (w_i, a_i) (p_i, c_i) \} \qquad (9.1)$$

- According to Condition 11, we add timeout transitions
    1. from $w_i$ to $a_i$, as $q_i \in S(w_i) \land (q_i, a_i) \in F_t$
    2. from $p_i$ to $a_i$, as $w_i \in S(p_i) \land (w_i, a_i) \in F_t$
    
    The set of timeout transitions, $T_t$ is therefore

$$T_t = \{ (w_i, a_i) (p_i, a_i) \} \qquad (9.2)$$

Two interesting conclusions rise from analyzing this algorithm in the multiple node failures cases [126]:

1. The only nonblocking node recovery strategy, the independent recovery, is resilient to the failure of a single mode.
2. A recovering node cannot determine whether it is safe to use independent recovery.

```
type
    ObjectJoint =
        record
                    (... see Program 17.1)
            ReplicaForum: set of ↑ObjectJoint ;
            VersionNumber: integer,unknown ;
            r, w: integer ;
            vote_r, vote_w: integer
        end

function Read (Obj: ↑ObjectJoint; item: DataBlockId) : Datablock ;
    var
        quorum: set of ↑ObjectJoint; best: ↑ObjectJoint;
    begin
        quorum := CollectReadQuorum ( Obj.ReplicaForum ) ;
        best := SelectReadRepresentative ( quorum ) ;
        Read := invoke (best.GetData.item )
    end
```
**Program 9.2** Weighted voting algorithm: read quorum.

## 9.2 Weighted Voting

The weighted voting algorithm for replicated data [57, 132] has set up the foundation of a family of quorum consensus algorithms. Out of its concepts, let us emphasize some major characteristics.

- Every passive object which includes or supplies data items is given a voting weight. Given a lower bound to the number of replicas of this object, there exists a lower bound to the number of votes. Let us denote the total number of votes as $v$.

- The data included within the object is marked by a version number, which maintains a global total order and thus allows identification of the most recent copy.

- Every executable object (e.g., a transaction manager) which uses the data included in the passive object, needs a quorum of $r$ votes to *read* from this object and $w$ votes to *write* to this object.

- The quorum restrictions are

$$r + w > v \quad \bigwedge \quad w > \frac{v}{2} \qquad (9.3)$$

The left restriction [57] ensures that there is a nonempty intersection between any read quorum and any write quorum. This property guarantees that any read quorum includes the most recent copy of the object. The right constraint [132] ensures that no two updates can occur in parallel.

```
function CollectReadQuorum (objects: set of ↑ObjectJoint) :
    set of ↑ObjectJoint ;
  var
    quorum: set of ↑ObjectJoint; i: ↑ObjectJoint;
    lastver, votes, limit: integer;
  begin
    quorum := ∅; lastver := votes := 0 ; limit := ∞ ;
    for ∀i ∈objects do
      break on votes≥limit ;
      if i.VersionNumber≠unknown then
        if i.VersionNumber>lastver then
          limit := i.r ;
          lastver := i.VersionNumber
        fi
        votes := votes + i.vote$_r$ ;
        quorum := quorum ⊕ {i}
      fi
    od
    CollectReadQuorum := quorum
  end
```

**Program 9.2**  Weighted voting algorithm: read quorum (continued).

Programs 9.2 and 9.3 describe procedures and functions employed in implementation of the weighted voting algorithm. The first program describes reading from replicated objects, and the second describes writing to replicated objects. It is assumed that the objects are constructed according to Section 1.2.3, each having a *joint* in which it contains its ID, a pointer to the object's body, its resource (and/or server) requirements, an access-control scheme, a time constraint for real-time object, and other features as described in Program 17.1. In addition, the joint contains a set of pointers to the joints of the replicas, a version number which reflects the last update, the size of the required read and write quorum, and the weight assigned to the vote of this specific object.

In the read quorum example we have chosen to describe only the Read function and the Collect-Read-Quorum function, in order to demonstrate the voting mechanism. However, the Select-Read-Representative function is also very important, but it is more application dependent. There, one may want to choose the object with the maximum weight, the object with fastest response, or the most recent copy.

The example of the Write procedure and the Collect-Write-Quorum function demonstrate the difference between the read and write. It is important to write to a quorum of recent copies only, avoiding updates

**procedure** Write (Obj: ↑ObjectJoint; block: DataBlock);
   **var**
      quorum: **set of** ↑ObjectJoint; $i$: ↑ObjectJoint;
   **begin**
      quorum := CollectWriteQuorum ( Obj.ReplicaForum ) ;
      **for** $\forall i \in$ quorum **do**
         UpdateVersion ( $i$.VersionNumber ) ;
         invoke ($i$.PutData ,block )
      **od**
   **end**
**function** CollectWriteQuorum (objects: **set of** ↑ObjectJoint) :
   **set of** ↑ObjectJoint ;
   **var**
      readQuorum, writeQuorum: **set of** ↑ObjectJoint; $i$: ↑ObjectJoint;
      lastver, writeVotes, readVotes, readLimit, writeLimit: integer;
   **begin**
      readQuorum := writeQuorum := ∅;
      lastver := readVotes := writeVotes := 0 ; readLimit :=
         writeLimit := ∞ ;
      { Try to form a quorum of current representatives }
      **for** $\forall i \in$objects **do**
         **break on** writeVotes≥writeLimit ;
         **if** $i$.VersionNumber≠unknown **then**
            **if** $i$.VersionNumber>lastver **then**
               readLimit := $i.r$ ; writeLimit := $i.w$ ; writeVotes := $i$.vote$_w$ ;
               writeQuorum := writeQuorum $\oplus$ $\{i\}$ ;
               lastver := $i$.VersionNumber
            **else**
               writeVotes := writeVotes + $i$.vote$_w$ ;
               writeQuorum := writeQuorum $\oplus$ $\{i\}$
            **fi**
         readVotes := readVotes + $i$.vote$_r$ ;
         readQuorum := readQuorum $\oplus$ $\{i\}$
      **od fi**
      { Exhausted forum without a quorum of current representatives? }
      **if** writeVotes<writeLimit $\wedge$ readVotes>readLimit **then**
         **for** $\forall i \in$readQuorum $\wedge$ $i \notin$writeQuorum **do**
            **break on** writeVotes≥writeLimit ;
            CopyCurrentVersion ( writeQuorum,$i$ ) ;
            writeVotes := writeVotes + $i$.vote$_w$ ;
            writeQuorum := writeQuorum $\oplus$ $\{i\}$
      **fi od**
      **if** writeVotes≥writeLimit **then** CollectWriteQuorum := writeQuorum
   **end**

**Program 9.3** Weighted voting algorithm: write quorum.

to obsolete representatives. All the writes can be done in parallel, once a quorum of current representatives has been established.

## 9.3 Consensus

Let us now consider the generalization of algorithms for reaching an agreement and maintaining consistency that can be used to solve the problem of consensus. In such a problem, each computation node is associated with a set of other nodes (forum) with which it communicates in order to develop a unified resolution. In order to minimize communication, proposed algorithms define a partial set (quorum) which is sufficient to develop a resolution. We now present different approaches which deal with quorum consensus:

1. A decentralized consensus scheme [78] uses finite projective planes for setting up communicating subsets to enhance the performance of the classical commit approach [125, 126] as well as to extend its application to agreement on the evaluation of associative functions or predicates.

2. A history based approach [63] generalizes the weighted voting algorithm [57] and enhances its robustness by merging histories of the nodes.

3. IO automata [59] extends the weighted voting algorithm [57] to include nested transactions and to accommodate transaction failures.

### 9.3.1 Decentralized commit and consensus

The following algorithm [78] applies to cases of a completely decentralized commitment scheme. In such a scheme, we assume that no single site controls the computation, and according to that assumption, the asymmetrical architecture of coordinator and cohorts is inadequate. The cost of a symmetrically decentralized commitment scheme has direct effects on communication overhead. For example, the scheme in Figure 9.5 in Section 9.1.2, demonstrates the performance of a symmetrical algorithm of $O(n^2)$. The following algorithm uses finite projective planes to set up intersecting subsets which intercommunicate and thus reduce the overhead.

> DEFINITION 27. A *finite projective plane* is a set of nodes and lines which obey the following properties:
>
> 1. Two distinct nodes are connected by one and only one line.
> 2. Two distinct lines share one and only one common nodes.

3. For every $m+1$ distinct nodes, no $m$ of them lie on the same line; $m$ is the *order* of this plane. □

Consider a finite projective plane with $n = m^2 + m + 1$ nodes and $n = m^2 + m + 1$ lines, for which there are $m+1$ nodes for each line and $m+1$ lines through each node. The order of this plane is $m$. Let us associate a line $L_i$ with each node $i$, such that $L_i$ shares $i$ with other $m$ lines. Therefore, $i$ receives messages from $2m$ nodes: the $m$ nodes that lie on $L_i$ and the $m$ nodes that are associated with the other lines which pass through node $i$.

Since $n = m^2 + m + 1$, we have

$$m = \frac{-1 + \sqrt{4n-3}}{2} \qquad (9.4)$$

In other words, the number of messages sent is $O(n\sqrt{n})$. However, we cannot always insure the existence of a finite projective plane with $n$ nodes. These cases embed an additional cost of maintenance of virtual nodes that achieves the symmetrical structure, but maintain the $O(n\sqrt{n})$ communication cost.

Figure 9.7 describes the consensus commit algorithm with two rounds of message exchanges. According to Condition 9 the protocol is nonblocking as it satisfies both requirements:

1. $\not\exists \sigma$: abort$\in C(\sigma) \bigwedge$ commit$\in C(\sigma)$.
2. $\not\exists \sigma$: $\sigma$=noncommittable $\bigwedge$ commit$\in C(\sigma)$. □

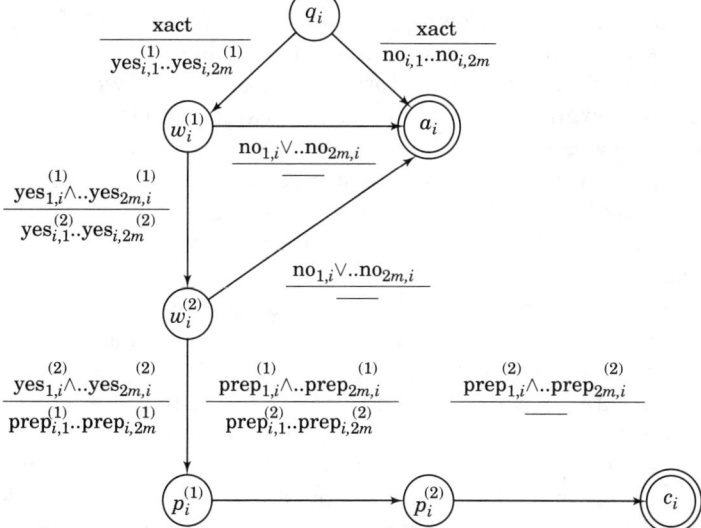

**Figure 9.7** Symmetrical consensus commitment protocol.

## 9.3.2 History-based quorum consensus

As discussed in Section 14.1.1, the history [64] of an object is the sequence of serial events which models its internal state. Since concurrent execution is allowed, an object's state is given by a trace of action executions, commit executions, and abort executions. The history is therefore a serialization of the concurrent trace of an object in a given ordering.

Based on this approach, the serial specification of an object is the set of permissible sequences of events of this object. The behavioral specification of an object describes the conflicts between operations that limit its concurrency and thus affect its trace. The following FIFO queue example emphasizes these specifications. The serial specification can be formalized as:

$$\#(EnQueue) \geq \#(DeQueue) \tag{9.5}$$

The behavioral specification reflects the fact that the returned value depends on the queue's state and is illustrated in the following conflict-table. EnQueue is always allowed, but Dequeue is in conflict with both EnQueue and Dequeue operations.

FIFO Queue Conflict-Table

|  | EnQueue /OK | Dequeue /OK |
|---|---|---|
| EnQueue/OK |  |  |
| DeQueue/OK | + | + |

Let us now consider the following history-based algorithm [63], in which each object maintains its local history variable. Failures can often lead to incomplete local subhistories, a disadvantage which can be dealt with by merging subhistories of replicated objects. In fact, in this algorithm we represent an object by its history rather than its "value." The algorithm associates with any action execution, a read quorum of $r$ objects to *read* from and a write quorum of $w$ objects to *write* to. Program 9.4 describes the algorithm for an object with a serial specification $S$.

The algorithm first collects a read quorum to assure a proper history of the item in concern. The function merge_and_test merges the particular subhistories which match the serial specification $S$. Thus, it is possible to create a more complete history $h_D$. If the latter is permissible, the required action is executed and the updated history is distributed to a write quorum. The quorum members are selected out of "object.ReplicaForum," a forum of objects which are pointed by the object's joint. The quorum sizes, object.action.$r$, and object.action.$w$ are action specific and they maintain the relation in Equation 9.3.

```
type
    ObjectJoint = record
                (... see Program 17.1)
        ReplicaForum: set of ↑ObjectJoint ;
        action₁ record
            r, w: integer ; end
        action₂ record
            r, w: integer ; end
        end
procedure ExecuteAction (action: ↑ObjectJoint.actionᵢ ;
    object: ↑ObjectJoint) ;
    var
        readQuorum, writeQuorum: set of ↑ObjectJoint; i: ↑ObjectJoint;
        writeVotes, readVotes, readLimit, writeLimit: integer;
        hD, historyᵢ: history;
        timestamp: time_interval;
    begin
        hD := ∅; readQuorum := writeQuorum := ∅;
        readVotes := writeVotes := 0; readLimit := object.action.r ;
        writeLimit := object.action.w ;
            { Form a read quorum }
        for ∀i ∈object.ReplicaForum do
            send request(JoinReadQuorum) od
        for ∀i ∈object.ReplicaForum ∧ i ∉readQuorum do
            break on readVotes≥readLimit ;
            if accepted historyᵢ then
                readVotes := readVotes + 1 ;
                readQuorum := readQuorum ⊕ {i} ;
                hD := merge_and_test(hD,historyᵢ,S)
        od fi
```

**Program 9.4** History-based quorum consensus algorithm.

The behavioral specification shows the relationships between different actions in an object. If an object has two different actions, say action₁ and action₂ as in Program 9.4, we can come out with an acceptability test of

$$object.action_2.r + object.action_1.w > v \qquad (9.6)$$

where $v = |object.ReplicaForum|$. This test assures the two quorums intersect. Thus, merging event(ExecuteAction) of an execution of action₂ includes the outputs of action₁. It is clear that a symmetrical constraint is required for including outputs of action₂ in events of execution of action₁:

$$object.action_1.r + object.action_2.w > v \qquad (9.7)$$

```
        if permissible($h_D$) then
            $h_D$ := $h_D \oplus$ event(ExecuteAction,timestamp) ;
            { Form a write quorum }
            for $\forall i \in$object.ReplicaForum do
                send request(JoinWriteQuorum) od
            for $\forall i \in$object.ReplicaForum $\wedge\ i \notin$writeQuorum do
                break on writeVotes≥writeLimit ;
                if accepted from $i$ then
                    writeVotes := writeVotes + 1 ;
                    writeQuorum := writeQuorum $\oplus\ \{i\}$
        od fi
            for $\forall i \in$writeQorum do
                send update($h_D$,timestamp) od
    fi
end
```

**Program 9.4** History-based quorum consensus algorithm (continued).

### 9.3.3 Quorum consensus with nested transactions

In the following section, we discuss an IO automata model [59] and an extension of the weighted voting algorithm [57] to include nested transactions and to accommodate transaction failures. The architecture proposed in this model employs DMs (data-manager automata) to retain state information and TMs (transaction managers) to physically access subsets of the DMs to accomplish read and write operations. In order to read, a TM accesses the DMs of a read quorum and chooses the value with the latest version, and in writing the TM, finds that the latest version so far and updates a whole write quorum DMs with the new version.

A system is modeled as a set of interfacing IO automata. Each automaton state transitions are associated with operation names. An IO automaton $\mathcal{A}$ is identified by

- $states(\mathcal{A})$: a set of states
- $init(\mathcal{A})$: the initial state of the automaton ($\subset states(\mathcal{A})$)
- $in(\mathcal{A})$: a set of input operations
- $out(\mathcal{A})$: a set of output operations ($in(\mathcal{A}) \cap out(\mathcal{A}) = \emptyset$)
- $steps(\mathcal{A})$: the transition relation of the automaton, defined as a set of triples $(s', \pi, s)$, where $s', s \in states(\mathcal{A})$ and $\pi \in in(\mathcal{A}) \cup out(\mathcal{A})$.

An execution of $\mathcal{A}$ is thus an alternating sequence $s_0, \pi_1, s_1, \ldots, \pi_n, s_n$, where $s_i \in states(\mathcal{A})$ $\pi_i \in in(\mathcal{A}) \cup out(\mathcal{A})$ and $\forall i : (s_i, \pi_{i+1}, s_{i+1}) \in steps(\mathcal{A})$. The subsequence which contains only the operations $(\pi_1, \pi_2, \ldots, \pi_n)$ is the execution *schedule*.

A serial system is a composition of a set of IO automata. The system primitives are transactions and basic objects and are organized in a tree structure. The root node ($T_0$) models the environment. The internal nodes create and manage subtransactions but do not manipulate the actual data. The leaves (called *accesses*) are the only nodes which actually access the data. The set of permissible transaction response values is denoted as $V$. Let $S$ be a system, and let $r$ and $w$ be the set of read quorums and the set of write quorums, respectively. The set $configuration(S)$ consists of the pairs $< r, w >$, where $r, w \in 2^S$. The set $legal(S)$ is a subset in $configuration(S)$, for which each element in $r$ has a nonempty intersection with each element in $w$.

Let us now consider the system primitives, nonaccessing transactions, and basic objects. The *in* set of a nonaccessing transaction $T$ consists of

- CREATE($T$)
- COMMIT($T', v$), $T' \in$ children($T$), $v \in V$
- ABORT($T'$), $T' \in$ children($T$)

The *out* set of a nonaccessing transaction $T$ consists of

- REQ_CREATE($T'$), $T' \in$ children($T$)
- REQ_COMMIT($T, v$), $v \in V$

CREATE starts the local execution of transaction $T$. $T$ sends REQ_CREATE ($T'$) requests to its children, which return COMMIT($T', v$) or ABORT($T'$) upon successful or unsuccessful completion, respectively. $T$ completes successfully with a REQ_COMMIT($T, v$) announcement to its parent. We note that a transaction is created only once and requests the creation of each particular child only once. In addition, a transaction receives neither repeating nor contradicting returns from any child, and receives no returns at all from a child whose creation has not been requested. It invokes no output operation before being created, nor after requesting commit.

The *in* set of a basic object automaton $X$ consists of

- CREATE($T$), $T \in accesses(X)$

The *out* set of a basic object automaton $X$ consists of

- REQ_COMMIT($T, v$), $T \in accesses(X)$, $v \in V$

Schedules of basic objects are restricted to alternating sequences of CREATE and REQ_COMMIT operations, starting with CREATE. Each pair of CREATE and REQ_COMMIT must refer to the same access transaction.

A serial scheduler controls communication between the primitives (transactions and basic objects) defining the permissible execution or-

- REQ_CREATE($T$) $\implies$ create_req($s$) = create_req($s'$)$\cup\{T\}$

- REQ_COMMIT($T, v$) $\implies$ commit_req($s$) = commit_req($s'$)$\cup\{(T,v)\}$

- CREATE($T$)
  $\wedge\quad T \in$ create_req($s'$) - ( created($s'$)$\cup$abortes($s'$) )
  $\wedge\quad$ sibling($T$)$\cap$created($s'$) $\subseteq$ returned($s'$)
$\implies$ created($s$) = created($s'$)$\cup\{T\}$

- COMMIT($T, v$)
  $\wedge\quad (T,v) \in$ commit_req($s'$)
  $\wedge\quad T \notin$ returned($s'$)
  $\wedge\quad$ children($T$)$\cap$create_req($s'$) $\subseteq$ returned($s'$)
$\implies$ committed($s$) = committed($s'$)$\cup\{(T,v)\}$
  $\wedge\quad$ returned($s$) = returned($s'$)$\cup\{T\}$

- ABORT($T$)
  $\wedge\quad T \in$ create_req($s'$) - ( created($s'$)$\cup$abortes($s'$) )
  $\wedge\quad$ sibling($T$)$\cap$created($s'$) $\subseteq$ returned($s'$)
$\implies$ aborted($s$) = aborted($s'$)$\cup\{T\}$
  $\wedge\quad$ returned($s$) = returned($s'$)$\cup\{T\}$

**Program 9.5** Serial scheduler for nested transactions.

der by running the transactions at a depth-first traversal of the transactions tree. It can abort a transaction whose creation has been requested but has not yet been created. Its *states* set is reflected by six variables: create_req, created, commit_req, committed, aborted, returned. Its *initial* set consists of

$$create\_req = \{T_0\}$$

and the other five variables are empty. The *in* set of the serial scheduler consists of

- REQ_CREATE($T$)
- REQ_COMMIT($T, v$).

The *out* set of the serial scheduler consists of

- CREATE($T$)
- COMMIT($T, v$)
- ABORT($T$).

Program 9.5 describes the rules obeyed by the *steps* set of the serial scheduler, a set which consists of the $(s', \pi, s)$ transition triples.

Replicas are modeled by DMs (data managers) which are basic objects with read and write accesses. Each read-write object $O$ has a set

- CREATE($T$), $T \in accesses(O)$
$\Longrightarrow$ active($s$) = $T$

- REQ_COMMIT($T, v$) $\land$ attribute($T$)=read
  $\land$ active($s'$)=$T$
  $\land$ $v$=data($s'$)
$\Longrightarrow$ active($s$)=nil

- REQ_COMMIT($T, v$) $\land$ attribute($T$)=write
  $\land$ active($s'$)=$T$
  $\land$ $v$=nil
$\Longrightarrow$ active($s$)=nil
  $\land$ data($s$)=$d$

**Program 9.6** Basic object with read/write accesses.

of transactions that access it $accesses(O)$. The *states* of a read-write object $O$ with a domain $D$ are defined by

- active: the current access to $O$
- data: an element of $D$

The *steps* of $O$ consist of the $(s', \pi, s)$ transition triples which obey the pre- and post-conditions described in Program 9.6. Each member $v$ of the domain $D$ is denoted as a pair (version,value).

The quorum consensus is implemented by TMs (transaction managers) which are nonaccessing nodes. There are two types of TMs: a read-TM and a write-TM. For a logical data item $x$, the read-TM ($T_r$) invokes accesses to multiple DMs for $x$ to review as a read quorum and to choose the latest version of $x$. The *in* set of $T_r$ consists of

- CREATE($T$)
- COMMIT($T', v$), $T' \in$children($T$), $v \in D_x$
- ABORT($T'$), $T' \in$children($T$)

The *out* set of $T_r$ consists of

- REQ_CREATE($T'$), $T' \in$children($T$)
- REQ_COMMIT($T, v$), $v \in D_x$

The *states* set of $T_r$ consists of four state variables: awake (boolean), data (a value of $D_x$), requested (a subset of accesses($x$)), and read (a subset of the DMs of $x$). In the *init* set, awake is false and requested and read are empty. Program 9.7 describes the $(s', \pi, s)$ transition triples of the *steps* set of $T_r$.

The write-TM ($T_w$) has the same *in* and *out* sets as $T_r$. The *states* set of $T_w$ consists of six state variables: awake (boolean), data (a value of

- CREATE($T$) $\implies$ awake($s$) = true

- REQ_CREATE($T'$) $\wedge$ attribute($T'$)=read
  $\wedge$ awake($s'$)= true
  $\wedge$ $T' \notin$ requested($s'$)
  $\implies$ requested($s$) = requested($s'$)$\cup\{T'\}$

- COMMIT($T', d$)
  $\implies$ read($s$)=read($s'$)$\cup\{T'\}$
  $\wedge$ **if** $d$.version>data($s'$).version **then** data($s$)=$d$

- ABORT($T'$) $\implies$ skip

- REQ_COMMIT($T, v$)
  $\wedge$ awake($s'$)= true
  $\wedge$ $\exists q \in$config($x$).$r$: $q \subseteq$read($s'$)
  $\wedge$ $v$=data($s'$).value
  $\implies$ awake($s$)= false

**Program 9.7** Transaction manager with read access.

$D_x$), read-requested and write-requested (subsets of accesses($x$)), and read, and written (subsets of the DMs of $x$). In the *init* set, awake is false and read-requested, write-requested, read and written are empty. Program 9.8 describes the ($s', \pi, s$) transition triples of the *steps* set of $T_w$.

The write-TM starts like the read-TM with invocations of read accesses, to verify the latest version of the data item. Then, it begins invoking write accesses. As presented in the REQ_CREATE($T'$) operation in Program 9.8, the version of the local data item is incremented for the access. Read commit is restricted to an empty write-requested set, since read accesses respond concurrently to write accesses after a quorum has been established. A late read access can return the write version index, and thereby increase it again for a later write access, and that is what the restriction prevents.

In order to accommodate transaction failures, one must cope with flexible reconfiguration of the system. In such cases, the read and write quorums are permitted to change dynamically during execution due to node or link failures. To accomplish that, we need an additional type of transaction manager, a reconfigure-TM, which preserves the following properties:

- It is a child of the user transaction, and thereby the proper atomicity restrictions are satisfied.

- It runs spontaneously as far as user transactions are concerned, and thus, a user transaction is not aware of its invocations.

- CREATE($T$) $\Longrightarrow$ awake($s$) = true

- REQ_CREATE($T'$) $\wedge$ attribute($T'$)=read
  $\wedge$  awake($s'$)= true
  $\wedge$  $T' \notin$ requested($s'$)
  $\Longrightarrow$ read-requested($s$) = read-requested($s'$)$\cup\{T'\}$

- COMMIT($T', d$) $\wedge$ attribute($T'$)=read
  $\Longrightarrow$ **if** write-requested($s'$)={} **then**
       read($s$)=read($s'$)$\cup\{T'\}$
       **if** $d$.version>data($s'$).version **then** data($s$).version=$d$.version

- REQ_CREATE($T'$) $\wedge$ attribute($T'$)=write $\wedge$ data($T'$)=$d$
  $\wedge$  awake($s'$)= true
  $\wedge$  $\exists q \in$ config($x$).$r$: $q \subseteq$ read($s'$)
  $\wedge$  $d$ =<data($s'$).version+1,value($T$)>
  $\wedge$  $T' \notin$ write-requested($s'$)
  $\Longrightarrow$ write-requested($s$) = write-requested($s'$)$\cup\{T'\}$

- COMMIT($T', d$) $\wedge$ attribute($T'$)=write
  $\Longrightarrow$ written($s$)=written($s'$)$\cup\{T'\}$

- ABORT($T'$) $\Longrightarrow$ skip

- REQ_COMMIT($T, v$)
  $\wedge$  awake($s'$)= true
  $\wedge$  $v$= nil
  $\wedge$  $\exists q \in$ config($x$).$w$: $q \subseteq$ written($s'$)
  $\wedge$  $v$=data($s'$).value
  $\Longrightarrow$ awake($s$)= false

**Program 9.8**  Transaction manager with write access.

- It is transparent to the user transaction, and thus, the user transaction is not aware of its returns.

Consistency maintenance can be achieved by extending the version specification to include configuration identification of the replica.

## 9.4 Approximate Agreement

In various problems, achievement of a unanimous agreement is difficult if not impossible. For example, collection of data from a set of sensors may never come out unanimously with a value. We generally consider agreement in such cases as having only small differences between the readouts. Furthermore, a noisy environment with a small number of faulty measurements which are unpredictable (i.e., entirely inconsistent and possibly malicious) is a common phenomenon in real-life systems. Therefore, the term *agreement* must be considered with an interpretation which is weaker than a unanimous decision [49]. Let us consider a computation graph in which several nodes are alternatives which compute the same value.

DEFINITION 28. An *approximate agreement* of a set of alternatives in a computation graph satisfies the following properties:

- For every two faultless alternatives in the computation graph which conclude with values $r$ and $s$, respectively,

$$\|r - s\| \leq \varepsilon \tag{9.8}$$

where $\varepsilon$ is a predefined small positive constant.

- For every faultless alternative which concludes with a value $r$, there exists two alternatives with initial values $x$ and $y$ respectively, such that

$$x \leq r \leq y \tag{9.9}$$

□

Approximate agreement protocols [49, 51] employ the iterative successive-approximation approach. In each round every node receives the current value held by itself and by other nodes and applies an approximation function to obtain its successive value. The value selected in each round depends solely on the values held by the nodes at the beginning of this round.

Let us consider the algorithm resiliency according to Definition 33, and verify what prevents the system from achieving agreement. If we consider a system with $m$ nodes, we can assume there are at least $m-n$ faultless ones. The approximate agreement is $n$-resilient to faults, if $m - n$ nodes satisfy the properties of Definition 28.

The value selection on each round is based on the following explanation. Let $V$ be a collection of values, bounded in the *range* $\rho(V)$, the interval [max($V$), min($V$)], with a *diameter* $\delta(V)$, (max($V$) − min($V$)).

DEFINITION 29. Refine$^k(V)$ is the multiset obtained by removing the $k$ smallest and $k$ largest values in $V$. □

DEFINITION 30. Select$_k(V)$ is the multiset obtained by choosing the smallest element of $V$ and every succeeding $k$'th element from a nondecreasing ordered collection of $V$. □

If $V$ consists of $m$ elements, select$_k(V)$ contains

$$\|select_n(V)\| = \left\lfloor \frac{m-1}{n} \right\rfloor + 1 \tag{9.10}$$

elements.

DEFINITION 31. A central tendency approximation function $f_{k,n}(V)$, where

1. $n \geq 0$: the number of faulty nodes
2. $k > 0$: a constant
3. $V$: a collection of values with $\|V\| > 2n$

is defined by

$$f_{k,n}(V) = mean(select_k(refine^n(V)))$$

□

In the following sections we present two approximate agreement algorithms [49]. The first one is intended for synchronous systems, for which the execution time and intercommunication delay are bounded. The second serves asynchronous systems, in which a slow node cannot be distinguished from a crashed one.

### 9.4.1 Synchronous approximation algorithm

Program 9.9 describes the synchronous approximation algorithm which employs the approximation function $f_{n,n}(V)$. The second $n$ in the subscript implies $n$ applications of refinement due to the fact that each node can receive at most $n$ faulty responses. The first $n$ in the subscript ensures its convergence since in every round the difference of two multisets in faultless nodes can contain no more then $n$ elements, those sent by faulty nodes. In each round, the diameter of the multiset is reduced by a factor of

$$\left\lfloor \frac{m - 2n - 1}{n} \right\rfloor + 1 > 2$$

as $m \geq 3n + 1$. The total number of rounds required is therefore

$$\lceil \log_{\lfloor \frac{m-2n-1}{n} \rfloor + 1}(\delta(V)/\varepsilon) \rceil$$

Summarizing the algorithm's resiliency brings forth the following condition.

CONDITION 12. If $m \geq 3n + 1$, then the synchronous approximation algorithm in Program 9.9 is $n$-resilient to faults. □

### 9.4.2 Asynchronous approximation algorithm

Program 9.10 describes the asynchronous approximation algorithm which employs the approximation function $f_{2n,n}(V)$. The second $n$ in the subscript implies $n$ applications of refinement due to the fact that each node can receive at most $n$ faulty responses, since in the synchronous case. The first $n$ in the subscript ensures its convergence,

**var**
    $v$: value; $V$: **set of** value; $H, h, m, n$: **integer**; $\delta(V), \varepsilon$: **real**;

**round** 1::
    **begin**
        **get**($v$) ;
        $V := \text{SynchExchange}(v)$ ;
        $v := f_{n,n}(V)$ ;
        $H := \lceil \log_{\lfloor \frac{m-2n-1}{n} \rfloor + 1}(\delta(V)/\varepsilon) \rceil$
    **end**
**round** $h$ $(2 \leq h \leq H)$::
    **begin**
        $V := \text{SynchExchange}(v)$ ;
        $v := f_{n,n}(V)$ ;
    **end**
**round** $H + 1$::
    **begin**
        **broadcast**($<v,$**concluded**$>$) ;
        **put**($v$)
    **end**

**function** SynchExchange($v$: value): **set of** value;
    **var**
        $V$: **set of** value; response(),default_value(): value;
    **begin**
        **broadcast**($<v,$**incomplete**$>$) ;
        $V := \{\emptyset\}$ ;
        $\forall \text{node}_i : i = 1, \ldots, m$ **do**
            **if** accepted_on_time response(node$_i$) **then**
                $V := V \oplus \{\text{response}(\text{node}_i)\}$
            **else**
                $V := V \oplus \{\text{default\_value}(\text{node}_i)\}$
        **od fi**
        **return** $V$
    **end**

**Program 9.9** Synchronous approximation algorithm.

as in each round there can be $2n$ elements in the difference of two multisets of faultless nodes: $n$ faulty values and $n$ faultless values received by one but not yet by the other. In each round, the diameter of the multiset is reduced by a factor of

$$\left\lfloor \frac{m-3n-1}{2n} \right\rfloor + 1 > 2 \qquad (9.11)$$

as $m \geq 5n + 1$. The total number of rounds required is therefore

$$\lceil \log_{\lfloor \frac{m-3n-1}{2n} \rfloor + 1}(\delta(V)/\varepsilon) \rceil \qquad (9.12)$$

**var**
    $v$: value; $V$: **set of** value; $H, h, m, n$: **integer**; $\delta(V), \varepsilon$: **real**;

**round** 0::
  **begin**
    **get**($v$) ;
    $V := \text{AsynchExchange}(v, 0)$ ;
    $v := \text{mean}(\text{enhance}^{2n}(V))$ ;
    $H := \lceil \log_{\lfloor \frac{m-3n-1}{2n} \rfloor + 1}(\delta(V)/\varepsilon) \rceil$
  **end**

**round** $h$ $(1 \leq h \leq H)$::
  **begin**
    $V := \text{AsynchExchange}(v, h)$ ;
    $v := f_{2n,n}(V)$ ;
  **end**

**round** $H + 1$::
  **begin**
    **broadcast**($<v,$**concluded**$>$) ;
    **put**($v$)
  **end**

**function** AsynchExchange($v$: value; $h$: bf integer): **set of** value;
  **var**
    $V$: **set of** value; $k$: **integer**; response(): value;
  **begin**
    **broadcast**($<v, h>$) ;
    $V := \{\emptyset\}$ ; $k := 0$ ;
    **while** $k \leq m - n \;\wedge\; \forall \text{node}_i : i = 1, \ldots, m$ **do**
      **if** accepted $h\_\text{response}(\text{node}_i)$ **then**
        $V := V \oplus \{h\_\text{response}(\text{node}_i)\}$ ;
        $k := k + 1$
    **od fi**
    **return** $V$
  **end**

**Program 9.10** Asynchronous approximation algorithm.

with an additional initialization round. Summarizing the algorithm's resiliency brings forth the following condition.

    CONDITION 13. *If $m \geq 5n + 1$, then the asynchronous approximation algorithm in Program 9.10 is $n$-resilient to faults.* □

A recovery of a node from a crash failure is somewhat complicated since the recovering node can join the algorithm in an outdated round. One way of overcoming that obstacle is to include "recovery" protocol in which nodes respond to the recovering node with their current round number. Another way is the use of history variables [51] as we have done in Section 9.3.2. The general scheme of such an approach for round $h$ ($1 \leq h \leq H$) follows. We note that a recovering node always starts from the round in which it crashed, and nodes which receive outdated messages ignore them.

1. Send $< h, me, history >$ to $\forall node_i : i = 1, \ldots, m$, where $history$ includes the local initial value and the messages $me$ received so far.
2. Wait for receiving $m - n < h', node_i, \ldots >$ messages for a round $h'$ where $1 \leq h' \leq H$ and $i = 1, \ldots, m$.
3. Update the local $history$ through various $h'$ states of the $m - n$ $node_i$s.

Then, the node applies the approximation function $f_{k,n}(V)$ on the local history and decides on the proper value. The nodes do not send or try to receive messages thereafter.

## 9.5 Byzantine Agreement

The solution to the Byzantine generals problem [80] deals with the behavior of a distributed system with faulty processing nodes which are unpredictable, that is, entirely inconsistent and possibly malicious. The nodes in this system are assumed to be organized in an hierarchical tree structure with a father and children relationship. The basic goals of this agreement protocol are summarized in the following conditions:

> CONDITION 14. All faultless children nodes follow the same algorithm. □

> CONDITION 15. If the father is faultless, then every faultless child agrees on the value sent by the father. □

Conditions 14 and 15 are called the interactive consistency conditions.

> CONDITION 16. A small portion of faulty nodes cannot cause faultless nodes to inappropriately interpret the correct value to agree upon. □

Some assumptions must be made regarding the communication links which interconnect the nodes in the system.

- Every message sent by a faultless node is properly received: link failures are excluded.
- Every received message has a properly identifiable sender: a faulty node cannot be disguised as a faultless one.
- An absent message can be detected: a faulty node cannot fail the system by just being silent.

**procedure** *UnSigned*( 0,father,children(father) )::
  **begin**
    **for** $\forall i \in$ { children(father)} **do**
      father **send** $v_{father}$ **to** $i$ **od**
    **for** $\forall i \in$ { children(father)} **do**
      **if** $i$ accepted($v_{father}$) **then**
        $v_i := v_{father}$
      **else**
        $v_i :=$ default_value
    **od fi**
  **end**

**procedure** *UnSigned*( $m$,father,children(father) ), $m > 0$::
  **begin**
    **for** $\forall i \in$ { children(father)} **do**
      father **send** $v_{father}$ **to** $i$ **od**
    **for** $\forall i \in$ { children(father)} **do**
      **if** $i$ accepted($v_{father}$) **then**
        $v_i := v_{father}$
      **else**
        $v_i :=$ default_value
    **od fi**
    **for** $\forall i \in$ { children(father)} **do**
      children($i$) := children(father) $\ominus \{i\}$
      *UnSigned*( $m - 1,i$,children($i$) )
      $v_i := f_{k,n}(\ \{v_j\},\ j \in$children(father) )
    **od**
  **end**

**Program 9.11** Byzantine Generals solution: unsigned messages.

### 9.5.1 Unsigned messages solution

The algorithm for nodes which use unsigned message communication is given in Program 9.11. The procedure UnSigned($m,\ldots,\ldots$) activates $n - 1$ executions of UnSigned($m - 1,\ldots,\ldots$) which activates $n - 2$ executions of UnSigned($m - 2,\ldots,\ldots$) and so forth. The decision on the value to be used is made by each node at each recursion level with the *majority* function $f_{k,n}(\{v_j\})$, where $j \in$children(father).

For up to $m$ faulty nodes, the algorithm UnSigned($m,\ldots,\ldots$) satisfies Conditions 14 and 15 if there are more than $3m$ fathers [80].

### 9.5.2 Signed messages solution

A requirement for a further increase in the robustness calls for making additional assumptions on the communication interconnection. These assumptions increase communication overhead, but protect the system from message falsification by faulty nodes.

**procedure** *Signed*( $m$,father,children(father) ), $m \geq 0$::
  **begin**
    **for** $\forall i \in \{$ children(father)$\}$ **do**
      $V_i := \emptyset$ (initially)
      father **sign_&_send** $v_{father}$ **to** $i$ **od**
    **for** $\forall i \in \{$ children(father)$\}$ **do**
      **if** $i$ accepted($v_{father}$) $\wedge$ $V_i = \emptyset$ **then**
        $V_i := \{v_{father}\}$
        **for** $\forall j \in \{$children(father)$\} \ominus \{i\}$ **do**
          $i$ **sign_&_send** $v_{father,i}$ **to** $j$ **od**
      **fi**
      **if** $i$ accepted($v_{father,j_1,...,j_k}$) $\wedge$ $v_{father,j_1,...,j_k} \notin V_i$ **then**
        $V_i := V_i \oplus \{v_{father,j_1,...,j_k}\}$
        **if** $k < m$ **then**
          **for** $\forall j \in \{$children(father)$\} \ominus \{i\}$ **do**
            $i$ **sign_&_send** $v_{father,j_1,...,j_k,i}$ **to** $j$ **od**
        **fi fi**
    **od**
    **for** $\forall i \in \{$ children(father)$\}$ **do**
      **if** no more messages for $i$ **then**
        $v_i := f_{k,n}(V_i)$
    **od fi**
  **end**

**Program 9.12** Byzantine Generals solution: signed messages.

- Faultless father nodes can sign their messages and forging without detection is impossible.
- Verification and authentication of a father's signature is feasible for every child node.

Program 9.12 illustrates the procedure Signed which implements an algorithm for communicating nodes which satisfy the above assumptions. The *selection* function $f_{k,n}(V)$ satisfies the following rule:

- If $V = \emptyset$, then $f_{k,n}(V)$=default_value.
- If $V = \{v\}$, a single element set, then $f_{k,n}(V) = v$.

For up to $m$ faulty nodes, the algorithm Signed($m,\ldots,\ldots$) satisfies Conditions 14 and 15 [80].

## 9.6 Concluding Remarks

Multiple resources used to obtain redundancy in threads or address spaces allocated for a computation, require mechanisms to resolve contradictions and ensure integrity. Most of today's distributed programming literature has treated this topic assuming the correct operation of these resources. We have directed the reader in this chapter through an environment in which faults may exist, however, agreement on commitment between concurrently executing processes is still required.

Committing to an action has been discussed with the goal of being able to complete a recovery if a crash occurs. Weighted voting has also been discussed with regard to commitment to action. Based on these two discussions, we have examined extensions which allow commitments and voting to be performed in a quorum. Approximate agreements have also been presented for cases where unanimous agreements are not necessary.

The chapter concludes with an examination of situations where the process of fault detection is uncertain. Such situations are defined as the Byzantine generals problem. In this section, we describe the process of reaching agreement under a faulty environment whose behavior is uncertain, and not necessarily as consistent as the fail-stop model.

This chapter extends the simple agreement mechanism of election, in which the computation nodes agree between themselves on the identity of a specific node to use with general agreement tools. The use of these tools establishes the basis of concurrently executing alternatives, which construct the strategies of modular redundancy as a mechanism for tolerating faults.

# Part 3

# Fault Tolerance

The third part of the book utilizes the conclusions of the first two parts, along with the tools presented in them, to examine in detail the problem of functioning under the presence of faults. This part begins with an examination of fault tolerance concepts and performability measures followed by an introduction of system modeling by elemental units. Next, sequential roll-back recovery and parallel modular redundancy schemes are presented in detail. In doing so, checkpointing schemes are analyzed for the sequential part while alternatives and replication schemes are examined for parallel or concurrent computation.

Once sequential and parallel schemes are established, we examine ways of treating exceptions in practical systems. Another topic covered in detail concerns the maintenance of consistency of distributed systems, and, in particular, strategies which support system partitioning due to failures. Special attention is then given to the resource allocation schemes which support damage containment and contamination control. Here, we emphasize the importance of safety requirements of systems along with discussions on ways to assure that they are satisfied in the presence of failures, with some compromise of reliability goals.

# Chapter 10

# Tolerating Faults

There are two major methods of increasing reliability with respect to faults in a system: fault prevention and fault tolerance. While preventing known faults is required, using fault prevention as the only method of dealing with faults has severe limitations. To begin with, removal of all faults from a system *a priori* is in most cases impossible. During the design and development we remove all known faults. However, we cannot remove those whose existence has not been revealed to us. Secondly, being satisfied that all faults have been prevented results in a stiff system with no provision for incorrect behavior. This may lead to a difficult maintenance when it is required, as the intervention constraints can disallow it. The third limitation of the fault prevention approach is that when things do go wrong, unpredictable delays are introduced. These delays can be critical in systems which cannot withstand deadline misses, and may be wasteful in systems in which penalties are employed.

Therefore, while prevention of known faults is appropriate, it must not be accepted as the only provision we have for an incorrect system behavior. In the following section, we review the basic definitions and concepts of fault tolerance, and then discuss various approaches of implementing the ability to withstand faulty subsystems in a system.

## 10.1 Fault Tolerance Concepts

### 10.1.1 Definitions

Let us begin by recalling the definitions that state what our system is, while we focus on the system whose correctness and reliability is of interest. Accordingly, we can then derive such a system's fault behavior.

In Section 1.2.1, a system is defined as an identifiable mechanism that maintains an observable behavior at its interface with its environment. The system's behavior is defined by a finite set of states and state transitions.

Each system can be modeled as a set of components which interact under disciplines derived from the design. Each component is, therefore, also a system. In order to emphasize the hierarchical structure of components in a system, we denote a lower level system as subsystem. Figure 10.1 depicts this hierarchical structure, demonstrating a system and its interfaces with the environment, subsystems, and subsystem level interfaces, and components within a subsystem (denoted by circles). The dotted arcs which leave the system (from $S_2$ and $S_3$) indicate limited observability into the system, supporting permissible interactions in addition to these through the the interface.

We distinguish between correct and acceptable behavior of a system. A correct behavior is that which conforms with the system specification. In order to produce such a behavior, let us recall the following.

In Definition 3, Section 1.3, we define an ASR (authoritative system reference) as a fictitious entity which produces a correct behavior of the system. Using this definition helps us define what an incorrect behavior of our system is, as we do in Definition 4 (see Section 1.3), which defines an error as a difference between the actual system behavior and that produced by an ASR.

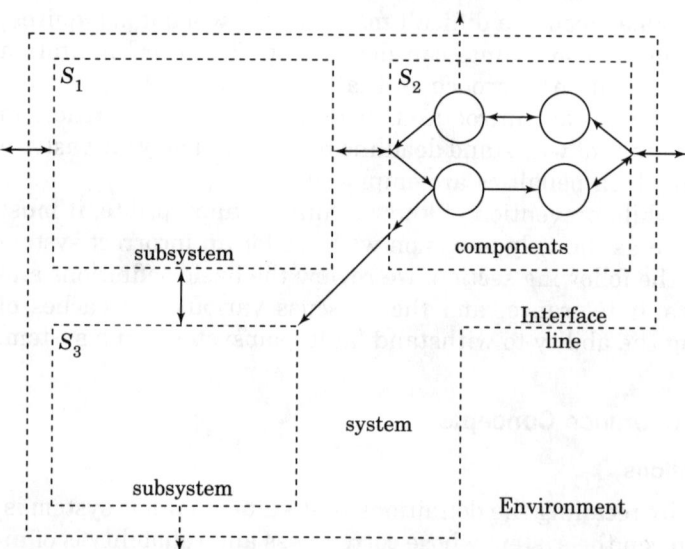

**Figure 10.1** System decomposition: subsystems and components.

We note that an error has a state, defined by the error's value, and is associated with an action in which the error was made. However, real life systems have observability limitations, that is, we may be able to detect the existence of an error not as the failure occurs but only when it is reflected in the system's behavior. Furthermore, the error we detect can be a propagated error and not necessarily the generated error, and in addition, a transient error is likely to be observable only within a limited time interval.

Definition 5 in Section 1.3 states that a failure is an event which corresponds to the first occurrence of the generated error. The action that has caused the failure originates in the source of the error. In a way this generated error source can be considered as a defect in the system, as is done by Definition 6 in Section 1.3 which defines a fault as a source of generated errors. Thus, a fault has not necessarily generated an error, but it has the potential of generating one.

### 10.1.2  Fault tolerance scenario

A process in a system is a sequence of actions that are generated when a program (or a set of programs) is executed. Each action causes an event to occur, thereby generating a state transition in the system. It can be an internal state, yet, an external state transition may be observed as the system behavior.

Let us now build a scenario which supports tolerating a fault through a process of recovery from a failure. Many fault tolerance mechanisms employ such a process sequentially, as a reaction to a failure. However, these actions are not necessarily sequential, and in the following chapters we describe some parallel approaches that achieve the same goals. Figure 10.2 illustrates this scenario.

**Error detection.** In many fault tolerance approaches, the detection of an error is the first step of the recovery mechanism. This step triggers the rest of the recovery according to its findings. The main objective of this step is to answer the question: "is the system state correct and in accordance with the specification?" A positive answer requires no

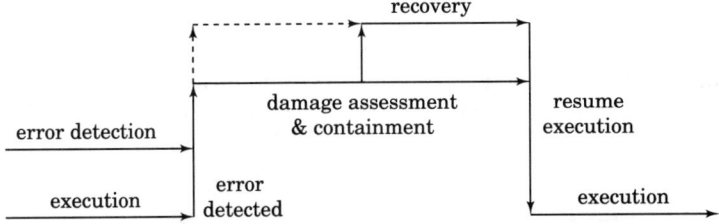

**Figure 10.2**  Fault tolerance scenario.

further treatment, and the regular activities can continue. A negative answer must invoke some maintenance mechanism that will assess the damage, narrow its influence, recover from the error, and eventually resume service.

Answering the above question is in fact a verification process, in which the system behavior is compared to its ASR. Therefore, this process must satisfy the following properties:

- It must depend only on the specifications. This property implies that the physical systems involved must support the required intersystem independence as well.
- It must be complete, or else it cannot cover all the reachable states of the system.

We note that achieving such properties results in a very expensive verification process. Thus, weakening the above requirements is often necessary for practical implementations.

**Damage assessment and containment.** When an error is detected, we assess its origin and the possibility of containment for activation of the proper maintenance activities. The origin should be identified for a correct replacement of the faulty element. However, for such an identification we need to know that the detected error is not a propagated error but rather a generated error. Furthermore, we want to limit the extent of error pollution in order to have simple and efficient maintenance mechanisms.

One solution is to construct every system operation from components called *atomic actions* [96, 120][1]. Each of these components have the following properties:

- *indivisibility*: either all the steps in the atomic action complete or none of them does.
- *serializability*: all computation steps which are not in the atomic action, either precede or succeed all the steps in the atomic action.
- *recoverability*: the external effects of all the steps in the atomic action either occur (the action completes) or not (none of the steps completes).

We note that such a solution must be a fundamental concept in the language used. It must support the atomic abstract data types, the

---

[1]See also Section 14.2 for further discussion of atomic actions.

encapsulation of the steps and their external effects, the locking mechanisms, and other mechanisms that avoid deadlocks and allow implementation of the required structural and logical design.

When a failure occurs in one of the steps of an atomic action, the error is detected and the action cannot complete. This results in a state usually called the *abort* state of the atomic action. This behavior achieves the required containment of the damage, and is due to the recoverability property of the atomic action. Applying this idea to wider and wider circles is feasible by using *nested atomic actions* (sometimes called spheres of control).

An example of an atomic action is the *transaction*, whose purpose is to maintain data consistency in a distributed environment. A transaction is a sequence of read and write commands sent by a user to a distributed database system. On a completion of a transaction, either all the sequence of read and write commands have been executed at the proper order, or none has been.

The transaction is used in various protocols of crash recovery of data base systems [84, 126]. Algorithms that support the transaction requirements have been given in Chapter 9.

**Error recovery.** The goal of any error recovery procedure is to bring the system to an error-free consistent state.

> DEFINITION 32. A *consistent* state of a system conforms with:
>
> - the system correctly reachable states[2] (the state has not necessarily been visited before).
> - the events history as reflected in the system behavior[3] (its interface).
>
> □

Let us distinguish between two approaches of recovery: backward and forward. The *backward recovery* restores the system to a prior state, which is assumed to be error free and is called a *recovery point*. The advantage of this approach is its simplicity: it provides a general solution and the only knowledge it requires is that the relevant prior state is error free. Consequently, this approach is frequently used, and we discuss it throughout the rest of this chapter.

---

[2]This requirement covers avoidance of contradictions and conflicts within the system.

[3]This requirement covers avoidance of contradictions and conflicts with external systems.

The *forward recovery* approach assumes that no previous state is known to be error free, and thus, it restores the system state, by some sequence of actions. However, in order to achieve a correct system state, there must be a good knowledge of what is wrong in the system current state. Having this knowledge, an appropriate correcting computation is activated (e.g., exception handler, voting) and an error-free state is achieved. In this approach the recovery point is the point in time in which the error-free state is subsequently made available. This approach is discussed in the following chapter. It is mostly used when there is no time for backward recovery, and instead other resources are made available.

**Continue service.** These phases support the detection of a fault and the reaching of a consistent and error-free state from which the system can recover. However, before resuming the system execution, some maintenance steps must be taken. The fault location must be properly isolated and identified, and the system should be repaired. Fault location is verified by diagnostic checks, which are generally carried out off line or by using a previously installed hardware feature called BIT (built in testing). Voting mechanisms provide another way of distinguishing the faulty element: the odd one is considered the most probable to be the faulty.

In order to repair the system a reconfiguration procedure is carried out. Full service can resume if we have a stand-by ("hot") spare which can take over the faulty system tasks. If one is not available, there are a number of ways to allow a *graceful degradation* in the system performance, supporting crucial activities only. For example, a backup mode in avionics control allows an approximate reference attitude when navigation is down, thus allowing a manual landing of the aircraft.

### 10.1.3 Hardware and software faults

Faults are categorized and classified according to various characteristics of their effects. Let us consider three of these classifications: the origin of occurrence of the failure, the predictability of the faulty behavior, and the permanence of the fault's effects.

**Origin of occurrence.** In view of our focus on computing systems we assume faults can occur both in the system's hardware and software components. Hardware faults can be generated by design faults (as overload and improper states) and by environmental stresses which cause physical degradation of materials. Software elements do not degrade physically with time, but their environment can change.

One cannot always distinguish hardware from software failures. We emphasize this diagnostics motivated view, mainly because hardware failures can produce identical faulty behavior identical to that generated by software. Memory failures are equivalent to software failures if they occur during instruction-fetch cycles of the processor, generating an erroneous execution of an instruction. A processor whose program counter is inappropriately altered produces an out-of-order execution of instructions, as does a software design error.

Therefore, we extend the definition of software errors to include both design errors and run-time errors, as found in hardware faults. Design errors are retained within the system until it is modified. Run-time errors occur during system performance periods, and can, in fact, cause system modification. Due to the above extension, we allow software errors to behave erratically as well as predictably.

**Predictability.** System behavior after failure can be either predictable or erratic. An example frequently given for a well defined behavior after failure is the model of fail-stop processes. This type of visible effect, due to a failure, can be summarized briefly in the following two rules [122]:

1. The processor stops executing upon error detection.
2. The failure causes a loss of the internal state and the content of the volatile storage.

This behavior is typically caused by faults such as loss of power, system halt due to CPU reset, or a message loss.

On the other hand, we find fault-generated behavior type which is totally unpredictable. This type of behavior occurs when a component of the system transfers conflicting information to different parts of the system. A general model of this problem is described in the Byzantine Generals Problem [80], which we discuss in Section 9.5.

**Permanence.** Permanent error sources include design faults, manufacturing faults, among others. The common denominator of these error sources is that the system includes an out-of-order mechanism which is continuously malfunctioning. In many discussions people tend to view software errors as permanent, especially in analyses that differentiate design faults from operational faults.

However, a permanent error can be generated during run time, and therefore does not necessarily originate in the system creation process. A component failure can lead to a permanent inability of the subsystem which includes this component. Transient errors, on the other hand, always occur during run time and fade away in a period which is short enough to justify the employment of recovery and maintenance mechanisms. Causes for such errors include communication noise and electromagnetic disturbances.

The issue of fault permanence has been extended from the simplified model of permanent and transient faults to that of a statistical model. High-speed integrated circuits have introduced a condensed organization of mechanisms that are capable of generating errors. Several terms have been used to characterize the statistical behavior of these error generators. One of these terms is the MTBF (mean time between failures) which is commonly used for comparing hardware designs for their resiliency to faults. The use of multiple solution versions in the software strengthens the need for a statistical consideration of error behavior of software components.

## 10.2 Recovery in Time and Space

We can distinguish temporal effects from physical effects of fault contamination. Physical effects of such a contamination, especially in permanent faults, result in an unusable system resource. Therefore, in order to recover from such a situation, we generally have to provide for an alternative resource. In case of a permanent fault, the problem and the solution exist in physical dimensions.

There are various temporal effects of faults which contaminate the system resource management and the system behavior.

- Operations are not completed in scheduled time, and thus resources are not released on time for other tasks to which they may have been previously committed.
- Various latency effects, especially in a hard real-time environment, can cause mission-critical risks.
- Invocation of time-out mechanisms usually changes the system's load balance, which is tuned to normal operation.

These effects mainly occur when repetition and retrials are used to recover from a fault, but they may also be found upon any failure detection. Repetitions of an execution carried out by the same resources are commonly used to achieve temporal redundancy, while the physical solution is given in terms of alternatives which are executed in parallel[4] by different resources. Figure 10.3 demonstrates this difference.

---

[4]Parallel executions do not necessarily start and end at the same time, except for temporally restricted systems.

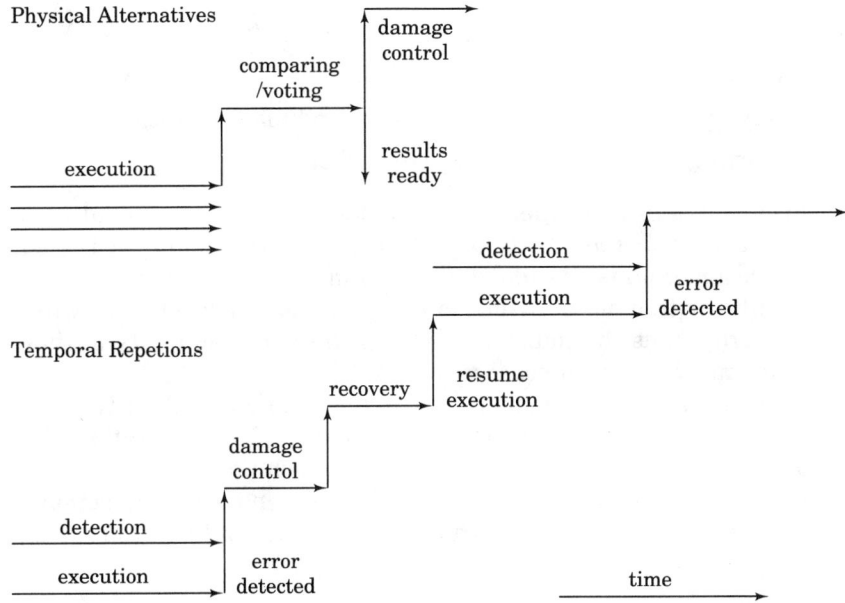

**Figure 10.3** Repetitions versus alternatives effects.

Temporal redundancy may be sufficient for recovery from transient faults, while physical redundancy is better suited for permanent faults. In a wide variety of transient faults in hardware and software, temporal redundancy is achieved by repetition of allocation and the execution of an object is adequate. An example of such a case is that of transient faults in communication which contaminate a bounded size of data.

A permanent fault in a hardware resource or link cannot be recovered by a repeated execution of the specific operation. However, a transient communication fault can be viewed as a time-limited contamination in a physical link. Such a fault exists only for a short period and has no after effects. Real-time constraints significantly narrow the ability to recover in the time domain, since a repetition can lead to a deadline miss. When temporal redundancy is used, time must be allowed for the execution of the replicas, making the utilization very low for normal execution.

## 10.3 Fault Detection Techniques

Fault detection requires mechanisms which test the functionality of system components. The goals of the inclusion of these additional mechanisms are:

- Determination of the occurrence of a failure.
- Identification and isolation of the fault.
- Supplying a starting point for recovery and maintenance.
- Supporting the masking of the error effects.

Inclusion of additional mechanisms generally adds additional error sources and is thereby likely to reduce system reliability. It is clear that once a mechanism is added, the system can be alerted by either its own faults or this mechanism's faults. Therefore, mechanisms which detect errors must be implemented such that the reliability of these mechanisms does not affect the system.

Errors in data management (i.e., data storage and data transfer) are much easier to detect compared to errors in data generation. The following section reviews some of the coding techniques which support detection of errors in data management. We then discuss detection of faults in data generation at both the component level and the system level.

### 10.3.1 Detection of errors in data management

Various error detecting codes are widely used today. A code is calculated with respect to a given data item, and attached to the data item. The data item and its code must allow detection of an error which can be found in either the item or the code. We review here only four major coding techniques for error detection: parity, checksum, residue, and cyclic codes. We avoid further investigation here since this subject mainly deals with a component level fault detection, rather than with system level fault tolerance. Readers who require further reading can find a considerable number of books and papers on this subject (e.g., [117]).

**Parity check.** Let $a_0, a_1, \ldots, a_{n-1}$ be $n$ bits of information, and let $p$ be the parity check bit computed for this information by

$$p = \left(\sum_{i=0}^{n-1} a_i\right) \bmod 2 \qquad (10.1)$$

Thus, the $n+1$ tuple $(a_0, a_1, \ldots, a_{n-1}, p)$ always contains an even[5] number of bits with a "1" value, and is therefore called *even parity code*.

---

[5]Equivalently, odd parity code is generated by $p = 1 + \left(\sum_{i=0}^{n-1} a_i\right) \bmod 2$

The parity code is widely used in binary systems, because it is very easy to implement. It is advantageous for serial data streams, since it can be calculated and checked "on line" through counting the "1"s as they arrive. Its major disadvantage is that it does not detect cases of even number of errors.

**Checksum code.** Let $A_0, A_1, \ldots, A_{n-1}$ be $n$ $k$-bit bytes of information, and let $C$ be the byte of the checksum code computed for this information by

$$C = \left( \sum_{i=0}^{n-1} A_i \right) \bmod 2^k \quad (10.2)$$

As mentioned for the parity check bit, the checksum code may not detect some cases of double and triple errors.

**Residue code.** The residue $R$ of an integer $A$ over a divisor $B$ is given by

$$R = A \bmod B \quad (10.3)$$

The pair $AR$ represents an interesting code word for arithmetic operations. Its properties are presented in the following example.

Let $A_1$ and $A_2$ be two integers with corresponding residues $R_1$ and $R_2$ over a divisor $B$. The following relations hold for these integers and residues:

$$(A_1 \pm A_2) \bmod B = (R_1 \pm R_2) \bmod B \quad (10.4)$$

$$(A_1 A_2) \bmod B = (R_1 R_2) \bmod B \quad (10.5)$$

Therefore, carrying out an arithmetic operation on the numbers can be checked by comparing the residue of the result over the same divisor to the residue of the result obtained by the same operation carried out on the corresponding residues. However, one must be careful not to overlook cases of overflow, for which these equalities may not hold.

The above residue property holds for addition, subtraction, and multiplication, but it does not hold for division. Another relation can be used in these cases [117]. Let

$$\frac{A_1}{A_2} = C + \frac{S}{A_2}$$

where $C$ is the quotient and $S$ is the remainder, with corresponding residues $R_C$ and $R_S$ over a divisor $B$. Then,

$$(R_2 - R_S) \bmod B = (R_C R_1) \bmod B \quad (10.6)$$

**Cyclic code.** A cyclic code is a parity code with an additional property that every cyclic shift of the code is also a code.

In other words if $(a_0, a_1, \ldots, a_{n-1}, p)$ is a code, $(a_1, \ldots, a_{n-1}, p, a_0)$ is also a code. Cyclic codes are often represented as polynomials, letting $(a_0, a_1, \ldots, a_{n-1})$ stand for

$$A(x) = a_0 + a_1 x^1 + \ldots + a_{n-1} x^{n-1} \qquad (10.7)$$

Having $x^i = 2^i$ and binary $a_i$s, together generate the value of the integer $A$. Let us further note that

$$\frac{A(x)}{B(x)} = C(x) + \frac{S(x)}{B(x)} \qquad (10.8)$$

where $C(x)$ is the quotient and $S(x)$ is the remainder, or in other words,

$$S(x) = A(x) \bmod B(x) \qquad (10.9)$$

From polynomial algebra we know that the degree of $S(x)$ is less than that of $B(x)$, and hence it is useful to serve as supplementary data for the information $A(x)$. If $A(x)$ is of degree $n$ and $B(x)$ is of degree $k$, then $S(x)$ is of a degree lesser than $n - k$.

Cyclic codes are generated by feedback shift registers which generate the remainder $S(x)$ for a given $B(x)$. Before code is generated, the register is initialized to zero. The generated word will be ready in the register after feeding it from the left as follows: starting with $a_{n-1}$, the information is entered with $n-k$ shifts and then the feedback shift register is shifted additional $n$ right shifts. The generated code now appears in the register. Attaching the generated code $s_{n-k-1}, \ldots, s_0$ to $A(x)$, yields the code word

$$a_{n-1}, a_{n-2}, \ldots, a_0, s_{n-k-1}, \ldots, s_0 \qquad (10.10)$$

Figure 10.4 describes a general scheme of implementing this shift register. The same feedback shift register, when fed by the code word (the

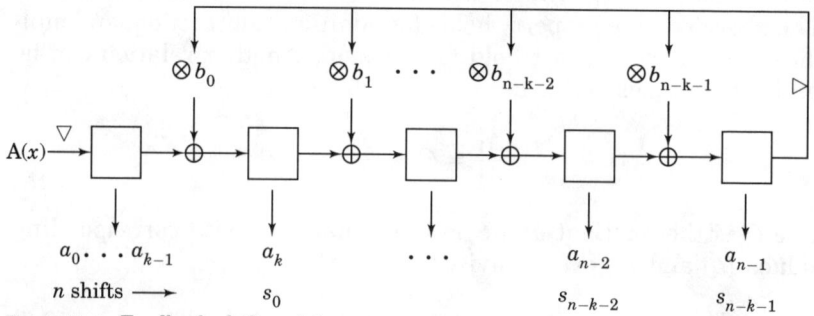

**Figure 10.4** Feedback shift register.

sequence of $A(x)$ coefficients followed by $S(x)$ coefficients as given in Equation 10.10), generates an error polynomial. If the code word is correct, the coefficients of the error polynomial are all zeros. Otherwise, their values include error identification information that can be used for error correction. A detailed discussion on this subject can be found in [113].

### 10.3.2 Component fault detection

The above coding techniques, along with their combinations, effectively yield methods of error detection in data management. They also adequately support fault detection at various component levels. The major advantage of the coding methods is their transportability which is demonstrated in the capability of attaching the codes to the information items. This property allows ascertainment of failure occurrence in the data paths of our system. Predefined sequences of data items and their attached codes can be transported through the communication subsystems and links to verify their proper functionality, pin-pointing faults where they exist.

It turns out that testing a complex system with external test equipment since very difficult, since the availability of information on the system internal states is always limited. Inclusion of local hardware and software modules, dedicated to tests and fault detection, helps to overcome this problem. For example, a sample of predefined sequences of data items and their attached codes can be stored in local ROMs and be transported during test sessions. Knowing what has been sent allows detecting faults through comparison with expected results, and embedded predefined faults in the data allow testing the detection mechanisms.

Built-in testing (BIT) capabilities are frequently implemented in reliable and safe systems. We can identify the following classes of widely used BIT elements:

1. Predefined pattern generators are used to feed the tested components with stimuli which cover a sufficient portion of the component state space. It not necessary to have an exhaustive pattern which exercises the module through its entire state space. Sometimes a randomly chosen pattern serves the system requirements. Selected patterns are frequently used to generate a *signature* of component states which can then be easily correlated with expected results to detect the presence or absence of faults.

2. Progress monitoring devices are generally used to identify illegal sequences, deadlock, or inactivity of components. Quiescence detection devices include time-out mechanisms and watchdogs, which

are devices programmed to wait for an operation-dependent period, and as this period elapses, they execute operation abortion.

3. Reference devices are used to measure, calculate, or generate a known value. Data acquisition systems employ such elements to feed selected entries in their input multiplexers. The result of measuring the reference input must match a predefined value. Computing elements often employ selftesting through invocation of a procedure which excercises CPU operations to test the ALU and register operations.

4. Check code generators and monitors are widely employed to generate and validate check codes for these data items.

The use of such local modules helps to identify component faults. System level control usually synchronizes the concurrent activation of these elements with the normal operations of the system.

Note that the built-in test modules can themselves have faults. A reference voltage can drop, a preprogrammed value can be inappropriately modified, and a parity generator can malfunction. Therefore, a proper analysis must be made before using the BIT results. BIT which exercises the BIT (sometimes denoted "BIT on BIT") may become too expensive and reduce reliability even more.

### 10.3.3 Detection tests at system level

Let us now discuss system level strategies concerning error detection in data or state. Due to the limited visibility available at the system level, comparison tests are needed in order to decide whether the data or state are acceptable. We distinguish between the following types of tests at system level:

1. Acceptance testing with respect to a known value.
2. Acceptance testing by comparison.
3. Voting.

These three types correspond to three recovery types: the recovery blocks, the N self-checking programming, and the N version programming, respectively.

The recovery blocks (RB) approach [119] carries out the same acceptance testing with respect to a known value after the execution of each alternative. It requires $n+1$ alternatives in order to withstand $n$ independent failures. The alternatives are executed sequentially, and so are the tests.

The N self-checking programming (NSCP) approach calls for testing by comparison of "hot" alternatives followed by result switching [85]. An example of this approach includes two alternatives for each NSCP component, and if comparison of the two fails, so does this component. It is clear that this approach requires $2(n + 1)$ alternatives in order to withstand $n$ independent failures. However, it delays results only for comparison and result switching. The acceptance tests in this approach can be executed concurrently, and can be categorized as one of the following

- Acceptance testing with respect to a measured (or calculated) value.
- Acceptance testing based on an indirect measurement.

The N-version programming (NVP) approach [28] employs a voting mechanism for error avoidance. It requires $2n+1$ alternatives in order to withstand $n$ independent failures. Instead of detecting a failure at an alternative, it processes all the alternatives and selects the results by voting (e.g., [57]). The error detection is implicit and refers to the alternatives not selected as a part of the quorum.

## 10.4 Performability Measures

Performance and reliability of fault tolerant systems often work at cross purposes with each other. System effectiveness can be regarded as the extent to which an application can tolerate the existence of faults and still achieve its performance objectives. This interpretation can be examined through system parameters as its *reliability* and *availability*. System effectiveness can be more precisely defined as the expected value of its worth as identified for a given benefit [101]. In this respect we consider the system worth as a random variable which maps the system's sample space to a real number set representing worth values.

System reliability is defined as the probability of not having intolerable faults. In other words, if a system is not resilient to a single fault, its reliability is the complement of the probability of having a fault in the system.

$$R(t) = 1 - P_1(t) \qquad (10.11)$$

However, if the system can tolerate faults, its reliability is improved significantly. Hence let us define what an $n$-resilient system is.

> DEFINITION 33. A system is *n-resilient* to *faults* if any $n$ faults in the system do not prevent the system from achieving any of its objectives.
> □

Thus, the reliability of an $n$-resilient system with independent failure modes can be expressed as

$$R_n(t) = 1 - P_{n+1}(t) = 1 - [P_1(t)]^{(n+1)} \qquad (10.12)$$

The system availability can be derived from its reliability and its maintainability. The frequency of faults in an $n$-resilient system is therefore derived from the expected value of the time of having $n+1$ failures in the system:

$$E_f(t) = \int_0^t \tau \cdot P_{n+1}(\tau) d\tau \qquad (10.13)$$

The mean time between failures (MTBF) can be written as

$$MTBF = E_f(\infty) \qquad (10.14)$$

If we derive from the system's maintainability characteristics its mean time to recover (MTTR), we can formulate its availability as follows:

$$A = \frac{MTBF}{MTBF + MTTR} = \frac{1}{1 + \frac{MTTR}{MTBF}} \qquad (10.15)$$

System performance is always associated with some sort of an accomplishment set. A model of the system performance can be considered a random variable [101], say $Y_S$,

$$Y_S : \Omega \to A \qquad (10.16)$$

where $S$ is a complete system (including the environment), $\Omega$ is its sample space, and $A$ is an associated accomplishment set. The simplest accomplishment set includes only two members: success and failure. An extended accomplishment set also includes partial successes that are associated with degraded performance. We can therefore denote $Y_S(\omega)$ as the accomplishment level which corresponds to the outcome $\omega$ of the system's underlying description.

Based on all this, one can define the system's *performability* at a given accomplishment level $a$ ($a \in A$), denoted $p_S(a)$, as

$$p_S(a) = P(\{\omega \mid Y_S(\omega) = a\}) \qquad (10.17)$$

Hence, evaluation of the reliability of a system can be regarded as the evaluation of the system's performability with the designated accomplishment subset associated with the system's success.

## 10.5 Modeling Fault Tolerant Systems

The scenarios and definitions of fault tolerance presented thus far in this chapter call for a uniform system model which encompasses functionality and constraints of objects in addition to fault tolerance. As such it must be an extension of the object-based model and include all specified requirements, constraints, restrictions, and conditions on its inter-object relations. The building block of this model is called an *elemental unit* (EU).

### 10.5.1 Elemental units and computation graphs

Each elemental unit consists of five logical entities, as depicted by Figure 10.5:

- Input data or invocation
- Input conditioning and testing
- Server or agent object
- Output acceptance test and status
- Output data or synchronization

This structure allows the modeling of each system as a graph, whose nodes are EUs interconnected by instantaneous events of synchronization or data transfer represented by directed arcs. It is important to notice that computation EUs as well as communication EUs are easily modeled by these building blocks. Furthermore, scope zooming is easily accomplished by the replacement of a subgraph by an elemental unit and vice versa, as depicted by Figure 10.6. However, the loss of observability in zooming is clearly noticed while intermediate monitoring and results are not available.

### 10.5.2 Fault handling modeling with elemental units

Event synchronization are supported by this model, as depicted by Figure 10.7 for synchronizing events $e$ and $f$. If, however, there is no need to synchronize several inputs to an elemental unit, we can show that their sequential nature imposes the required precedence accordingly. For example, if events $a$ and $c$ in Figure 10.7 are asynchronous, then their effects are synchronized only after being processed by $A$ and $C$ accordingly.

The failure in the elemental unit model always occurs in the execution of an EU and is detected by either the input condition testing

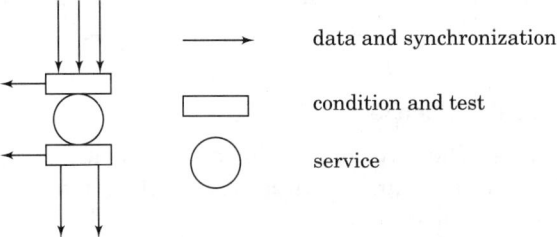

**Figure 10.5** Elemental unit of a fault-tolerant system model.

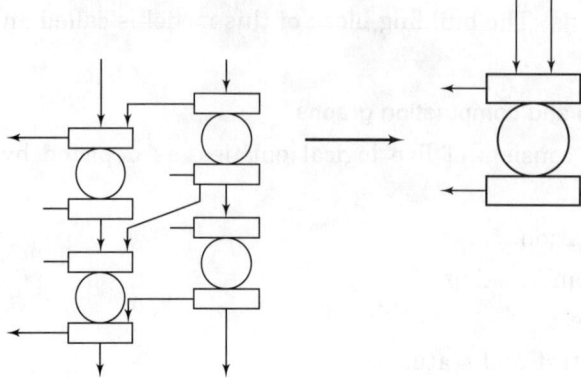

**Figure 10.6** Scope zooming from subgraph to elemental unit.

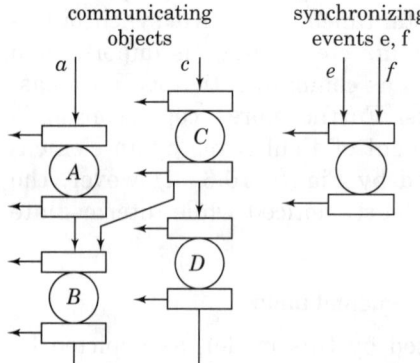

**Figure 10.7** Synchronization and communication elemental units.

or the output acceptance test. A fault detected at a specific application EU invokes a specific fault handler linked to the arc leaving the specific test. Fault handlers can be user defined or system defaults, but as the fault always begins at an EU level, detection always begins at this level as well. However, it is not essential to have a private fault handler for each EU, as shown by Figure 10.8. Moreover, it is demonstrated how fault handler1 and fault handler3 further invoke fault handler4 for encompassing both fault types.

### 10.5.3 User view of elemental unit model

In Section 1.2.4, we have identified five phases in the composition of a computerized system: design, compilation, integration, allocation, and execution. During the design phase, a user specifies all the objects in the system and their interrelations. Then, in the compilation phase, each of the system modules is coded and compiled with two products:

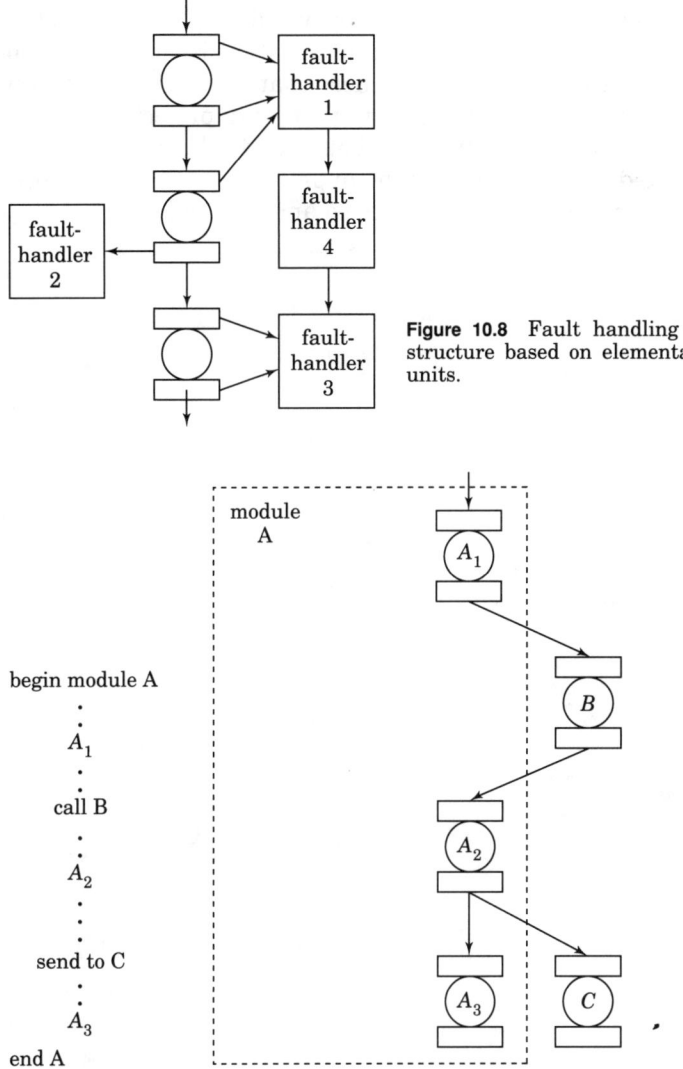

**Figure 10.8** Fault handling structure based on elemental units.

**Figure 10.9** User view of elemental unit model.

the EU structure and the machine-dependent service representation. If we consider the example depicted by Figure 10.9, module A is prepared with three internal EUs, two input arcs (invocation and return from B), and two output arcs (calling B and communicating with C). However, the directed arcs are not connected, because B and C do not exist in the scope of the compilation phase.

During the integration phase, the EU graph is created, as depicted by Figure 10.9. Incoming and outgoing arcs which have to be connected are syntactically and semantically verified. If there exists a need for remote communication, a communication EU (an agent object) is inserted in the graph. We note here that constraints (e.g., mutual exclusion, precedence, timing projections) are projected between objects in this phase. In addition, resiliency and robustness requirements are treated here, generating the appropriate redundancy forks and joints to interconnect the alternatives and replicas.

# Chapter 11

# Roll-back Mechanisms

## 11.1 Roll-Back Mechanisms

A roll back mechanism establishes recovery points in states through which the system has already been. In addition, it uses tests to detect a failure, and upon a detection, the recovery mechanism is activated. Let us first examine the tests through which an error is detected in roll-back recovery schemes.

### 11.1.1 Acceptance tests

Acceptance tests aim at but are not able to answer the question: "Is the system behavior correct and in accordance with the specification?" Instead, they verify that the system behavior is acceptable. Within this compromise there exists a hidden consent: some of the errors may stay undetected, since we test only for the anticipated errors.

An acceptance test is therefore an assertion[1] on the anticipated system state, which returns the logical values TRUE or FALSE. The acceptance test must have no side effects on the system behavior. It must be simple and reliable, such that even though it does not detect all the faults, each fault that it detects does actually exist.

Let us now consider some frequently used candidate categories for acceptance tests.

---

[1] It can be regarded as a function.

- *Replication checks.* Replication of either data and/or mechanisms that are required for and used in a given computation provides a very powerful means for maintaining resiliency to faults. Replication generally employs either matching or majority selection rules, which must be very simple and reliable with respect to the system. These checks, however, are very expensive in resource consumption, and require special care for consistency maintenance to retain a required resiliency.

- *Timing checks.* This category is regularly found in real-time systems, in which timing constraints are imposed on all computations. These checks usually employ absolute or interval timers to invoke the detection mechanism. They result in the detection of activities which failed to satisfy a specified time bound. The checks are very powerful and simple to implement, making them very popular for computer embedded systems.

- *Reasonability.* Many computation results are range bounded in nature by some constraints. For example, it is physically impossible to have water in liquid state under one atmosphere pressure at a temperature higher than one hundred degrees centigrade. Thus, the knowledge of such range bounds can be used as the basis for tests of the reasonability of computation results. Another example is the knowledge of continuity properties of a controlled phenomenon. Based on previous results, we can establish an expected range, according to which we can compare the computation result and decide if it is reasonable. Assertions derived from the system specification can be used as reasonability range bounds as well.

- *Reversed checks.* In some cases, it is possible to formulate a reverse mapping of the system state transitions. Say, for example, that we started a computation at a system state $\alpha$, and that the computation performed a specified mapping $f$ that resulted in a system state $\beta$. If it is possible to formulate $f^{-1}$, then we can verify that $\beta \xmapsto{f^{-1}} \alpha$.

- *Structural tests* are very common in database and communications. The knowledge of an expected structure of data, can be used to verify the existence of some types of faults in the system. For example, the length of a message is in most cases bounded, the size of records in a database can be semantically defined, and the distinction between pointers and data in an abstract data type can be verified.

- *Low level tests.* This category generally employs tests which are embedded in the hardware. Examples are parity generators and detectors, error detection and correction codes, and cyclic redundancy checks (CRC).

- *Diagnostics checks.* This category is intended for isolating the location of a fault once an error is detected. Many diagnostic tests are based on a failure mode analysis of the system, according to which the most probable fault is selected.

### 11.1.2 Recovery blocks

Once a test has detected an error, a recovery scheme must be activated. Let us examine a fundamental sequential roll-back recovery scheme which is based on a programming structure called a *recovery block* [119].

The recovery block structure includes an acceptance test that must be ensured and alternatives with user defined priority: primary executed first, secondary only if the primary failed, etc.

Figure 11.1 illustrates the execution of a recovery block scheme. First, a recovery point is established. Then the primary alternative is executed, followed by an acceptance test. If the test is successful, the block is exited. Otherwise, upon a failure in the test, the system state is restored to the recovery point and the second alternative is executed. If all the alternative tests fail, the block fails.

Program 11.1 describes the recovery block programming structure.

### 11.1.3 Nested recovery blocks

Let us examine the case of a block failure. What if all the alternatives fail to satisfy the acceptance test? Generally, problems of this type are solved by means of a defined hierarchy: recovery blocks can be parts of a higher level recovery block, while the highest block is the whole system. This hierarchical structure employs *nested recovery blocks*. Program 11.2 describes this programming structure.

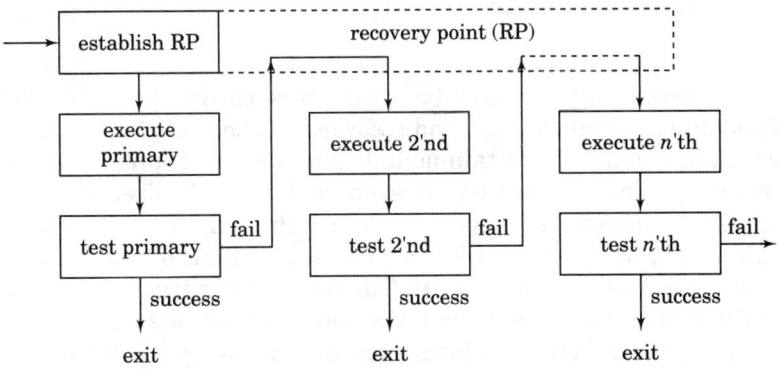

**Figure 11.1** Recovery block.

**begin**
    **ensure** acceptance test ;
    **by** primary alternative ;
    **else by** 2'nd alternative ;
    $\vdots$
    **else by** $n$'th alternative ;
    **else** error ;
**end**

Program 11.1  Recovery block.


RB1: **begin**
    **ensure** RB1 acceptance test ;
    **by begin**
        primary alternative RB1 ;
        $\vdots$
    RB2:  **begin**
            **ensure** RB2 acceptance test ;
            **by** primary alternative RB2 ;
            $\vdots$
            **else by** $m$'th alternative RB2 ;
            **else** error ;
        **end**
    **end**
    $\vdots$
    **else by** 2'nd alternative RB1 ;
    $\vdots$
    **else by** $n$'th alternative RB1 ;
    **else** error ;
**end**

Program 11.2  Nested recovery block.

Nesting recovery blocks must be organized such that the principles of indivisibility, serializability, and recoverability are kept. Otherwise damage assessment and containment do not allow an appropriate rollback recovery. These principles are even further emphasized when we have parallel processes that communicate with each other. The upper example in Figure 11.2 describes an implementation of two parallel processes in a recovery block [119]. The horizontal parallel lines represent the fork and join points, and the downward arrows describe the process progress directions. This structure of recovery block for a set of processes is called *conversation*.

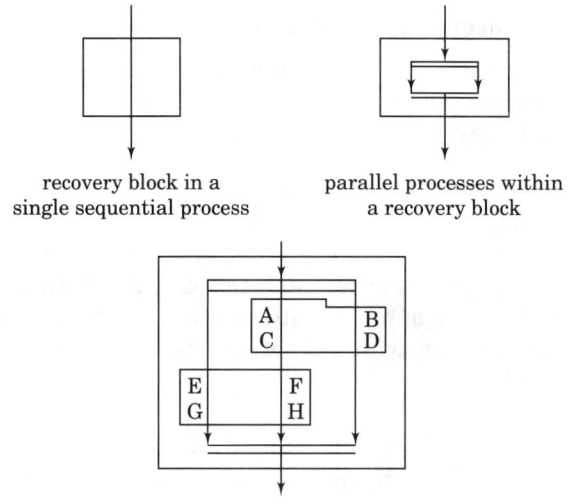

Figure 11.2   Process interactions in recovery blocks.

### 11.1.4 The domino effect in roll-back recovery

Let us consider a system which consists of multiple processes that communicate with each other only via messages. Let each of the processes have its own error detection mechanism and establish its own recovery point. Once a process detects an error, it rolls back to its recovery point. However, since these processes communicate with each other, others are forced to roll back too, this creates an uncontrolled rolling back that resembles the domino effect [119].

The above effect occurs due to following reasons.

- There is no coordination of the recovery block structures of the processes that have inter-dependencies due to their interaction.

- There exists a cycle of error propagation due to which each process can cause another process, which interacts with it, to roll back.

The avoidance of such an uncontrolled rolling back is achieved in the establishment of system *consistent states* which serve as recovery points. Let us emphasize here that by the term consistent state we require consistency of the whole system. A consistent state is defined in Definition 32, and it allows achieving an error-free state that leads to no contradictions and conflicts within the system and its interfaces. Thus, all communication events between processes are taken into account, maintaining their order of occurrence as reflected at the system interfaces. However, we must enforce some restrictions on the communication system to support such a consistency.

- Communication delay is negligible and can be considered zero.
- Communication maintains a partial order of data transfer. All messages sent between a particular pair of processes are received at the destination in the order they were sent.

We can determine consistent states statically or dynamically. The static approach is a language based approach and the consistent state is determined at compile time. At that time a *recovery line* is set which is a set of recovery points, one for each process, to which the processes roll back. The dynamic approach utilizes stored information about communication and recovery points to set up a recovery line only after an error occurs.

**A static approach.** An example of a static determination of a recovery line is the previous conversation. However, parallel processes that communicate with each other raise some limitations due to possible nested recovery blocks. The lower example in Figure 11.2 describes the forking of three parallel processes. The middle and the right processes enter a conversation at points A and B, respectively. Reaching their respective termination of the conversation, points C and D in the figure, each process must satisfy its respective acceptance test and neither is able to proceed until the other has done so.

We note that conversations between different processes must enforce the damage containment principles, and thus the middle process in Figure 11.2 enters its second conversation with the left process (points F and E, respectively) only after the previous conversation has ended.

Let us summarize the conversation properties.

1. No communication is allowed across the conversation boundaries.
2. All processes exit the conversation simultaneously.
3. On entry to the conversation, each process establishes its recovery point.
4. On exit from conversation, each process executes an acceptance test. If any process fails, all processes recover; if none fails, all exit the conversation.
5. Conversations can be nested.

We note that the properties that define exiting the conversation enforce a wait state after an acceptance test.

**A dynamic approach.** Let us first consider the case in which we have asynchronous communication. The only knowledge on ordering of events comes from causality relations between them. Knowing that

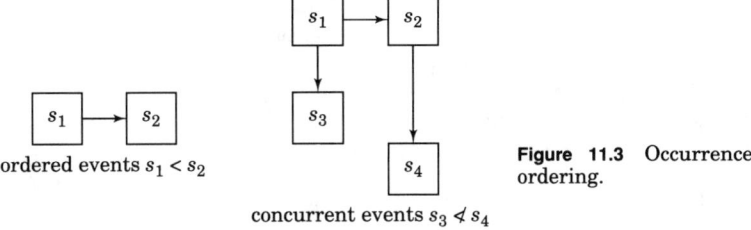

**Figure 11.3** Occurrence ordering.

$s_1$ causes $s_2$ allows the partial ordering $s_1 < s_2$ as described in Figure 11.3. An example of such states can be the case of $s_1$ transmitting a message and $s_2$ receiving that message. However, causality does not allow ordering of concurrent states, as illustrated in the right side of Figure 11.3. Although we know that $s_1 < s_2$ and $s_1 < s_3$ and $s_2 < s_4$, we can not determine the ordering of $s_3$ and $s_4$. Thus, when we come to examine a dynamic approach of setting up consistent states in a distributed system, we must take into account the need to restore the proper ordering as well.

One way to dynamically determine a consistent state to which we can roll back originates in the state space of the system. We describe here the principle of such methods, and the reader can find detailed implementations of models and recovery protocols based on Petri nets and other state descriptions (e.g., [100]).

Let us define the following *occurrence graph* of a system. The states of the elements in the system are described by labeled squares in the graph. The active states, those which are currently occurring, are described as ovals.[2] Causality relations between states are described by directed arcs, which define the ordering as well as the transition from one state to another. The states which are connected by the directed arcs form the occurrence graph. In order to distinguish consistent states from the others, the outcoming arcs of consistent states are doubled. The occurrence graph is constructed during execution, recording the states with the proper causality relations.

Using this model, the system in Figure 11.4 has three consistent states of its elements, $s_1$ $s_2$ and $s_6$, and two active states, $s_7$ and $s_8$, when an error is detected on the transition caused by $s_5$. Once the error is detected, all active states which are going to be influenced by the roll-back recovery must be deactivated. Then, we must find a subgraph of the occurrence graph, whose removal ends in a set of consistent states. This removal represents the roll-back recovery, which

---

[2]In a Petri net formalism it can be regarded as places and their marking.

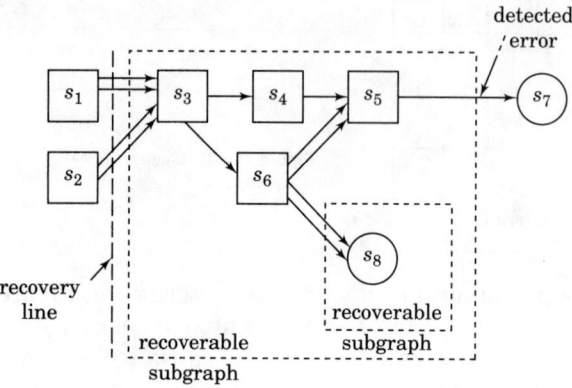

**Figure 11.4** Occurrence graph with recoverable subgraphs.

must retain the proper ordering of events up to this consistent state. In order to be able to find such a subgraph, let us define the following subgraph.

> DEFINITION 34. A *recoverable subgraph* is a subgraph of the occurrence graph which has no outgoing arcs to other subgraphs of the occurrence graph and has only doubled incoming arcs. □

In the example in Figure 11.4, we describe two recoverable subgraphs: one which consists of $s_8$, and the second which consists of $s_3$ to $s_6$ and $s_8$.

Each recoverable subgraph introduces a possible recovery line: the set of consistent states whose outcoming arcs are the incoming arcs of the subgraph. Thus, restoring the system to the state defined by the elements' states on the recovery line has the following properties.

- All restored elements are in consistent states.
- No causal relation from the rest of the system to the restored elements exists.

The recovery process is therefore regarded as a sequence of

1. deactivation of active states in the minimal (if possible) recoverable subgraph, and
2. removal of this subgraph by restoring the element's states to the recovery line of that subgraph.

When an error is detected in the transition in which we try to reach $s_7$ in the example in Figure 11.4, we first deactivate $s_7$ (obviously we did not activate it due to the fault) and deactivate $s_8$. Then, we trace back the occurrence graph to isolate a recoverable subgraph which includes the causes to the faulty transition. This subgraph introduces the illustrated dashed line as a recovery line, and thereby the restoration of consistent states $s_1$ and $s_2$. We note that the rest of the system, which may have other concurrent activities, is not involved.

## 11.2 Checkpointing

Two recovery mechanisms are used in establishing recovery points in a roll-back recovery: the *audit trail* and the *checkpoint*. The process of saving the essential data for a complete system state restoration is commonly called a *checkpoint*, and the log of transactions which have modified the system state since the last checkpoint is called an *audit trail*.

The comparison of recovery processes based on these two mechanisms is depicted by Figure 11.5. Each time the system state is modified, a permanent record is updated to show this modification. This log of activities can be used to "undo" all the modifications (in reverse computing) up to a consistent error-free state. We note here that not all modifications can be "undone," and therefore additional mechanisms to handle such modifications are required. A record of all the data whose validity is essential for the restoration of the recovery point is periodically stored in a nonvolatile memory. In some cases the data is a representation of the entire system state. On recovery, restoration of this entire system state produces a consistent error-free state.

There are various strategies for setting checkpoints in the application processes. Some employ randomly selected points, others set a checkpoint after the completion of a specified number of successful transactions, and others choose to maintain a specified time interval between two checkpoints. Models of these strategies have been compared (e.g., [108]) considering their effects on the system availability. In the following, we set up a framework in which analytical expressions can be stated and checkpointing performance can be evaluated.

### 11.2.1 Analytical models for checkpoint performance

Let us define an analytical model which will help us analyze the performance of the checkpoint mechanism we employ in the system. Let us

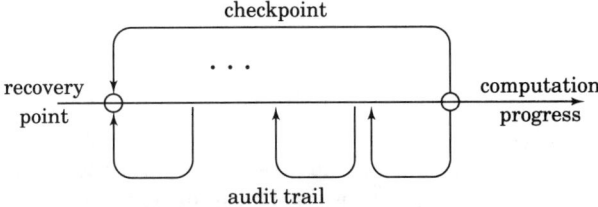

**Figure 11.5** Audit trail and roll-back recovery.

assume we have a system of $n$ nodes with equal processing potential. Each node, say $i$, processes an incoming stream of transactions which arrive at an average rate of $\lambda_i$ transactions per second and require an average processing time of $1/\mu_i$ seconds in a nonfaulty node. As long as the node is nonfaulty, it is assumed to be in state $\mathcal{N}$ (normal).

We model the failures that occur in a processing node $i$ in a Poisson process with rate $\gamma_i$. The recovery process, which later in the section will be described in various formulations, requires $R_i$ seconds. Once an error occurs, the node changes its state to $\mathcal{F}$ (failed), in which it spends on average $R_i$ seconds.

Recovery is assumed to be executed according to a whole-system state which was stored at the last checkpoint, and an audit trail which records the requests since that last checkpoint. In order to support this roll-back recovery scheme, an overhead must be paid. We assume that a checkpoint is performed each $T_i$ seconds, and it takes on average $\tau_i$ seconds to perform a checkpoint, and $\rho_i$ seconds to restore a checkpoint. While performing the checkpoint, the node is assumed to be in $\mathcal{CP}$ state.

We note that the only state transitions to (and from) $\mathcal{F}$ and $\mathcal{CP}$ are allowed from (and to) $\mathcal{N}$. The reader must note that in the rest of this section, we omit the node index when we deal with the single node case.

**Audit trail and checkpointing.** Let us start our checkpoint analysis with some simple models of recovery schemes [38]. Let us consider a single node system, which has a constant arrival rate of transactions, $\lambda$.

First, we want to express the system overhead, or in other words, the time it spends in states $\mathcal{F}$ and $\mathcal{CP}$. Since it is convenient to express it in time intervals of $T$ seconds, we can compute the overhead of this interval, $O_T$, and then divide it by $T$ to get the overhead per second. Due to the $\mathcal{CP}$ state, $O_T$ obviously gets $\tau$ seconds, since a single checkpoint is performed each $T$ seconds. Let us now find the time we spend in the $\mathcal{F}$ state.

The expected recovery time in our single node system, $t$ seconds after a checkpoint was performed, can be described by

$$R(t) = \rho + \frac{\lambda}{\mu} t \qquad (11.1)$$

where $\rho$ represents the restoration time of the checkpoint, and $\lambda \cdot t$ transactions from the audit trail require $\mu$ seconds to process each.

Let us constrain the system to allow no errors during the recovery phase of a previous error. Then, let $C(x)$ be the expected recovery cost in the time interval $(x, T)$. When we consider the time interval

$(t - \Delta t, t)$, the expected recovery cost function shows

$$C(t - \Delta t) = \gamma \cdot \Delta t \cdot (R(t) + C(t)) + (1 - \gamma \cdot \Delta t) \cdot C(t) \quad (11.2)$$

The left term in the sum represents the probability of having an error in the $\Delta t$ interval, and the right term of having none in that interval. In other words, we can say

$$\frac{d}{dt}C(t) = -\gamma R(t) \quad (11.3)$$

Recalling that $C(T) = 0$, since that is the time we update the checkpoint copy, and integrating Equation 11.3 yields the overhead of the $\mathcal{F}$ state for the checkpoint interval

$$C(0) = \gamma \rho T + \frac{1}{2}\gamma\frac{\lambda}{\mu}T^2$$

Thus, the $\mathcal{CP}$ state and the $\mathcal{F}$ state contribute together the following overhead

$$O_T = \tau + \gamma\rho T + \frac{1}{2}\gamma\frac{\lambda}{\mu}T^2$$

and the overhead per second

$$O_1 = \frac{\tau}{T} + \gamma\rho + \frac{1}{2}\gamma\frac{\lambda}{\mu}T \quad (11.4)$$

Minimizing $O_1$, yields

$$T_{opt} = \sqrt{\frac{2\tau\mu}{\gamma\lambda}} \quad (11.5)$$

Figure 11.6 plots the overhead $O_1$ and $O_T$ versus the intercheckpoint time $T$. Removing the above constraint and allowing errors during the recovery phase of a previous error changes the definition of $R(t)$ as defined in Equation 11.1 to

$$R(t) = \frac{1}{\gamma}(e^{\gamma(\rho + \frac{\lambda}{\mu}t)} - 1) \quad (11.6)$$

However, Equation 11.3 holds for this case too. Description of this case and others are given in [38].

**System availability.** Let us now extend the above system model to a queueing model, with a Poisson transaction arrival rate and transaction recovery which is proportional to the intercheckpoint period and the transaction size [54]. The transactions are served in "first come

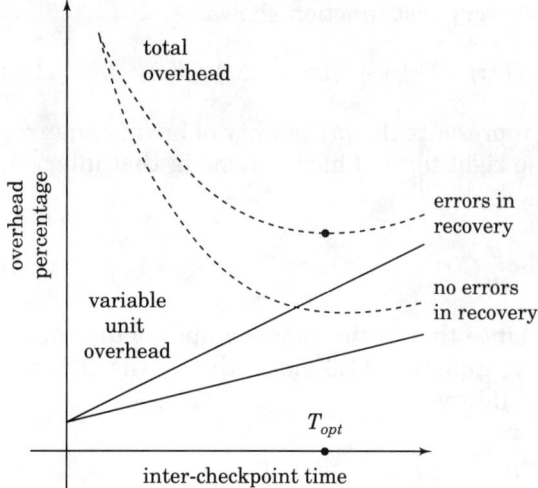

**Figure 11.6** Overhead versus intercheckpoint time.

first served" sequential order. Let us also assume that performing the $k$th checkpoint consumes $T_k$ seconds, a process with a mean of $\tau$ seconds.

Figure 11.7 describes the queueing model of this system in its three allowable states: $\mathcal{N}$ the normal queue with the Poisson arrivals with rate $\lambda$ and the exponential service with rate $\mu^{-1}$, $\mathcal{CP}$ the periodic checkpointing performance which consumes an average of $\tau$ time units, and $\mathcal{F}$ failed state with a Poisson failure rate of $\gamma$. We assume now that the transaction recovery time is proportional to the transaction "size" and thus to its service time $\mu$ and to the expected time between checkpoints $T$. Therefore, we model the transaction recovery time as $k\mu T$, where $k$ is a constant.

The system state can be defined by two random variables $N$ and $Q$, which represent the queue size and the server state, respectively. These two variables constitute a stochastic process, which is a continuous time Markov chain. Let us define the system *availability*.

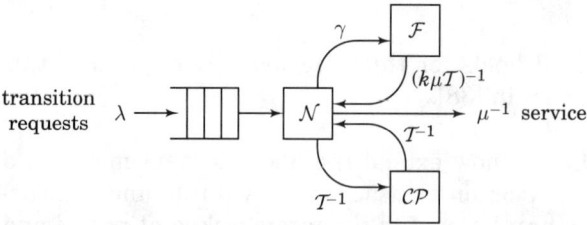

**Figure 11.7** Queueing model for checkpointing and recovery.

DEFINITION 35. The system *availability* equals the steady state probability of the system to be in state $\mathcal{N}$. □

Let us define its stationary joint probability distribution

$$p(n,q) = \lim_{t \to \infty} Pr(N = n, Q = q)$$

where $n \geq 0$ and $q \in (\mathcal{N}, \mathcal{F}, \mathcal{CP})$, and for which the generating function is obtained by

$$G_q(x) = \sum_{n=0}^{\infty} p(n,q)x^n, \quad |x| \leq 1$$

Applying the ergodicity conditions [76, pp 30] and solving the equilibrium equations for state transitions [54], yields the following results.

- Stationary probability distribution for the above model exists if and only if

$$\frac{\lambda}{\mu} < \left(1 + \gamma k \mu T + \frac{\tau}{T}\right)^{-1} \quad (11.7)$$

- The steady-state probability of the system to be in state $\mathcal{N}$ is

$$G_\mathcal{N}(1) = \left(1 + \gamma k \mu T + \frac{\tau}{T}\right)^{-1} \quad (11.8)$$

From the above result, we can obtain the optimal intercheckpoint interval that maximizes the system availability:

$$T_{opt} = \sqrt{\frac{\tau}{\gamma k \mu}} \quad (11.9)$$

**Additional load considerations.** We can define the system availability slightly differently from that given in Definition 35, without changing its properties. This definition is better from the application point of view, and it gives some new insights to the system properties [55].

DEFINITION 36. The system *availability*, $\alpha$, is the mean portion of time in which the system is in state $\mathcal{N}$. □

Considering the above results, as stated in Equation 11.8, they agree with this definition. For every time period $T$ in which the system is in state $\mathcal{N}$, it spends additional time in establishing checkpoints, $\tau$, and recovering from failures, $\gamma T \cdot k\mu T$. However, the above model for error recovery is too simple, ignoring the audit trail part which is proportional to the time passed since the last checkpoint. A better model of that part is defined in Equation 11.1.

Adopting this model, and summarizing the total time in all the system states, while it spends $T$ seconds in state $\mathcal{N}$ yields

$$T + \tau + \gamma \int_0^\infty dF(y) \int_0^y R(x)dx$$

where $F(y)$ is the general distribution function of the time spent in state $\mathcal{N}$ between two consecutive transitions to state $\mathcal{CP}$. Substituting Equation 11.1 gives the total time

$$T + \tau + \gamma\rho T + \frac{1}{2}\gamma\frac{\lambda}{\mu}T^2$$

and thus the availability is

$$\alpha = \left(1 + \frac{\tau}{T} + \gamma\rho + \frac{1}{2}\gamma\frac{\lambda}{\mu}T\right)^{-1} \quad (11.10)$$

This availability is also the steady-state probability of having the system in state $\mathcal{N}$, denoted $\Pi_\mathcal{N}$.

Let $k$ be the portion of transactions that have been reprocessed after failure. We can update Equation 11.1 to a better measure of the recovery cost

$$R(t) = \rho + k\frac{\lambda}{\mu}\Pi_\mathcal{N} t \quad (11.11)$$

where $\rho$ represents the restoration time of the checkpoint, and $k\lambda \cdot t$ transactions from the audit trail require $\mu$ seconds to process each, assuming the processor is in state $\mathcal{N}$.

Using this model of recovery yields

$$\alpha = \Pi_\mathcal{N} = \frac{1 - \frac{k\gamma T\lambda}{2\mu}}{1 + \frac{\tau}{T} + \gamma\rho} \quad (11.12)$$

The checkpoint interval that minimizes the availability is then

$$T_{opt} = \frac{\tau}{1+\gamma\rho}\left(\sqrt{1 + \frac{2\mu(1+\gamma\rho)}{\gamma\lambda k\tau}} - 1\right) \quad (11.13)$$

In a heavily loaded system

$$\frac{2\mu(1+\gamma\rho)}{\gamma\lambda k\tau} \gg 1$$

and therefore

$$T_{opt} \approx \sqrt{\frac{2\tau\mu}{(1+\gamma\rho)k\gamma\lambda}}$$

In a lightly loaded system

$$\frac{2\mu(1+\gamma\rho)}{\gamma\lambda k\tau} \ll 1$$

and therefore

$$T_{opt} \approx \frac{\mu}{\gamma\lambda k}$$

**Multisystem considerations.** We can now extend our system to a multiple node system, and extend its performance with the following properties. When a node is not functional, that is, not in state $\mathcal{N}$, its transaction requests are transferred to other nodes. However, in order to update itself with the state of other nodes and avoid transfers to nodes which are in $\mathcal{F}$ state, each node has a test state $\mathcal{T}$. It enters this state on average every $1/\theta_i$ seconds and spends there on average $\Theta_i$ seconds. We note that transitions to state $\mathcal{T}$ are allowed only from state $\mathcal{N}$.

The overall rate of transactions, $\Delta_i$, which arrives in a node is therefore greater or equal to $\lambda_i$, including those transferred from other nodes. In order to compute $\Delta_i$, there is an important quantity that must be derived according to the system transfer policy: $p_{ij}$. This is the mean portion of transactions at node $i$ transferred to node $j$, given that node $i$ is in state $\mathcal{F}$ and node $j$ is not. This property, the matrix $p_{ij}$, depends totally on the system architecture and its routing properties. Another property needed in order to determine $\Delta_i$ is the mean portion of time node $i$ is in state $\mathcal{F}$, denoted $f_i$. We show an iterative approach to evaluate this parameter.

The mean rate of transactions coming from node $j$ to node $i$ is therefore

$$\Delta_j f_j p_{ji}(1-f_i)$$

and thus

$$\Delta_i = \lambda_i + \sum_{\forall j \neq i} \Delta_j f_j p_{ji}(1-f_i) \qquad (11.14)$$

and accordingly

$$\lambda'_i = \Delta_i - \sum_{\forall j \neq i} \Delta_i f_i p_{ij}(1-f_j) \qquad (11.15)$$

Hence, for a sufficiently long observation period $L$, we can state that at node $i$

$$L = \alpha_i L + \frac{\Theta_i}{\theta_i}\alpha_i L + \frac{\tau_i}{T_i}\alpha_i L + \gamma_i R_i \alpha_i L \qquad (11.16)$$

The terms in the right side of the equation represent the time spent in states $\mathcal{N}$, $\mathcal{T}$, $\mathcal{CP}$, and $\mathcal{F}$, respectively.

From this equation we can extract the system availability

$$\alpha_i = \left(1 + \frac{\Theta_i}{\theta_i} + \frac{T_i}{T_i} + \gamma_i R_i\right)^{-1} \quad (11.17)$$

and as we have stated previously, the ergodicity conditions and the solution of the equilibrium equations for state transitions [54], require

$$\forall i: \quad \alpha_i > \frac{\lambda_i}{\mu_i}$$

Let us examine a suggested approach to optimize the selection of $\theta$ and $T$, the intervals for test and checkpoint, respectively [56]. First, let us assume that our recovery model reflects a behavior such that the higher the load in a node is and the longer the checkpoint period is — the longer the recovery takes. In other words,

$$R_i = c_i \cdot \frac{\lambda_i' \cdot T_i}{\mu_i \cdot \alpha_i} \quad (11.18)$$

where $c_i$ is a constant. Substituting it in Equation 11.17 gives

$$\alpha_i = \left(1 + \frac{\Theta_i}{\theta_i} + \frac{T_i}{T_i} + \gamma_i c_i \frac{\lambda_i' T_i}{\mu_i \alpha_i}\right)^{-1}$$

Maximizing $\alpha_i$ over $T_i$ gives

$$T_i^{(opt)} = \frac{\gamma_i c_i \lambda_i'}{\mu_i}\left(1 + \sqrt{1 + \frac{1 + \frac{\Theta_i}{\theta_i}\mu_i}{c_i \gamma_i \tau_i \lambda_i'}}\right) \quad (11.19)$$

The second step is to add the testing overhead to our recovery cost. To do that, we may assume the following model.

$$R_i = c_i^{(1)} \cdot \frac{\lambda_i' \cdot \theta_i}{\mu_i \cdot \alpha_i} + c_i^{(2)} \cdot \frac{\lambda_i' \cdot T_i}{\mu_i \cdot \alpha_i} \quad (11.20)$$

Using the availability Equation 11.17 and the previous $T_i^{(opt)}$ from Equation 11.19 yields the update availability

$$\alpha_i = \frac{1 - \gamma_i \frac{\lambda_i}{\mu_i}(c_i^{(1)}\theta_i + c_i^{(2)}T_i)}{1 + \frac{\Theta_i}{\theta_i} + \frac{T_i}{T_i}} \quad (11.21)$$

Optimizing this availability with respect to the selected test interval gives

$$\theta_i^{(opt)} = \frac{c_i^{(1)}\gamma_i\lambda_i'\left(1 + \sqrt{1 + \frac{(T_i^{(opt)} + \tau_i)(\mu_i - c_i^{(2)}\gamma_i\lambda_i' T_i^{(opt)})}{c_i^{(1)}\gamma_i\lambda_i'}}\right)}{\mu_i - c_i^{(2)}\gamma_i\lambda_i' T_i^{(opt)}} \quad (11.22)$$

**begin**
    1. choose initial $f_i$ such that $f_i < 1 - \frac{\lambda_i}{\mu_i}$ ;
    2. **while** no convergence **do**
        2.1. $\forall\, i,j$ nodes:
            2.1.1. compute $p_{ij}$ ;
            2.1.2. compute $\Delta_i$ :
                solve $\Delta_i - \sum_{\forall j \neq i} \Delta_j f_j p_{ji}(1 - f_i) = \lambda_i$ ;
            2.1.3. compute the effective arrival rate :
                $\lambda_i' = \Delta_i - \sum_{\forall j \neq i} \Delta_i f_i p_{ij}(1 - f_j)$ ;
        2.2. optimize availability :
            2.2.1. use Equations 11.19 and 11.22 to compute $T_i^{(opt)}$ and $\theta_i^{(opt)}$ ;
            2.2.2. use Equation 11.21 to obtain new $\alpha_i$ ;
        2.3. compute a new set of $f_i$
            $f_i' = \gamma_i R_i \alpha_i$
        2.4. test convergence of $f_i$
    **od**
**end**

**Program 11.3**  Optimal availability algorithm.

Stability conditions require

$$\forall i : \quad \frac{1}{T_i^{(opt)}} > \frac{\gamma_i \lambda_i' c_i^{(2)}}{\mu i}$$

We can summarize this suggested approach in an algorithm that optimizes the availability with respect to the selected checkpoint and test periods. The algorithm is described in Program 11.3.

## 11.3 Concluding Remarks

In this chapter, we considered roll-back mechanisms and models for evaluating their performance. The roll-back mechanisms which call for repeating a portion of the computation, require additional time. Clearly, such mechanisms can be used in real-time systems only if provisions are made for this additional time.

    Checkpoint techniques are very common in transaction processing systems. The analysis of the performance of these schemes is based on stochastic models of Poisson arrivals and exponential time distributions. While these assumptions are made for mathematical tractability, the results obtained are often found to be reasonable for practical systems.

# Chapter 12

# Modular Redundancy

Roll-back recovery schemes do not always establish a system design which meets the fault tolerance goals. Hard real-time jobs are one example in which a retry solution may not work, because these jobs have hard deadlines which must be met. Furthermore, it is impossible to overcome design faults by just running them again and again. For this type of fault, we need other versions of our proposed solution which are created independently.

In this chapter we discuss a fault tolerance approach that relies on parallel alternatives and forward recovery as a response to failures. The terms that pop up in association with this type of solution are redundancy, replication, and back-up modes. We do not discuss the degraded performance after failure, but only the methods that keep the system resilient to it.

## 12.1 N-Version and Modular Redundancy

Roll-back recovery schemes that assume no design faults can be regarded as redundancy implemented in *time*. In this approach, we reactivate a process which we suspect to be faulty using the same resources employed during the failure. If we want to employ different resources upon recovery, we require *space* redundancy as well. This physical redundancy can be activated while the primary computation is active, thus consuming no additional time. A necessary mechanism in this case is a decision algorithm that can select the proper output of the set of redundant computations. However, if we activate it only after the primary fails, it is equivalent to temporal redundancy.

Another approach is that of using *information* redundancy. In this third approach, the reactivated software element is a different version of the computation. Having N versions of a program[1] can show resiliency to N independent design faults, when the versions are independent modules.

The fundamental requirements of the activation of redundant processes are:

1. consistent initial states of primary and redundant computations, and
2. a reliable decision algorithm.

Let us examine how these requirements satisfy the requirements of various design-fault recoveries. Let us divide design faults into two categories. *Autonomous faults*, sometimes called simplex faults, are faults in a computation module that do not influence the correctness of other computation modules. There can be a single or a multiple autonomous fault and for all of them, the criterion holds. The good damage containment of this type of fault, which confines the fault to a given locality only, implies that a consistent activation of all the redundant modules and an effective decision algorithm are sufficient for recovery.

However, for the second category of design faults, the *related faults*, the criterion is not sufficient. Detecting multiple faults of this category can lead to an indistinct recovery that is required. Say that we have a generated error detected at one module, and a propagated error detected at another. The decision algorithm cannot distinguish a generated error from a propagated one, and therefore, both are treated as generated errors. In this case, if each of the errors requires a different solution from the decision algorithm, we may end up with a false decision. Furthermore, if the decision making is based on voting, an error that spreads in the redundant modules distorts the algorithm's results.

The N-version programming approach attempts to minimize related errors through design diversity. By isolating each design version from another, this approach tries to achieve minimization of identical error causes. The quality of this isolation has, therefore, a major role in the quality of the results obtained by this approach. When this isolation is achieved, the approach has some advantages. For example, a simple and efficient decision algorithm can be used, no assumptions

---

[1] The term NVP stands for N-version programming.

are required regarding the cause of the faults, and we accomplish a forward recovery scheme. Unfortunately, there are also some disadvantages to this approach. Isolation achievement is very expensive and in many cases unachievable, the amount of resources that are required is massive, the synchronization of versions is difficult, and common environment factors (e.g., communication) can still introduce similar errors.

Information redundancy does not necessarily have to be based on the isolation of design solutions. Modular redundancy as a forward recovery scheme has many advantages without isolating design of redundant modules. Figure 12.1 compares the recovery block approach to the modular redundancy approach, demonstrating the parallel nature of modular redundancy, and the sequential nature of recovery blocks.

Let us define our model of modular redundancy. Our redundant modules are schedulable entities. Each fault tolerant module corresponds to a set of processes that form its redundancy set. Each member of a redundancy set of modules has the same set of specifications, is physically independent of the rest of the members, and has its own distinct input and output communication channels. The independence restriction excludes the use of shared variables, and we limit this model to a message passing system.

Communication with the redundancy set follows the following rules:

1. an incoming message to the redundancy set is replicated to all the members' input channels, and

2. an outgoing message from the set is obtained by a majority vote on all the messages sent on the members' output channels.

In addition, each modular redundancy design must obey the following conditions [99]. First, we have to guarantee a consistent conflict resolution in the redundancy set.

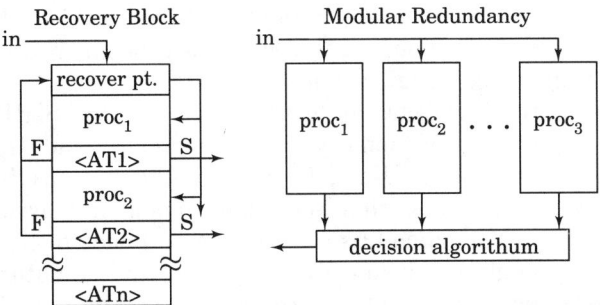

**Figure 12.1** Recovery block versus modular redundancy.

CONDITION 17. All the members of a redundancy set must resolve any nondeterminism in an identical manner. □

In order to support the proper serializability of the individual execution of each of the members in the redundancy set, we need to impose ordering restrictions on the command and control communication.

CONDITION 18. All the members of a redundancy set must service external requests in an identical order. □

Not only requests and control messages influence execution control of the processes, but also some of the incoming data is used to do that as well. For example, message content can be used as the guard for a particular command in the process. Therefore, we need another condition to assure proper serializability.

CONDITION 19. All the members of a redundancy set must process incoming data items which control process execution in an identical order. □

We can summarize these conditions as being the requirements for consistency in conflict resolutions, serializability of communication, and serializability of processing control data.

Let us consider an example of a system architecture which is based on modular redundancy principles: the SIFT (Software Implemented Fault Tolerance) computer system [135].

## 12.2 SIFT

The SIFT system was designed to be used as the central computer of advanced commercial aircraft. This computer system contains approximately twenty tasks which can communicate with each other. These tasks perform the control functions of the aircraft. The tasks use information supplied by the aircraft sensors, and they output data to operator displays and to the aircraft actuators. The reliability requirements of the mission critical tasks, require a failure rate that does not exceed $10^{-9}$ per flight hour, enforced the use of hardware redundancy with the technology available in the 1970s.

Figure 12.2 describes the hardware architecture used in the SIFT system. It consists of $n$ processor–memory pairs, denoted $P_i$ and $M_i$ in the figure, which we call nodes. Each $P_i$ can write to $M_j$ only if $j = i$, and can read from any $M_j$. A triple redundant bus is employed to interconnect the nodes. The same bus connects these nodes to the input output system, through which sensor, transducer, and actuator data items are transferred. Computations are carried out by tasks, which are SIFT's schedulable entities. Tasks in SIFT have restrictions

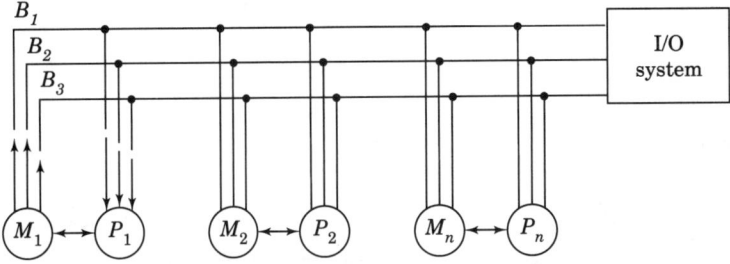

**Figure 12.2** SIFT hardware architecture.

on resource requirements, such that the requirements of a single task do not exceed the resources available at a single node.

Each task is replicated for redundancy, creating multiple versions of it. The versions are distributed among the nodes. Resources are allocated to tasks such that redundant versions do not reside on the same node with their equivalents. The resource allocation is controlled by a *global executive* task, which is also distributed among some nodes to provide the proper degree of fault tolerance.

Since the system is a closed one (no more tasks are expected to appear), the initial allocation is static. However, if a failure occurs, a reconfiguration process is carried out by the global executive. This process changes the state of only one node at a time, generating the required tables and schedules for the tasks this node will be expected to execute.

A *local executive* controls the execution in each node. It carries out the local task rescheduling, their loading, and their error status reporting.

Each SIFT module is defined by abstract data structures, called *v-function*, and a set of operations that manipulate these structures, called *o-function*. This approach allows determination of the module state by its v-functions. The restoration of a task that failed is therefore an activation of a new version of its o-functions after an agreement is reached on its v-functions.

Tasks may require data items from each other. On such data transfers, each version of the receiver locally votes on the data items sent by all the sender versions. Different busses are used for communication with different senders, to avoid bus fault dominance. The local voting procedure uses majority rules for conflict resolution, and discrepancies are logged for later analysis by the global executive.

Being a closed system, the SIFT system is more restricted than the modular redundancy described above. It enforces an *a priori* planned schedule of tasks for each node. Since most of the aircraft tasks are periodic ones, SIFT tasks are dispatched according to their periodicity.

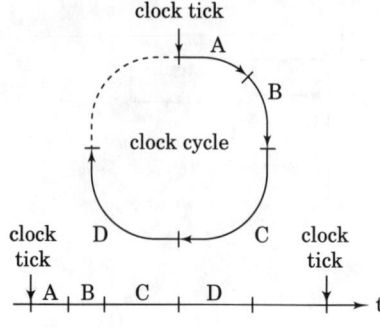

**Figure 12.3** SIFT clock ticks.

Figure 12.3 describes the cyclic nature of the SIFT dispatcher. The rate of the clock ticks defines the resolution limit of the SIFT dispatcher. In Figure 12.3, we see four tasks, A to D, at a given node. Each clock tick initiates a dispatch cycle in which we can schedule all four tasks, or any subset of them.

Progress of versions should be synchronized, to reduce wait times when a vote on incoming data is required. On such a closed system, this type of progress control is a part of the schedule plan. If such a plan is not prepared *a priori*, the receiver version must start receiving with the first sender and continue until the last sender reaches that point. This "wait until receiving last" time is optimized by the plan.

The logical hierarchy that appears in every node in the SIFT system is described in Figure 12.4. The system uses some circular lists of schedule plans and bus assignments for every clock cycle. In this approach, bus-to-task relations are switched all the time, providing a good database for fault allocation. These features are represented by the bus connection and circular list modules in the above figure.

The reader/writer/voter module in Figure 12.4 controls intertask data transfers. Output data is sent by a "write" o-function, and input data is received by a "read" function which is both o-function (generating a task state change) and a v-function (returning data). The "read" operation supplies a returned value according to a vote on several data items that correspond to the different redundant versions of the data supplier.

The support of periodic tasks must provide means for every task instance to use data generated by its precedent instance, and by the precedent instances of those with which it communicates. Since SIFT does not enforce a rendezvous-like synchronization, it must provide double buffering for the write function. This buffering is further restricted by not allowing tasks to write to other tasks, but only to read from them. SIFT makes use of the odd/even iteration numbers for controlling the double buffers. All the system nodes have synchronized clock ticks, with synchronized odd/even status. Thus, reading from a

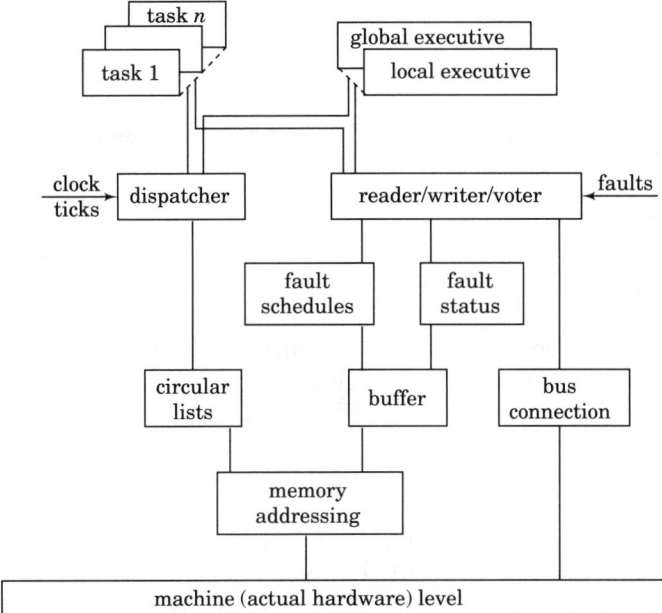

**Figure 12.4** SIFT logical hierarchy.

previous iteration, while an odd iteration is executing, refers to the even iteration buffer in every node.

Only the reader/writer/voter module of nodes that run the global executive are allowed to access the fault schedules. These schedules contain schedules that the system gets as a function of the detected fault. When the global executive decides on a fault detection, it updates the fault status module.

This system is a good example of the use of modular redundancy to obtain high degree of fault tolerance.

## 12.3 Replicas

Replication of active and passive objects is a special case of using parallel versions as modular redundant modules. This approach is widely adopted for fault tolerance or in distributed databases. This type of relationship requires the grouping of objects of the same type into a meta-object to which the rest of the objects may refer to as an entity. We distinguish between the grouping of a set of executable objects (a *troupe* [41]) and the grouping of a set of non-executable objects (a *pack* [92]).

Fault tolerance of non-executable objects is often achieved by maintaining replicas of the objects and forcing the operations on all members of each pack to obey the consistency conditions defined in Section

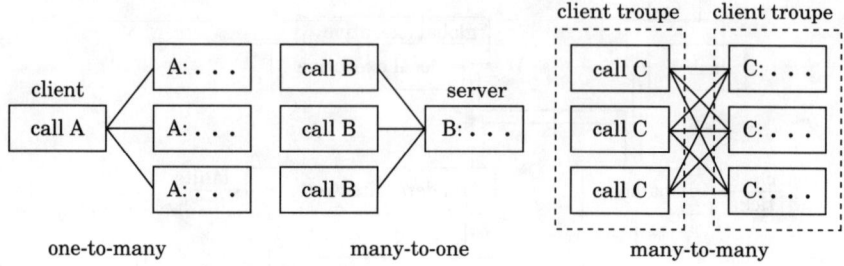

**Figure 12.5** Replicated procedure calls.

12.1. A property which ensures that concurrent execution of operations on a passive object and its replicas is equivalent to a serial execution on the passive object is called *one-copy serializability*. We deal with it in Chapter 14. Let us now describe invocations of replicated executable objects, and in particular, let us examine the case of replicated procedure calls [41].

Three types of procedure call replication are shown in Figure 12.5. The *one-to-many* type uses duplicated servers. The single client includes a call for a procedure (call $A$) which calls three replicas of $A$. The *many-to-one* type uses duplicated clients which use a single server ($B$). Each of the clients includes a procedure call statement (call $B$), referring to the same server. The *many-to-many* type is a combination of the two, with multiple clients and multiple servers. Each of the client's troupe includes a procedure call to the server troupe (call $C$), and each member of the server troupe is a duplicated server ($C$).

## 12.4 Alternatives

An executable object has a set of resource requirements and a set of service requirements [92]. A service requirement can be executed by another executable object instance, which can be chosen out of a set of alternatives.

Figure 12.6 represents an example of multiconnected alternatives to execute the client $R$. There are three server alternatives, each of which can do the job, and they are marked as 1, 2, and 3. The dotted lines in the figure represent the alternative (or) relations.

In order to execute server alternative 1, we may choose subservers 1.1 and 1.3 or 1.2 and 1.3. Here, 1.2 serves as an alternative for 1.1. However, in both cases 1.3 is necessary along with subserver 1.1.1 which serves 1.1 and 1.2. Therefore, although 1.1 serves as an alternative for 1.2, the fact that both rely on services of 1.1.1 shows it is not a full back-up alternative.

If, on the other hand, we choose alternative 2, we still need server 1.3 as before, but it is required in conjunction with server 2.1. Server

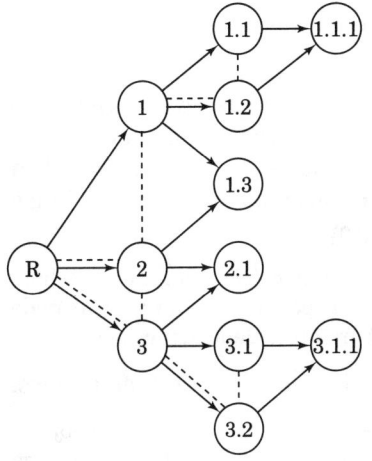

Figure 12.6 Multiconnected alternatives.

3 requires services of 2.1 and services of either 3.1 or 3.2. Both 3.1 and 3.2 rely on services of 3.1.1.

Hence we can summarize this example as follows:

$$R \Rightarrow 1 \wedge (1.1 \vee 1.2) \wedge 1.1.1 \wedge 1.3$$
$$\bigvee 2 \wedge 1.3 \wedge 2.1$$
$$\bigvee 3 \wedge (3.1 \vee 3.2) \wedge 3.1.1 \wedge 2.1 \qquad (12.1)$$

Analysis shows that if we run the three alternatives, two failures (in 1.3 and 2.1) can make it inoperative.

In order to obtain quantified conditions of the resiliency of a system to faults, let us define our system properties more formally. Let each executable object instance $p$ have a set of resource requirements $\{R_i^{(p)} : 1 \leq i \leq k\}$ and a set of service requirements $\{S_i^{(p)} : 1 \leq i \leq n\}$, forming its *dependency set*, which we denote as $DS_p$. Restricting $p$ with a time constraint $TC_p$ applies a projected time constraint to each member of its dependency set. Each projection is a result of the temporal relation between $p$'s execution and its requirements. A service requirement can be executed by another executable object instance, which can be chosen out of a set of alternatives. Hence, we can define the dependency set as follows.

DEFINITION 37. The *dependency set* of an object $p$ with a time constraint $TC_p$ is

$$DS_{p,TC_p} = \{\, \{< R_i^{(p)}, TC_{R_i^{(p)}} >: 1 \leq i \leq k\} \,,\, \{S_i^{(p)} : 1 \leq i \leq n\} \,\}. \qquad (12.2)$$

where

$$S_i^{(p)} = \{< s_j^{(p)(i)}, TC_{S_i^{(p)}} >: 1 \leq j \leq M_i^{(p)}\}$$

and $M_i^{(p)}$ is the number of service alternatives of service requirement $S_i^{(p)}$. □

Let us consider a graph that models the dependency relations between an object $p$ and its requirements. We denote each relation by a directed arc from an object to a member of its dependency set. If a member of the dependency set is another object $q$, then $q$'s dependency graph is a subgraph of $p$'s dependency graph.

> DEFINITION 38. The *dependency graph* of an object is a graph in which the object is represented as a node, and directed arcs connect this node to the dependency graphs of the members of its dependency set. □

We can also define the set of members in each subgraph as follows.

> DEFINITION 39. A *reachability set* of an object $p_i$ is the set of all objects $p_k$, such that there exists a finite path from $p_i$ to $p_k$ in the dependency graph of $p_i$. □

The allocation resiliency that we are interested in is of the type that can guarantee a certain connectivity of our computation graph. Thus, we want to define resiliency to monotonic faults as well as to transient faults.

> CONDITION 20. A dependency graph of object $p$ is $n$-*resilient* to *monotonic faults* if it contains at least $n + 1$ distinct alternatives whose intersection with each other contains at most the node $p$. □

Let us now examine how a dependency graph demonstrates resiliency to transient faults, recalling that we allow multiple executions of instances of the same object as temporal redundancy.

> CONDITION 21. A dependency graph is $n$-*resilient* to *transient faults* if it is $k$-resilient to monotonic faults, and each of its $k + 1$ alternatives, $0 \leq i \leq k$, can have $\tau_i$ timely disjoint instances, such that
>
> $$\sum_{i=0}^{k} \tau_i \geq n \qquad (12.3)$$
>
> □

These conditions quantify the resiliency of a system to monotonic and transient faults. The conditions already take in to account the constraints one is likely to face in real-time system applications.

## 12.5 Dynamics of Replicas and Alternatives

The dynamics of invocations and gathering the results of alternative and replicated objects requires some mechanisms of arbitration. A system's computation graph can contain an object which invokes a set of alternatives for a particular service and that all these invoked servers can invoke yet another service. This last service has neither replications nor alternatives, and thus it should react to all these invocations as if it were a single invocation. In [104] such arbitrators are introduced: *forker*s for distributing and propagating an invocation to alternatives and *joiner*s for merging invocations coming from several alternatives to a single invokee. This scheme is described in Figure 12.7.

The left side of the figure demonstrates a forker that distributes the invocation of object A to servers $S_1$, $S_2$, and $S_3$ and a joiner which merges the servers' invocations of object B. The right side of the figure shows a consistent-link formation (the dotted line) where the alternatives $A_1$, $A_2$, and $A_3$ invoke through a single link the alternatives $B_1$, $B_2$, and $B_3$. If we exclude $S$, this formation ensures link consistency for $\{A_i\}$ invocation of $\{B_i\}$. This property of a back-to-back joiner and forker pair is a very strong characteristic for fault tolerant links, because it maintains the resiliency of $\{B_i\}$ even after a fault in $\{A_i\}$.

The use of arbitrators does not significantly affect the complexity of the fault tolerant algorithms in the system, since the number of forkers and joiners inserted in the computation graph is bounded by $O(n)$. Each regular computation node can be extended to include at most a forker and a joiner as its service access points (see Section 1.2.3).

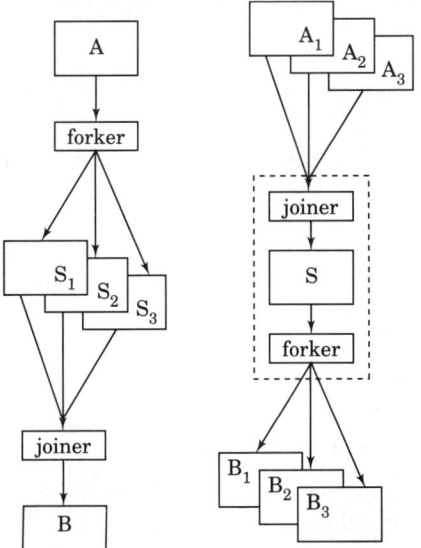

**Figure 12.7** Forkers and joiners as links and objects

### 12.5.1 Forker

A forker distributes the invocation message to a known set of alternatives. In order to properly accomplish its tasks, the forker needs knowledge of the alternatives' proper location, access restrictions, and operational status. The standard procedure of a forker is as follows.

1. It accepts an invocation at time $t$.
2. It checks the invocation legality and reasonability (logical as well as temporal).
3. It sends the correct invocation to all the members of the alternatives set at time $t + \Delta t$.
4. Upon an incorrect invocation, it invokes the exception/error handler.

We note that the forker plays an important role in the establishment of a semantic link to each of the alternatives. The critical role of such bounded semantic links is defined in Section 18.3.

### 12.5.2 Joiner

A joiner gathers invocation messages which are expected from a given set of alternatives. The standard procedure of a joiner is as follows.

1. It receives the invocations from all the members of the alternatives set at their arrival time $\{t_i, R_i\}$.
2. It selects an invocation $R$ based on the accepted $\{R_i\}$ and propagates it at time $t_f + \Delta t$, where

$$t_f = \max(T_{max}, \max_i\{t_i\}) \qquad (12.4)$$

3. If invocations fail to arrive before $T_{max}$, it invokes the exception/error handler, and $R$ is based only on the subset of the expected invocations: the accepted ones.

We note that the way in which we base $R$ on the incoming invocations can be any of the agreement methods introduced in Section 9. Hence, it plays an important role in the recovery from failures in a redundant architecture.

### 12.5.3 Resilient computation graph using elemental units

The dynamic nature of forking and joining is suitable for the construction of resilient computation graphs based on the computation graph of an application. The model of elemental units (EU) presented in

Section 10.5.2 identified their appropriateness for use in the construction of a computation graph for an application. When we need the tolerance of a given number of faults we can construct a resilient computation graph by replication of the elemental units. Let us consider an example.

Figure 12.8 illustrates modeling redundancy with elemental units. The example computation is a simple sequential execution of objects $A_1$, $A_2$, and $A_3$. The figure depicts three resilient elemental unit graphs. The global redundancy has a redundant branch of $A'_1$, $A'_2$, and $A'_3$ linked to the basic branch on invocation of the first and on exit of the last. The total redundancy has replicas $A'_1$, $A'_2$, and $A'_3$, where each object replicates its corresponding partner. The partial redundancy illustrates replication of $A'_2$ only.

The notion of having global redundancy, total redundancy, and partial redundancy gives the user flexibility in structuring the application to tolerate faults. The primary advantage of total redundancy is that after a failure, a degree of resiliency is restored if the necessary hardware on which the subsequent elemental unit is to run is still functioning. The graph connectivity provides a good measure for qualifying the resiliency achieved.

Note that the resilient computation graph requires the use of forkers and joiners which can automatically be added in the integration phase. Clearly, the user must retain control of the technique used by the joiner in determining the single message to send after it receives all the incoming messages. For example, it can be appropriate for one

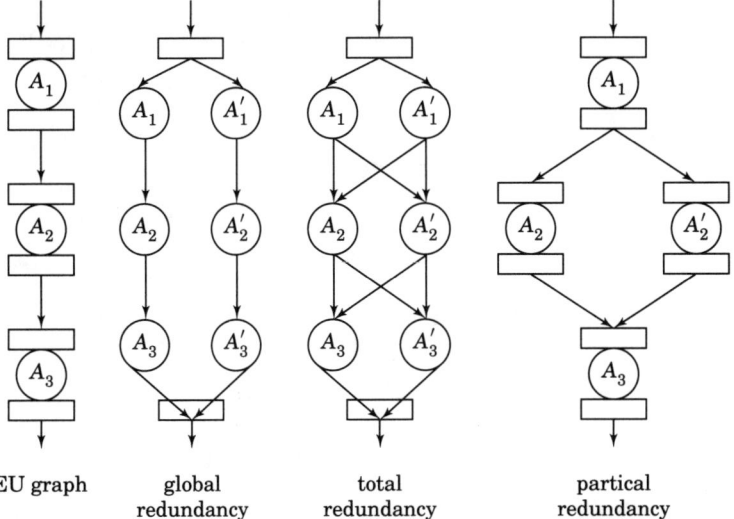

**Figure 12.8** Modeling redundancy with elemental units.

application to use the first message arrived and discard all the rest, while other applications can insist that voting or a quorum decision be used.

## 12.6 Concluding Remarks

Concurrently executing alternatives show promise as a solid basis for forward recovery schemes. This approach has been analyzed in this chapter for both replication and N-version redundancy. The fault tolerance goals are examined to satisfy particular conditions of the dependency set, constructing a powerful tool for verification based on the dependency graph of alternatives.

However, in addition to static verification of the goals, the dynamics of forking (and joining) the invocations (and results) of alternatives is an important implementation issue. We have specified various properties of these forkers and joiners. Implementation of these dynamic elements as a part of a fault-tolerant operating system constitutes a transparent redundancy control feature to applications.

Applications can require different resiliency levels for various subsets of their objects. Thus, merging and distributing invocations through the communication subsystem can vary between one subset and the other. Leaving the forking and joining in the application domain unnecessarily complicates all aspects of an application. We show later how an allocation scheme allows the operating system to provide means for joiners and forkers to be incorporated transparently to an application.

**Chapter**

# 13

# Handling Exceptions

Exceptions are detected and treated at various levels of the system. At the lowest level, the hardware and machine primitives level, most systems employ interrupts and traps to signal the occurrence and the handling of exceptions. Among these primitives we find several that support event detection, others that support identification of the source, and several that support event masking and discarding.

Operating system and language support provide upper level mechanisms, such as handlers and servers which implement the system policies at exception occurrence. Such mechanisms include the processor's interrupt and trap handlers, rescheduling and reconfiguration servers, and communication support tools. Services of various system resources are generally included as well.

At the application level, users implement their application-specific exception support. User defined servers support application-specific irregularities, and in large applications even priority management of these services.

The objective of exception handling is the formation of an appropriate response for unpredictable events that can occur during execution. We note that such irregularities can involve design flaws, timing errors, and even singularity points which should have been avoided (e.g., division by zero). However, in a fault tolerant system with some alternatives executed redundantly by overlapping tasks, we can assume local restriction of the irregularity and "hope for" a global solution construction. The overall system may, therefore, be able to recover from this local fault.

## 13.1 Interrupts and Traps

Detection of unpredictable irregularities during execution generally interrupts the execution thread to avoid contamination and to allow corrective actions. Upon detection of such an irregularity, the resource manager receives an interrupt request. As the urgency and criticality of the request are not known *a priori*, especially with respect to other pending requests, most processors implement priority and queueing models to support an appropriate resource management process.

Interrupting the execution thread of a single processor system generally involves the following steps:

1. The singularity source (or detector) signals its interrupt request.
2. The processor completes its current instruction.
3. All interrupts are disabled.
4. The processor acknowledges the interrupting source with the highest priority.
5. The latter identifies itself to the processor.
6. The processor masks this interrupting source, disabling it from requesting additional service until this service is complete.
7. The processor stores the volatile parameters of the interrupted thread, including the program counter of the interrupted thread. It then loads the parameters of the interrupt service, including its start address.
8. Interrupts are enabled.
9. Execution is resumed, serving the interrupt request.
10. The service ends by unmasking the interrupt requesting source, enabling its requesting capability.
11. Once service is over, the interrupted thread parameters are retrieved from storage, and execution of this thread continues.

We note that the interrupted thread can be the service of an interrupt request with a lower priority. This case of interrupt nesting is not always permissible, and operating system features use criticality parameters to enable or disable interrupting sources while servicing another.

Multiple processor systems are more complicated and in most cases require interrupt support in their interprocessor communication (IPC). The reason originates in the architecture which can allow service of requests by processors which do not necessarily participate in the request detection. Therefore, an interrupt request accepted locally by a processor may be linked to a consecutive IPC initiative, to be forwarded to another processor. The forwarding sequence must include a minimal handshake process of acknowledgement by the destination processor, followed by a delivery of the detailed request.

### 13.1.1  A general model

Asynchronous external events can be detected by a computer program if it is associated with an accessible indication flag. A simple polling mechanism supports such a detection, interrogating each of these flags to determine event occurrence. However, polling is very inefficient and has very slow response characteristics. Furthermore, it is impractical to interrogate all the indicators all the time. It is unreasonable to suspend CPU execution unless one knows an indicator has been turned on. Therefore, our general model assumes the processor is not interrupted unless an indicator has been turned on.

Figure 13.1 describes a simple interrupt structure. Each of the event detectors (or external devices) produces its interrupt request into a single channel, indicated as *intr*, which is an integration of all the requests. Once the CPU detects the existence of a request, it polls all the indicators to identify the request's source. The polling is done through the control and the address/data buses.

Before the control is transferred to execute the service routine, the current state of the CPU is saved in the memory. This activity is frequently performed by the operating system's interrupt handler. Upon completion of the service, this state is restored and execution resumes. Generally, state restoration is performed by an "interrupt return" command (IRET), with which every service routine must conclude.

**Priority of requests.** It is very impractical to poll all the indicators in order to determine which is the source. Normally, we do not want to serve the first one to be identified but rather the most crucial one.

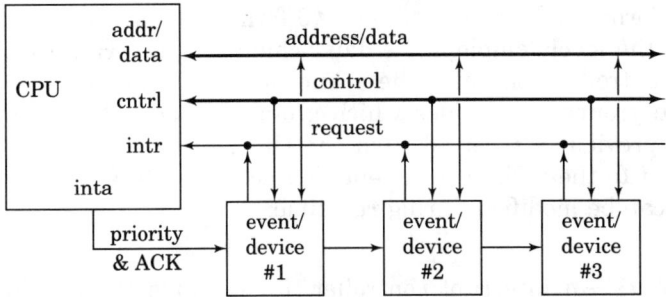

**Figure 13.1** Single interrupt request, daisy-chain prioritized acknowledge.

Therefore, it is reasonable to employ a scheme in which the most critical interrupt source will be the first to identify itself when interrogatation of indicators takes place.

One such scheme is the daisy chain described in Figure 13.1. The signal is passed from one indicator to another. Device #1 has the highest priority, and if it has no pending request it passes the signal to device #2, and so on. The first one to have such a pending request, is currently the indicator with the highest priority, and it reacts to the acknowledge signal by identifying itself by putting its address on the address/data bus.

This priority scheme is inflexible, and is based solely on the hardware wiring. Run-time modification of neither the priority nor the scheme is allowable, and thus this scheme by itself is usually inadequate.

Another issue which is related to the priority scheme is the capability of interrupt nesting, or the interrupting of an interrupt service routine. It is imperative to block interrupts when we interrogate the indicators. Still, we want to be able to interrupt the execution of a service routine of a lower priority interrupt by a higher priority interrupt. Furthermore, this nesting capability assures that upon servicing an interrupt, it is blocked from re-interrupting its own service. Thus, the interrupt request indicator of a particular event/device is cleared upon starting its service, and it is marked "in service" as long as the service did not complete.

**Vectored interrupts.** An extended scheme of communicating the interrupts to the CPU is the interrupt vector scheme. The CPU is informed of the pending requests by a vector and not by a single bit. The number of bits in the vector (say $v$) defines the number of priority levels which the CPU discriminates ($2^v$). The processor can then dynamically tune itself to block interrupts below certain threshold.

Figure 13.2 describes the generation of such a vector, indicated as *ivec*. There are three interrupt priority levels in this figure: devices #1, #2, and #3 form one level, #4, #5, and #6 form another level, and so on. Each of the levels employs the daisy-chain scheme. However, the interrupt controller generates the acknowledge (marked *inta*) to only one of the groups. It decides which group it is according to a scheme it has previously received from the CPU. Furthermore, the vector generated to the CPU, *ivec*, is encoded according to a priority scheme which can be modified during run time.

**Interrupt controller.** An interrupt controller, for example the one in Figure 13.2, supports the CPU in the following interrupt activities:

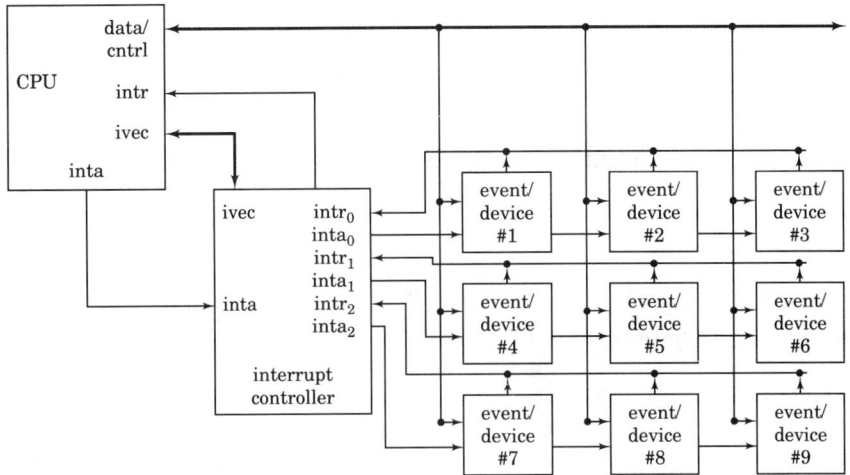

Figure 13.2  Multiple interrupt requests, daisy-chain acknowledge.

- handling multiple requests
- dynamic masking of requests
- dynamic prioritizing, including altering the priority of various groups
- nesting interrupts, including clearing indicators, marking the in-service events, and filtering threshold by CPU commands.

These features constitute a generalized hardware support for the CPU to handle the multiple-interrupt system structure.

**High level synchronization with interrupts.** Interrupts are used for synchronization of activities, as well as the detection of irregularities. It is therefore improper to anticipate that the whole service required by an interrupt will be reconstructed as an interrupt service routine. Catching the indication from the hardware is simple, and is generally performed by the interrupt service routine. However, the interpretation of the event for a process performed, by the system software, is executed at a higher level.

Interlevel synchronization is done by various means.

- Flags are employed to support proper interpretation of interrupt generated in the same locality of the handler.
- Messages are used for the same task that employs the flags, when the generator resides remotely of the handler.
- A queue is used in the presence of multiple sources, such as the scheduler case.

**Traps.** Traps are actually software generated interrupts. There are events which are detected during run time that do not need additional processing in order to detect irregularities. Some examples are:

- division by zero (integer and floating)
- arithmetic overflow
- index overflow
- string (or array) size out of bounds
- operands type mismatch.

Detection of any one of the above events is a consequence of the execution of a particular software instruction. The system handles the trap condition in the same way as it handles the interrupt condition. However, as the trap condition is generated within the CPU, the identification of the condition is simpler.

### 13.1.2 Real-time performance

Interrupt driven systems are frequently used in real-time applications, especially in control systems. The system reacts to each interrupt that triggers the system with an interrupt service process that is defined by the system designer. This service process is invoked by the operating system through its interrupt handler. The handler identifies the arriving interrupt and passes control to the appropriate service process.

The interrupt service definition in a real-time system must contain a time constraint [24]. The designer of the system defines the maximal frequency within which each interrupt is allowed to occur. A deadline for the service of each interrupt is distinctly set, along with computation requirements. The system designer also specifies conditions on start times. Hence, each of the interrupt service processes is expressible as the time constraint. However, an additional mechanism is required to ensure that a particular service process is to be executed if and only if the interrupt which it serves has already occurred. In other words, the time constraint of an interrupt service process depends on the occurrence of an *event*, and not only on the initial timing condition.

The requirement to guarantee deadlines in a real-time system implies that the scheduling mechanism must take interrupts into account. Most scheduling mechanisms ignore the effect of services given to interrupts. A simple interrupt handler serves an interrupt from the moment it occurs, preempting the object currently being scheduled for execution. In such a preemption, one might cause a deadline miss, which might be an unrecoverable fault. Further, there might be enough time to execute the service process after the currently executing process completes, without any preemption at all. These reasons justify a selective preemption approach. We can use the object *state*, which we can evaluate and change, in order to deal with interrupt driven processes without contradicting a guaranteed deadline.

**Interrupt driven objects.** An interrupt-activated executable object can have two states: *idle* and *active*. Let the state of such an object be *idle* as long as the interrupt it serves has not occurred. The prescheduler (if employed) and the process of allocating resources must take the interrupt service requirements into account. This way, the operating system may be able to guarantee meeting the deadline of an interrupt service while maintaining guarantees given to already *active* objects. Therefore, *idle* and *active* objects are all considered executable, regardless of their states.

The on-line scheduler applies a state-evaluation procedure to the object which is most likely to use the resource next. If it is an *active* object, then this object receives control of the resource. We must postpone the execution of an *idle* object, even though it may be more favorable from the timing point of view. However, we can advance an *active* object scheduled for a later time. To do it, we must be able to carry out the execution ahead of time without violating the postponed object's time constraint. In this case, the *active* object receives control of the resource. Figure 13.3 describes such a case, demonstrating that $P_1$ can execute before serving the interrupt. Note there is no rea-

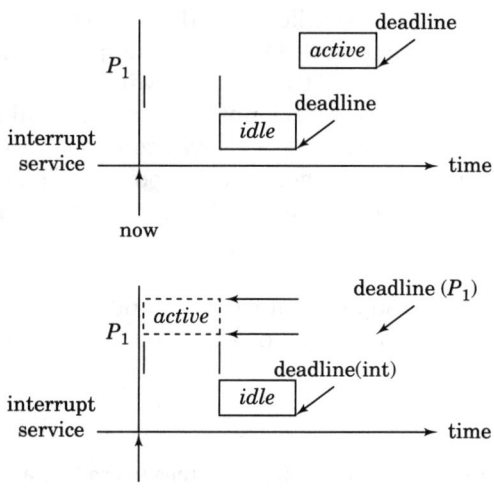

**Figure 13.3** Interrupt service without preemption.

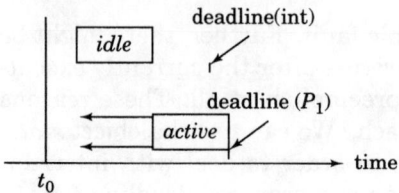

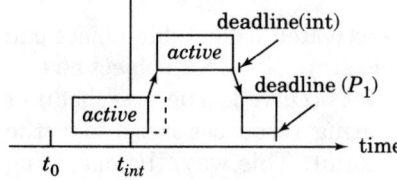

**Figure 13.4** "First-deadline" scheduling with preemption.

son to preempt $P_1$, even if the interrupt occurs during the execution of $P_1$.

The interrupt handler invokes the proper *state-change* procedure according to the interrupt source. This procedure changes an *idle* object to an *active* one. Both the *state-evaluation* procedure and the *state-change* procedure are very short in their execution time. Each represents a constant overhead cost for all the devices at a specific processing node.

However, it is wasteful to apply such a very safe approach. So far, we have allowed advancing execution of active objects only if they can complete without violating the time constraint of the postponed interrupt server. Hence, we want to utilize the slack[1] of an idle object, as long as it is idle and we do not violate its time constraint. Figure 13.4 gives such an example. $P_1$ executes although the interrupt service deadline is earlier, until there is a conflict with the service time. Upon conflict detection the service preempts $P_1$. One of the problems with this approach is the overhead cost of the context switching. An additional problem is the depth of nested preemption allowed, while guaranteeing satisfaction of all the constraints. One way to solve these problems is to selectively allow such switches at checkpoints. Thus, we allow preemption only at a finite number of points in the execution trace of an object.

**Interrupt handler example.** The system's interrupt handler is a hardware-invoked server. An occurrence of an interrupt event trig-

---

[1]Time slack is defined as $\max(d_{Id} - c_{Id}, 0)$ where $d_{Id}$ is the time to deadline and $c_{Id}$ is the computation time.

Upon Interrupt_Define(*device,server*) ::
    **begin**
        **if** ($\exists x : \exists < device, x > \in Interrupt\_Map$) **then**
            **signal** exception(ambiguous definition)
        **else**
            $Interrupt\_Map := Interrupt\_Map \cup < device, server >$ ;
            $server$.user $:= server$.user $\oplus \& < device, server >$
        **fi**
    **end**
Upon Interrupt_Remove(*device*) ::
    **begin**
        **if** ($\exists x : \exists < device, x > \in Interrupt\_Map$) **then**
            $server$.user $:= server$.user $\ominus \& < device, server >$ ;
            $Interrupt\_Map := Interrupt\_Map\ (\cup)^{-1} < device, server >$
        **else**
            **signal** exception(nonexistent device)
        **fi**
    **end**
Upon interrupt occurrence (*device*) ::
    **begin**
        interrupts **disable** ;
        **if** ($\exists server : \exists < device, server > \in Interrupt\_Map$) **then**
            $server$.state := active ;
            **mask**(*device*)
        **else**
            **clear**(*device*) ;
            **signal** exception(nonexistent device)
        **fi**
        interrupts **enable**
    **end**

**Program 13.1** Interrupt handler example.

gers this invocation. Such a server needs three service access points (SAPs) to perform the tasks listed below. In the description of these SAPs we use the following notation:

- $\cup$ is an insertion algorithm
- $(\cup)^{-1}$ is a deletion algorithm
- $\ominus$ is a set subtraction operator
- $\oplus$ is a union operator
- $\& < device, server >$ is a backward link from the object's joint to the node.

1. *Interrupt Definition.* When a hardware device signals that it needs its service, the system must recognize this service identification in

order to invoke the proper service. An invocation of this entry of the interrupt handler defines and updates the mapping from device identifiers to the server objects. Often, systems manipulate data structures in order to perform the mapping. Other systems use a dedicated memory in the hardware vector interrupt driver. We denote this data structure as the *Interrupt_Map*.

The Interrupt_Map must support a very efficient mapping mechanism, being a part of the interrupt overhead. This action interrupts the execution of a guaranteed time constraint, even when we do not invoke a particular interrupt server. When we have a large number of interrupt sources, lookup tables become inefficient. In such cases, one can implement the mapping data structure with an ordered and balanced tree, whose nodes are the tuples

$$< device\_identifier, object\_server >.$$

When one invokes the Interrupt_Define entry, supplying the proper tuple, we expect the interrupt handler to insert the node to the *Interrupt_Map*, as described in Program 13.1. For an example of an insertion algorithm, see [77, Chapter 6.2.3].

2. *Interrupt Removal.* The converse of the above operation is the removal of a service from the recognized list of services. When one invokes this entry, indicating which device the handler must remove, we expect the interrupt handler to remove the node from the *Interrupt_Map*. An implementation example of these requirements is given in Program 13.1.

3. *Interrupt Response.* The interrupt service is not within the responsibilities of the interrupt handler. The pre-scheduler and the allocator reserve resources and account for the server's requirements. This server executes after the interrupt occurrence, according to the on-line scheduler's decision. Therefore, the only activity required from the interrupt handler in a response to an interrupt occurrence is to record the event occurrence. The occurrence recording must reflect an active state of the server to the on-line scheduler. It also must acknowledge the hardware in order to clear the request and to allow other requests to arrive. A possible implementation that achieves these goals is given in Program 13.1.

The response handler is homogenous in the sense that all the interrupts receive the same response. Its execution load is predictable and deterministic, and can be derived from the implementation of the Interrupt_Map.

Let us consider an operating system with a core resident kernel and a boot mechanism that loads it. When the boot mechanism loads the kernel, it loads a basic Interrupt_Map with it, which contains services for the devices of the system. Users can define new services for existing devices, and expand the system with additional devices. We take care of mutual exclusion in accessing Interrupt_Map by making sure that the users cannot access the Interrupt_Map directly. Only the handler's object manipulates this data structure, which is done on behalf of one user at a time.

Let us now consider cases that do not agree with our assumptions. These are cases such as those in which the system clock's granularity is extremely fine and the sum of responses to interrupt occurrences is not negligible. In such cases one should compensate for the maximal number of possible interrupt occurrences with a continuous overhead cost of the system. In order to guarantee meeting a job's time constraint we have to advance its required deadline accordingly. Let each response require $\delta$ time units (assuming homogeneity). Let the maximum number of interrupts in the system be $n$, and let each interrupt $i$ occur at a frequency $f_i$. For a time constraint with a frequency $f_k$, one must advance the deadline by $\sum_{i=1}^{n}(f_i/f_k)\delta$. This way we can guarantee meeting the deadline.

## 13.2 Reaction to Exceptions

The invocation of some mechanisms is required upon exception detection. A scenario of these activities follows.

1. Detect the exception.
2. Come out of the exception detection with some diagnostics.
3. Report and cease local execution.
4. Produce a special value or a boolean (e.g., returning $-1$ in Pascal).
5. Invoke a corrective action.

We designate corrective actions performed by the object which contains the fault as *local*. If, however, a local action is insufficient, a *remote* action must follow.

Let us assume that when a detection is permissible, the detector can *signal* the remote handler that is able to react to that exception. In order to allow remote reactions, the system must define which object is to handle a signaled exception before its occurrence. In addition, the system must impose a constraint on resumption or preemption of execution of the object where exception is detected. We note here that

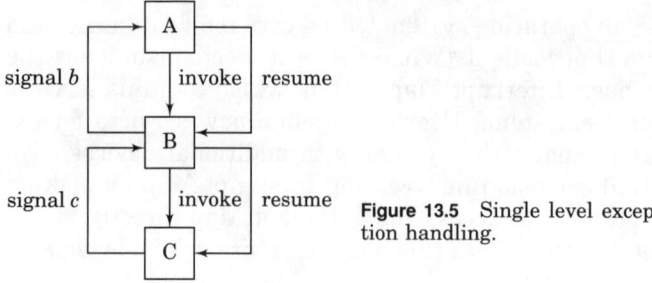

**Figure 13.5** Single level exception handling.

one must specify a default state of object in order to resume execution. Default values by their nature are application dependent.

Two signal mechanisms can be used:

1. A single level mechanism, when upon exception detection by the invoked object, this object signals the invoking object. We find examples of these mechanism type in Ada[2] [23, Chapter 11].

2. A multilevel mechanism, as in Mesa[3]. Here, execution can resume at a higher level than the one which generated the signal.

### 13.2.1 A single level approach

A single level mechanism is characterized by having the object which detected an exception signal, signal the object which invokes it. The invoking object, upon being signaled, is responsible for resuming the execution of the invoked object which has aborted due to the exception. Figure 13.5 illustrates this hierarchy of resumption, as found in CLU [95] and Ada.

**Signal generation.** The exception is signaled from the procedure level upon its detection. The signals which a procedure invokes are declared at the procedure statement. For example, consider the division procedure in Program 13.2.

Detected exceptions are signaled (Ada uses the term "raised") by local or concurrent execution threads. The exception handling of the procedure scope is defined by the "exception" statement. The "divide_by_zero" exception is raised and treated locally. The "unrepresentable" and "failure" exceptions are signaled to the caller level and

---

[2] Ada is a trademark of the U. S. Department of Defense (Ada Joint Project Office).

[3] Mesa is a trademark of Xerox Corporation.

```
procedure div(x,y:integer) return (real) signaling
    (divide_by_zero,unrepresentable,failure)
  begin
    if non_integer(x) or non_integer(y) then
        signal unrepresentable
    else if y=0 then
        signal divide_by_zero
    else
        return x/y
    fi
  exception
    when unrepresentable =>
        signal unrepresentable
    when divide_by_zero =>
        return -1
    when others =>
        signal failure
    end
  end div
```

**Program 13.2**  Exceptions in division procedure.

are to be dealt there. The term "others" is the inclusion of all other exceptions defined for the current scope.

**Backward error recovery.** The roll-back recovery described in Chapter 11 can also be described by the single level exception scheme. Let us recall the recovery block of Program 11.1, which has the following structure.

```
begin
        ensure < AT >
        by primary alternative
        else by secondary alternative
        else error alternative
end
```

The acceptance test $< AT >$ must be satisfied for the primary alternative to succeed. The secondary alternative is an exception handler for this level, and the error alternative should inform the upper levels of the malfunctions at this level.

Program 13.3 describes a pseudo implementation of roll-back by exceptions signalling. The primary, secondary, and error alternatives execute concurrently, with the secondary waiting for the primary to terminate or to signal. The error causes the secondary to wait or to signal. The error alternative signals the failure to the upper level.

```
primary::
    begin
        init(cache) ; /* save state for roll-back */
        do primary algorithm ;
        if not < AT > then
            signal not_accepted
        else
            discard(cache)
        fi
    end
secondary::
    begin
        ...
    exception
        when not_accepted =>
            restore(cache) ; /* roll-back */
            do secondary algorithm ;
            if not < AT > then
                signal error
            else
                discard(cache)
            fi
    end
error::
    begin
        ...
    exception
        when error =>
        discard(cache) ;
        signal failure
    end
```
**Program 13.3**  Roll-back by exception.

### 13.2.2  A multilevel approach

A multilevel mechanism, allows execution to resume at a higher level than the one which generated the signal. Such a mechanism is employed by the Mesa support on Xerox machines.

**Signal origination.**  The signalling mechanism takes over the polling roles, and generates the exception declaration to be caught by a handler. The signal generator does not aim its declaration for a specific handler. The operating system and the run-time support take care of the destination selection.

Figure 13.6 describes multilevel exception handling in which a signal InputTooBig "climbs" up in the calling hierarchy to find the nearest handler for service. The calling stack shows the calling sequence

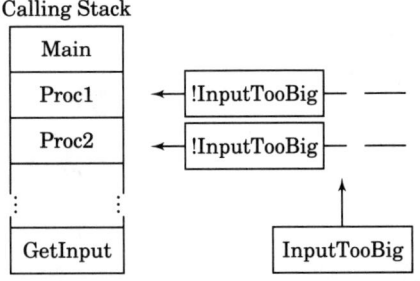

**Figure 13.6** Multilevel exception handling.

down to the GetInput procedure, in which the signal InputTooBig is generated upon a test.

GetInput::
**begin**
    ...
    **if** InputNumber> ··· **then**
        **signal** InputTooBig
    **fi**
    ...
**end**

This can be expressed in the calling procedure as

$n \leftarrow$ GetInput(!InputTooBig$\Rightarrow$RETRY)

marking the signals to be caught.

**Catching a signal.** We can distinguish between two ways of activating a signal catch.

1. The signal catch (!) is defined in the procedure call statement, following an exclamation mark after the parameters list.

$rln \leftarrow GetInput(\ param1, param2, \ldots !InputTooBig \Rightarrow RETRY,$
$InputTooLittle \Rightarrow begin \ldots end)$

The RETRY handler is therefore activated within the calling level.

2. The signal catch is activated at run-time by "enabling" the catch.

Proc1::
    **begin**
        ...
        Proc2 ;
        **if** InputCase ... **then**

```
            enable InputTooBig ⇒
                begin
                    GetMoreMemory ;
                    RESUME
                end
            do ...
        fi
        ...
end
```

The RESUME command activates the InputTooBig signal generator, which is at a lower level than this handler. Therefore, the signal generator is expected to support such a resume entry.

**Rejecting a signal.** A signal can be rejected in several ways:

1. an explicit reject by a REJECT command
2. implicitly, by not being caught
3. implicitly, by falling off the enablement "end".

Otherwise, the signal is caught.

**Multilevel roll-back.** In a multilevel roll-back, there are some cases in which we want to roll back to the signal generator level, and some cases in which rolling back to the handler level is preferable.

1. We have seen that RESUME takes us back to the signal generator level for exception recovery.
2. We need a CONTINUE command to resume execution at the level of the handler which caught the signal.

**Clean-up.** Resuming execution at the level of the handler which caught the signal by a CONTINUE command causes some problems. At lower levels, such as the one which generated the signal, we may have left open channels, resources which have not been released, queues which have been blocked, and other loose ends.

A special UNWIND command deals with the unfinished processes. However, the cleanup process is used for a regular exit, as well as for a stack pop-up during the CONTINUE recovery.

### 13.2.3 Comparison

The multilevel mechanism is more powerful, but it contains some drawbacks. When execution resumes at a higher level, we need a "clean-up" of all the lower level activities that we have preempted. In addition, this may contradict our objectives of autonomy and containment. Letting the exception affect a distant activity at a higher level may spread the effects of the exception. For example, in hierarchical systems a higher level affects more objects than a lower one.

Another drawback is the degree of unpredictability that results from this approach. Real-time systems, for example, cannot allow using loosely controlled activities for clean-up and garbage collection.

## 13.3 Exception Handling Model

The behavior of each software module can be viewed as a set of states and a set of state transitions. Say that the program $P$ started at a state $s'$ and resulted in $s$. Program $P$ supports the service $post(s', s)$ as its specification.

### 13.3.1 Standard and exception domains

The module behavior can be categorized [42] in two complementary domains: the specified behavior is the module's standard domain and the faulty behavior is the exception domain.

> DEFINITION 40. The *standard domain* (SD) of a program $P$ are the set of states $s'$ for which $P$ terminates normally in states $s$, such that $post(s', s)$.
>
> □

Consequently, the corresponding definition of the exception domain follows.

> DEFINITION 41. The *exception domain* (ED) is the set of states which do not belong to the standard domain.
>
> $$ED \equiv \neg SD$$
>
> □

The exception domain is further divided into an anticipated exception domain (AED) and an unanticipated exception domain (UED). Figure 13.7 describes the relations between these domains. The anticipated exception domain is dealt with by the exception handlers.

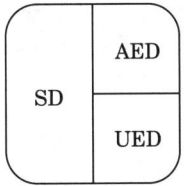

**Figure 13.7** Standard and exception domains.

The unanticipated exceptions can be detectable, as they are invocations of the module in states which belong to neither SD nor AED. However, being unanticipated they do not have an exception-specific recovery process. Invocations at this domain ($s' \in AED$) may lead to nonterminating or blocked states, or even to states ($s$) which contradict the module specification $post(s', s)$.

### 13.3.2 Modeling exception handling with elemental units

The model of elemental units (EU) presented in Section 10.5.2 for treating detected faults is suitable for handling exceptions as well. The activation of an elemental unit requires that the input values be checked for correctness. If an exception condition occurs before the activation of an elemental unit, it can be detected by the input monitor. If an exception condition occurs during the execution of an elemental unit, the output monitor will detect its occurrence. Both the input and the output monitors have the ability of generating a message. The exception condition can be signaled to some other unit through this mechanism.

The input and the output monitors of an elemental unit check the syntax and the semantics of the information they receive. Clearly, these checks must be defined by the programmer so that they are the most appropriate for the functions being carried out by the elemental unit. For example, the values for input variables can be checked for being within range. Similarly the output values can also be checked to be within range. When the input values are outside the range an exception condition is detected and a message is generated.

The exception message contains information which is very similar to any other message and thus can be handled by an elemental unit. This design permits the use of elemental units for the receipt of the exception messages. The Exception Handling elemental unit, on receiving a message, can take the appropriate action for handling the exception condition. This organization permits the use of a uniform mechanism for the handling of the exception conditions in a uniform way. An example of this arrangement is shown in Figure 13.8.

### 13.3.3 Dynamic adjustment of exception handling EUs

Each elemental unit carries out a user defined function and, for the given input variable values, generates the output values. The checking of the input variables is done by the input monitor and the checking of the output variables is done by the output monitor. Clearly, one way of defining the checks for the input and the output monitors is to do them statically at the compile time. For most cases, such static tests will have to be rather loose.

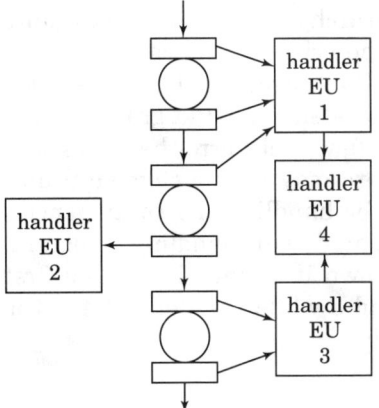

**Figure 13.8** Exception handling structure based on elemental units.

For many functions carried out by elemental units, the output values are related to the input values of the variables. For such elemental units the ranges and other checks to the output variables may be better defined after the input variable have been observed. This may require some computations to be carried out but is likely to yield a much tighter check on the output values. This conforms to the principle of detecting any error or exception condition as early as possible.

The dynamic adjustment of the monitors of an elemental unit requires that additional computations be carried out as a part of the input monitoring function. One way to generate the output checks is to use a much simpler algorithm which may yield approximate values but which are within bounds. These bounds can be used as checks for the output monitor. Another advantage of using a simple algorithm for generating the bounds is that we may be able to obtain a default value which, if required, can be sent to the next elemental unit to be executed under exception conditions.

### 13.3.4 Error reporting hierarchies

As previously noted, the input and the output monitors of an elemental unit send messages to another elemental unit when they detect the presence of an exception condition. Clearly, the monitors can be organized to send regular status or other messages also. The elemental unit receiving the messages from the monitors of an elemental unit is its *handler EU*. A handler EU may be connected to several elemental units and therefore, can take into account the information it obtained from all these units in handling the exception condition reported by one or more of these units.

Since an error handler is also an elemental unit, it can be assigned an error handler as well. In this way an exception handling hierarchy can be defined. At the base of this hierarchy are the elemental units forming the elemental unit graph of the original application. As we go up in the hierarchy, we have first level error handing EUs, each monitoring a few EUUs of the application elemental unit graph. At the next level, the error handlers monitor a few level 1 error handlers, and so on. At the top of this hierarchy there may be one elemental unit which provides the human interface for handling the contingencies which cannot be handled by the automated error handling structure. An example of such a hierarchy is shown in Figure 13.8. The first level handlers are handler EU 1, 2, and 3 while handler EU 4 is the second level handler.

## 13.4 Concluding Remarks

In this chapter we considered the problems associated with the treatment of run-time exceptions. The basic tools used were described for hardware and software interrupts and an architecture was presented for the multilevel handling of exceptions. A consistent way of modeling the structure of exception handling is through the use of elemental units.

We must recognize that the occurrence of exceptions and their handling can compromise the integrity of a system. Especially for mission critical systems, it is essential that the exception domain be fully addressed as a part of the design. This is very critical for systems having hard real-time constraints.

**Chapter**

# 14

# Consistency

A system must support mechanisms to assure its consistent behavior. The tasks of such mechanisms, as mentioned in Chapter 10 Definition 32, is the avoidance of contradictions and conflicts within the system itself and with external systems with which it interacts.

In this chapter, we consider aspects of concurrency control, define and analyze atomicity and transactions, describe partitioning effects, and broadcasting mechanisms, all in order to support the maintenance of consistent system states.

## 14.1 Concurrency Control

Multiple computing resources linked for a single computation is a common phenomenon today. In such systems, the concurrent execution of the various parts of the computation which are geographically distinct is inevitable. A concurrent system, as defined in Chapter 5, has a program with multiple threads and multiple address spaces. If the address spaces do not intersect, it is a distributed program, however, if they do, memory is shared. In both distributed and shared memory systems concurrency control is extremely significant to the achievement of the system's correctness and performance.

The basic element of activity in any mechanism is called an *action*. However, actions performed by complex systems are not totally basic. Generally, actions are constructed of primitives which are the generic elements of activities. The primitives are indivisible, but the actions are not. Therefore, in concurrent systems one must consider the effects of executions of "half" actions in perfectly functioning systems and even in cases of failure occurrences. In distributed database systems inconsistent data may result if half actions are allowed to occur.

Another aspect which must be considered is the control of the sequential execution of the actions. Say that an action is constructed of some primitives, each of which is separately invoked. Let us consider a case such that due to latency effects, the order of invocation requests for the same primitives may be different for two processors which work on two copies of the same database. Clearly, the results can come out different. Hence, serial ordering must be imposed on action executions. One way to ensure serial ordering is through mutual exclusion. However, mutual exclusion is too restrictive: why not allow parallel "read" actions when no write is executed and objects are not to be modified?

#### 14.1.1 Serializability criteria

We want the result of the execution of an application concurrently with other applications to be as if it were produced by *some* serial ordering. Let our system be composed of objects (see Section 1.2.3), each of which can be invoked to perform an action, and respond. Each event in this system is therefore described by the $(invocation, response)$ pair, and we order events precedence ($\succ$) by their responses. The *history* [63] of an object is the sequence of serial events which models its internal state. Since concurrent execution is allowed, an object's state is given by a *trace* of action executions, commit executions, and abort executions. The history (say $h$) is therefore a serialization of the concurrent trace (say $H$) of an object in an order (say $\succ$).

The relation of ordering events by precedence and the serialization is very precise. In fact, we want the ordering relation to be a serialization relation. Thus, a concatenation of an event to a history which consists of a serialized sequence of events (each of which precedes this concatenated event) produces a permissible history.

The serialization of a trace $H$ is a reordering process that results in a history $h$. Let $H$ contain two subtraces $H_1$ and $H_2$, such that $H_1 \succ H_2$. Let $h_1$ be a subsequence of events in $h$ associated with $H_1$, and accordingly $h_2$ be likewise associated with $H_2$. The reordering constructs $h$, such that

$$\forall H_1, H_2, h_1, h_2: \quad H_1 \succ H_2 \Longrightarrow h_1 \succ h_2 \qquad (14.1)$$

Hence, a *serializability criterion* can be defined as follows:

> DEFINITION 42. A trace $H$ is *serializable* if its serialization $h$ is a permissible history. □

When we use multiple copies (replicas) of the same object, serializability gets an additional dimension. A commonly used criterion for serializability is the condition of *one-copy serializability*, which is given in Condition 22. This condition refers to the correctness of updates applied to replicas, where the order in which the read and write primitives are executed must be carried out as if we have a single copy of the data.

> CONDITION 22.  An action is *one-copy serializable* if the concurrent execution of the action's primitives on replicated data items is equivalent to a serial execution of this action on a single item. □

One-copy serializability and adherence to the permissible history require mechanisms to support them. The following locking mechanisms have been introduced to overcome conflicts between concurrently executing actions.

### 14.1.2 Locking

Detection of conflicts and contradictions in a system can be avoided if we take them into account at design time. In a simplistic model of read and write accesses performed concurrently on the same data item, multiple read actions do not impose any serialization conflict. However, conflicts in a multiple access environment occur whenever a write is carried out:

Conflicts Table

|       | read | write |
|-------|------|-------|
| read  |      | +     |
| write | +    | +     |

Let us consider a lock with the following properties:

- A lock is required for each entity.
- We distinguish between a read lock and a write lock to allow an independent access rule at least for the read-read case.
- A process must acquire a lock before an access is initiated.

**Single phase locking.** Let us examine if a lock is a sufficient tool to ensure serializability. We can define a single phase lock as a lock obtained for every write access, like the one in the multiple-readers single-writer case. The following example demonstrates how the lock fails to ensure serializability in the multiple-readers multiple-writers case.

Consider the example depicted by Figure 14.1. There are two actions $T^1$ and $T^2$, where $T^1$ consists of the read/write sequence

$$R^1(x)\ W^1(x)\ R^1(y)\ W^1(y)$$

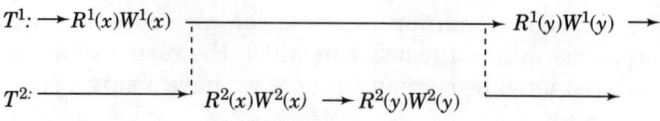

Figure 14.1 Serializability in single phase locking.

and $T^2$ of the sequence

$$R^2(x)\ W^2(x)\ R^2(y)\ W^2(y)$$

According to serializability criteria, the result expected after a sequence ordered $T^1T^2$ is $\{x_2y_2\}$. After a sequence ordered $T^2T^1$ the expected result is $\{x_1y_1\}$. However, Figure 14.1 shows that a single phase lock may produce a result of $\{x_2y_1\}$.

**Two phase locking.** In order to overcome this nonperformance, we can carry out actions in two phases, a "growing" phase that includes acquisition of all the locks needed for the whole sequence and a "shrinking" phase that includes release of all these locks. This approach is depicted by Figure 14.2.

Important properties demonstrated by this approach include the following.

- An action is not allowed to acquire any lock after releasing a lock.
- All requests for locks are made during the growing phase.
- All releases are made during the shrinking phase.

We note that such a mechanism may lead to a deadlock, and therefore requires a nontrivial approach in governing the acquisition phase.

One can describe the serializability relations in a graph representation. Let the nodes represent the actions and let the edges represent the dependency between processes for any $RR$, $RW$, $WR$, and $WW$ relation for an entity. Let each node in the graph represent an action and let a directed arc between any two nodes represent a precedence dependency based on the serialization order ($\succ$). The edges are therefore marked by the entity identifiers. Figure 14.3 demonstrates such

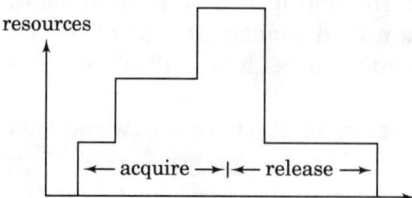

Figure 14.2 Two phase locking.

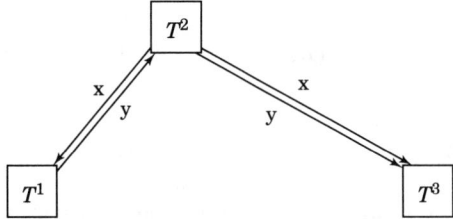

**Figure 14.3** Graph representation of locking dependencies.

a graph. Action $T^1$ succeeds both $T^2$ and $T^3$ in the serialization order for entity x, while for entity y, $T^1 \succ T^2 \succ T^3$.

> CONDITION 23. Let the *serializability graph* include nodes which represent actions, interconnected by labeled directed arcs representing precedence relation on entity accesses according to a serialization order. Let the actions acquire and release locks in a two phase locking mechanism on each of the entities, implementing this serialization order. If the graph is *acyclic*, then the log is *serializable*. □

Having an acyclic serializability graph means there exists a topological sort of this graph which represents the sequence of actions. Hence, there is an equivalent serialization order of the concurrent trace. If for each entity, the read set $\mathcal{R}$ and the write set $\mathcal{W}$ satisfy the relation $\mathcal{W} \subset \mathcal{R}$, then there exists a simple serialization graph.

## 14.2 Atomicity and Transactions

Atomicity of actions is needed in order to support the sequence

{ consistent state } atomic action { consistent state }

both for impeccable and for faulty executions of the action. The atomicity ensures that only the initial and final states are observable by the environment. Thus, we allow an atomic action to behave only in an "all or nothing" manner, discouraging any effects of partial executions.

> DEFINITION 43. An *atomic action* is an action with the following properties:
>
> - It is *indivisible*: either all the primitives which construct it complete execution or none of them does.
> - It is *serializable*: all the computation steps which are not in the atomic action, either precede or succeed all the steps in the atomic action.
> - It is *recoverable*: the external effects of all the steps in the atomic action either occur (the action completes) or not (none of the steps completes) leaving no side effects of any partial execution. □

The indivisibility and serializability properties ensure atomicity of an execution that is properly carried out. The recoverability property ensures consistent states in cases of improper execution, or in other words upon failures.

Data consistency plays a major role in database systems, in which data items can be replicated and accessed by numerous consumers of these items. Actions carried out on a database, performing operations on different data items in a certain order are called *transactions*.

> DEFINITION 44. A *transaction* is a set of database read and write actions and a partial order on them, which is observed as a high level atomic action. □

The definition of a transaction being observed as a high level atomic action implies its indivisibility, serializability, and recoverability requirements.

### 14.2.1 Transaction model description

Three sequential steps are identified in a transaction

- *Read* ($R$) *and Write_of_Intentions* ($W_i$). In this step users submit their intentions for reading or altering data of a particular object.
- *Certification*. In this step serializability criteria are checked.
- *Validation*. If the transaction is certified, or in other words if it is found to be serializable, the written intentions become effective. Otherwise, this step is not performed.

The validation step of the transaction must be performed atomically.

In various approaches, the certification step is a part of one of the other steps. Pessimistic approaches execute it when intentions are declared, and optimistic ones just prior to validation. We have chosen to model it separately in order to support both pessimistic and optimistic solutions later in this chapter.

**Concurrency conflicts.** Transactions can execute concurrently using the same database. Consequently, the following table summarizes the possible conflicts between two transactions accessing the same object. There are four serialization orders depending on the intentions, out of which three raise potential conflicts between the accessing transactions.

Transaction Conflict-Table

| Order | $RR$ | $RW_i$ | $W_iR$ | $W_iW_i$ |
|---|---|---|---|---|
| Potential Conflict Raised | | + | + | + |

These potential conflicts become real once the first transaction is validated.

**Distributed and concurrent certification.** The distributed nature of databases calls for the distribution of the certification of transactions. If a transaction involves objects which reside in different sites, only a collective certification is able to certify this transaction. Assuming sites do not have global knowledge, the following algorithm is self-evident.

1. *Local certification.* According to the local knowledge, the site checks for feasible conflicts. Its result, say $OK_i$ and its constraints are passed to the next step.

2. *Global certification.* This algorithm step collects $\forall_{j \in T} OK_j$ and produces $OK_{glob}$, which it passes to $\forall j \in T$.

3. *Local validation.* Every site which receives $OK_{glob}$, validates or rejects the transaction.

Clearly, the global certification step can be executed centrally in a site whose knowledge includes global knowledge,[1] or by means of a distributed algorithm when global knowledge is unavailble.

The problem of concurrent certification [2] further complicates the situation but it is essential for improving parallelism. We can categorize concurrent certification methods in two classes:

1. Methods in which the serialization order corresponds to the chronological order of certifications.

2. Methods in which the serialization order forms a total order.

The first class requires some restrictions on the chronological order in order to maintain serializability. In particular, the precedence relations between certifications should not violate the serializability constraints. On the other hand, the second class (e.g., timestamped transactions) do not restrict the chronology of certifications, but enforce total order on the transactions.

### 14.2.2 Approaches to recoverability

The objective of all recovery procedures is to allow correctness after recovery from a failure.

---

[1]For example, the site which initialized the transaction.

- If the already executed actions had no effect on their environment, we can *undo* them.
- If the already executed actions had an effect on their environment, we must *compensate* the environment.
- If the already executed actions had an effect on their environment and compensation is improbable, we must apply some *correction* mechanism which will transform the system back to a consistent state.

Let us consider some of the tools for recovery from failures, and how they deal with the undoing, compensating, and correcting of inconsistent effects on the environment.

**Checkpoints.** Checkpoint mechanisms are discussed in Section 11.2, where procedures for saving the essential data for a complete system state restoration establish recovery points for a roll-back recovery. In general, roll-back strategies are founded on undoing local failure results which have no effect on their external environment. Hence, between checkpoints, we must manage an audit trail to identify:

- what is to undo,
- what is to redo,
- what is to restart.

Management of a stable log of completed actions allows distinction between the three.

When an action is rolled back, other actions which have used data produced by it need to be rolled back as well. Therefore, the stable log must maintain information on such dependency relations.

The desire to increase efficiency demands that user submission of transactions as well as transaction processing be executed concurrently with the checkpointing process. In distributed database systems, such schemes require properties of noninterference [127] to support global consistency of the database. Furthermore, in real-time application, noninterference of the checkpointing scheme by transaction processing is essential for obtaining predictable execution times.

**Two-phase commit.** Commit protocols are discussed in Section 9.1, are used to establish agreement which results in only two observable states: abort and commit. These states are irreversible, and therefore commit protocols are excellent devices for establishing recovery points.

The two-phase commit protocol [125, 126] assures the establishment of these recovery (commit) points, but it is a blocking protocol. A three-phase protocol solves this problem. The intermediate states of these protocols are assigned with failure and timeout transitions, as defined previously in Conditions 10 and 11. These rules are sufficient to assure the protocol resiliency for a single node failure, maintaining the system with only two observable states, as well as recovering to these states only.

**Combine two-phase locking with two-phase commit.** In distributed database systems, it is common to ensure serializability through a local two-phase locking approach along with inter-node two-phase commit protocol. Hence, this combination requires that every lock acquired for a particular transaction be released only after a stable commit or abort decision. A local serializability graph can therefore represent the two-phase locking, and a global serializability graph can represent the commit protocol.

However, since conflicting transactions cannot be concurrently executed because they need the same locks, different transactions will be executed in different nodes during a given time interval. This may affect serializability to a point of inability to commit. Say that we have a sequence of two transactions $A$ and $B$ to be executed in sites 1 and 2. If $A_1$ is going to be the first in site 1, then the locks for $A$ are acquired by site 1, and site 2 can only execute $B_2$. Therefore, upon completion, site 1 cannot commit without the consent of site 2 on $A$, and site 2 cannot commit without the consent of site 1 on $B$. Hence, they cannot release the locks, and deadlock will occur if nothing else is done.

**Integrity constraints.** There are various ways to support recoverability through detection of violation of data integrity.

- Semantic constraints: In various systems, semantic links establish the relationships between objects in the form of invocations which are syntax and semantics restricted, as defined by the peer protocol and the interfaces. The link's logical properties determine the invocation type, as determined by the parameters transferred from the invoking object to the invoked object. The invoked object gets the parameters directly from the link. The semantic restrictions allow the detection of integrity violation, and an intelligent selection of a default replacement.
- Application dependence: Some of the system properties always emerge as an *a priori* knowledge which is application dependent. For example, in a bank account a transaction is always restricted to a positive balance.

### 14.2.3 Approaches to indivisibility

**Two-phase locking.** As discussed above, a two-phase locking requires that all requests for locks be made during the growing phase and that all releases be made during the shrinking phase. A monotony property is imposed allowing no action to acquire locks after releasing a lock. From a fault tolerance point of view, it is important to commit before one starts releasing the locks. Otherwise, upon the failure of a transaction in its release phase, one needs to roll back all other transactions that used the lock used by the failing transaction.

Since all requests for locks are made during the growing phase (only then when we start releasing) does the question of lock granularity arise: should we lock the whole database or just a single record? The more restrictive the system is, the less concurrent the computations are. On the other hand, over simplification and an increased granularity can be compensated for by hierarchy in locks that yield levels of isolation. For example, consider an adjustable use of the following levels of lock isolation:

1. No uncommitted action can alter any object.
2. No uncommitted action can alter any part of the object.
3. No other action can alter the object that this action addresses currently.
4. No other action can alter any object that this action has addressed.
5. This object is not aware of other actions at all.

**Timestamping.** Timestamps are widely used for ordering and serialization purposes. In [120], we find a *possibility* to which one writes before the commit. Every action (read or write) comes with a timestamp. The action "selects" a timestamp from a monotonic pseudo-time generator. Every object carries two timestamps:

1. read-time: the timestamp of the action that read it last
2. write-time: the timestamp of the action that wrote it last

This approach is characterized as follows:

- An action which attempts reading and finds a "younger" write-time invokes an action restart.
- An action which attempts writing and finds a "younger" read-time invokes an action restart.

For example [120], consider the following sequence of a transfer between two bank accounts $B_1$ and $B_2$.

1. Create a *possibility* Q and link it to the monotonic pseudo-time generator.
2. Select pseudo-time $t_1$, and read($B_1$,$t_1$,Q,balance).
3. Select pseudo-time $t_2$, and write($B_1$,$t_2$,Q,balance−amount).
4. Select pseudo-time $t_3$, and read($B_2$,$t_3$,Q,balance).
5. Select pseudo-time $t_4$, and write($B_2$,$t_4$,Q,balance+amount).
6. Try to change the state of Q from "unknown" to "committed."

Since the pseudo-time generator is monotonic, we know that $t_1 < t_2 < t_3 < t_4$, which is in accordance with the above characteristics. Once $B_1$ has been written with the timestamp $t_2$, any other action with a timestamp later then $t_2$ which attempts reading $B_1$ waits until the end of the last step or until its abortion.

We note that deadlock is avoided by this approach, however, starvation is not.

## 14.3 Partitioning

Partitioning of a distributed system is a fragmentation of the system into isolated subsystems, each called a *partition*. If the system contains a distributed database, or duplicated elements of a database, partitioning may lead to uncoordinated updates of replicas. In a distributed system with alternative servers (see Section 16.2.2) partitioning may lead to inconsistent states of the alternates.

An interesting classification of recovery strategies is presented in [45]: pessimistic and optimistic strategies. The pessimistic strategies make worst case assumptions about what can go wrong, while the optimistic strategies assume that inconsistencies seldom occur and thus are left to be resolved after they are detected. [64] assumes optimistic strategies are based on the premise that "it is more effective to apologize than to ask permission."

The cost associated with the pessimistic strategies is that of their rigidity, due to which activity is frozen even when it is not required to be. On the other hand, optimistic approaches have the cost of detecting a conflict and the associated cost of resolving it.

The differences are emphasized in the following tables [64], which describe account type actions and their conflicts. The account actions include three types of events: deposit which is granted, withdrawal which is granted, and withdrawal which generates an overdraft. When we consider these actions in a concurrent environment, we find various conflicts. For example, should a withdrawal be approved when other withdrawals are still to be processed, or should serializability be imposed on these processes?

In an optimistic approach, we first execute the action, and before we confirm it we certify it while observing its already known results. Therefore, successful withdrawals do not depend on deposits. On the other hand, unsuccessful withdrawals do depend on deposits which can increase the balance to a confirmation level. Successful withdrawals also depend on each other as they can be accumulated to an overdraft level. Certifying after execution considers conflicts between events, rather then between actions.

Optimistic Conflict Table

|               | deposit /OK | withdraw /OK | withdraw /over |
|---------------|-------------|--------------|----------------|
| deposit/OK    |             |              |                |
| withdraw/OK   |             | +            |                |
| withdraw/over | +           |              |                |

In the pessimistic approach, we consider worst case results before we execute. Therefore, a withdrawal has a conflict with any action, because the withdrawal results are unknown. The pessimistic approach, therefore, considers conflicts between invocations, as shown by the following table.

Pessimistic Conflict Table

|               | deposit /OK | withdraw /OK | withdraw /over |
|---------------|-------------|--------------|----------------|
| deposit/OK    |             | +            | +              |
| withdraw/OK   | +           | +            | +              |
| withdraw/over | +           | +            | +              |

### 14.3.1 Pessimistic strategies

**The asymmetrical approach.** Asymmetry indicates that not all the nodes are equivalent. Only one leading copy, the coordinator, is responsible for the data updates. It forwards updates to its cohorts only after updates are locally processed. Upon partitioning, only the partition which contains the coordinator, and hence the leading copy is given access.

As in any asymmetrical approach, this approach is extremely sensitive to the coordinating site failure. If we want to improve its robustness, upon coordinator failure we can pass the leading role with a token to one of the cohorts, following some election algorithm. However, such a solution is also very sensitive, but this time due to a token loss.

**Accessible copies approach.** The Accessible Copies algorithm [50] uses a very inexpensive reading access to data items. Its basic rules are as follows.

1. A data item is *accessible* for a read or a write action primitives only if a majority of its copies belong to the partition.
2. Reading an accessible data item is carried out by accessing the *nearest* copy within the partition for reading.
3. Writing an accessible data item is carried out by accessing *all* copies within the partition for writing.

If properly implemented, these rules impose one-copy serializability such that writing is carried out either to all the copies or to none. Furthermore, it is important that when a writing failure occurs, all nodes be immediately notified. Otherwise, one-copy serializability can be violated.

**Agreement.** Systems with replicated elements or with concurrently executing alternatives give ground to on-line resiliency. However, the existence of multiple threads, which must join at some point to allow deterministic and coherent actions throughout the system, calls for the use of agreement protocols such as the ones discussed in Chapter 9.

Voting algorithms are discussed in Section 9.2, based on [57]. Each replica of the data object is given a voting weight, and the total voting of the replicated population is $v$. Read and write access are quorum restricted: an external object needs a quorum of $r$ votes to read from the data object and $w$ votes to write to this object. The quorum restrictions, as formulated in Equation 9.3, provide important consistency measures when partitioning occurs. The constraint of

$$r + w > v \qquad (14.2)$$

assures that there can be no case of an object modified by one partition and read by the other. The constraint of

$$w > \frac{v}{2} \qquad (14.3)$$

assures that no modification is carried out on the same object at different partitions.

Quorum consensus algorithms are discussed in Section 9.3 associating each computation node with a set of other nodes (forum) with which it communicates in order to obtain a unified resolution and a partial set (quorum) which is sufficient to obtain a resolution. The weighted voting algorithm can be generalized based on history traces [63] to serve as a consensus algorithm.

### 14.3.2 Optimistic strategies

As noted above, optimistic strategies allow concurrent execution of transactions without synchronization, cerifying and validating serializability and resolving conflicts only upon conflict detection.

**Relaxing serializability constraints.** The most optimistic strategy proposes an extreme relaxation of the serializability constraints, up to totally disregarding one-copy serializability. Certification and validation are performed concurrently, but do not interfere with the read and write actions. Once a conflict is detected, backward tracing is performed to "clean" all contaminated effects of the detected conflict, using one of the approaches of recoverability introduced above.

**Precedence graph approach.** Recalling that execution of a transaction within a partition is serializable, each partition maintains transaction history. Let the transaction history of partition $i$ be (in serialization order)

$$T_{i1}, T_{i2}, \ldots T_{in}$$

A precedence graph [45] can be constructed, in which the nodes represent the transaction within the partition and the edges represent relations between the transactions.

- Dependency edges ($T_{ij} \Rightarrow T_{ik}$) represent in-partition producer-consumer relations: $T_{ik}$ reads a value of a data item produces by $T_{ij}$.

$$WriteSet(T_{ij}) \cap ReadSet(T_{ik}) \neq \emptyset \quad j < k \quad (14.4)$$

- Precedence edges ($T_{ij} \rightarrow T_{ik}$) represent in-partition order: $T_{ij}$ reads a value later changed by $T_{ik}$ in the same partition.

$$ReadSet(T_{ij}) \cap WriteSet(T_{ik}) \neq \emptyset \quad j < k \quad (14.5)$$

- Interference edges ($T_{ij} \rightarrow T_{lk}$) represent out-of-partition producer consumer relations: $T_{ij}$ reads a value of a data item written by $T_{lk}$ in another partition.

$$ReadSet(T_{ij}) \cap WriteSet(T_{lk}) \neq \emptyset \quad i \neq l \quad (14.6)$$

All these edges stand for precedence in occurrences. Hence, an acyclic precedence graph between partitions ensures a consistent database state. In other words, an acyclic precedence graph indicates the existence of a serial history which describes the transactions of all the partitions.

For example, consider the precedence graph presented in Figure 14.4 ([45]). Near each transaction, the readset is denoted above the

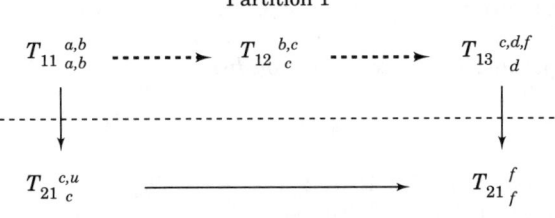

**Figure 14.4** Acyclic precedence graph: no partitions' conflicts.

line and the writeset below the line (i.e., $T_{1,2}$ reads $bc$ and writes $c$). This example demonstrates a serializable partitioning, with serialization order
$$\{T_{21}, T_{11}, T_{12}, T_{13}, T_{22}\}$$

**Conflict based validation.** An optimistic approach is described in [64], where transactions execute without synchronization. However, before commitment they are validated. A conflict based validation uses conflicts between events for validation which is carried out on top of a two-phase commit protocol during the first phase of the commit. Each transaction which receives a "prepare" message is validated locally. If all the cohorts validate, the coordinator sends a timestamped "commit" message. Thus, no extra messages are required in addition to those used for the commit protocol.

Let us recall our definitions from Section 14.1.1, and define the following types:

- HIST ≡ a set of sequences of events
- TRANS ≡ a set of transaction ID's
- EV ≡ a set of events
- TIME ≡ a set of logical clocks

Based on these types, the following variables are constructed:

- *State*: { HIST } (initialized empty)
- *Intention_list*: { TRANS } → { HIST }
- *Op_lock*: { EV } → list of TRANS members (initialized none)
- *Committed*: list of TRANS members
- *Aborted*: list of TRANS members
- *Clock*: { TIME }
- *First*: { TRANS } × { EV } → { TIME }
- *Last*: { EV } → { TIME }

For transaction $A$ to execute event $e$ ::
**begin**
    **if** $(A \notin Commited) \wedge (permissible(State \oplus Intention\_list(A) \oplus e))$
    **then**
        **execute** event $e$ ;
        $Clock := Clock+1$ ;
        $Intention\_list(A) := Intention\_list(A) \oplus e$ ;
        $Op\_lock(e) := Op\_lock(e) \cup \{A\}$
    **fi**
**end**
For transaction $A$ to commit::
**begin**
    **if** $(A \notin Commited \cup Aborted) \wedge$
            $e \in Intention\_list(A) \wedge e \succ_o e' \Longrightarrow Op\_lock(e') - \{A\} = \emptyset$
    **then**
        **execute** *commit* ;
        $Clock := Clock+1$ ;
        $State(e) := State(e) \oplus Intention\_list(A)$ ;
        $Op\_lock(e) := Op\_lock(e) - \{A\}$
    **fi**
**end**

**Program 14.1** Forward conflict validation algorithm.

The object which contains and handles the data is modeled by its permanent state (*State*) and a transition rule (*Intention\_list*) applied to this state upon commitment. The object handles lists of committed and aborted transactions, and has access to a set of logical clocks.

[64] has proposed and verified two validation methods. A forward validation method is established when each transaction confirms that its results are not invalidated by other concurrent transactions. A backward validation method is established when each transaction confirms it does not invalidate the results of other concurrent transactions.

Program 14.1 describes a forward validation algorithm in which the object uses *optimistic lock*s for concurrency control. A transaction which attempts to execute an event receives such a lock for that event, denoted *Op\_lock*. Only if there is *no event* in the intentions list of transaction $A$, whose *Op\_lock* is held by another transaction (and thus conflicts with $A$), will transaction $A$ be validated. This condition is given by the following equation

$$e \in Intention\_list(A) \wedge e \succ_o e' \Longrightarrow Op\_lock(e') - \{A\} = \emptyset \qquad (14.7)$$

where $\succ_o$ is the conflict relation. The locks are released on commitment or abortion.

For transaction $A$ to execute event $e$ ::
**begin**
    **if** ($A \notin Commited$) $\wedge$ (permissible($State \oplus Intention\_list(A) \oplus e$))
    **then**
        **execute** event $e$ ;
        $Clock := Clock+1$ ;
        $Intention\_list(A) := Intention\_list(A) \oplus e$ ;
        $First(A,e) :=$ **if** $First(A,e) = \perp$
            **then** $Clock$
            **else** $First(A,e)$
        **fi**
    **fi**
**end**
For transaction $A$ to commit::
**begin**
    **if** ($A \notin Commited \cup Aborted$) $\wedge$
        $e \in Intention\_list(A) \wedge e \succ_o e' \implies First(A,e') > Last(e')$
    **then**
        **execute** $commit$ ;
        $Clock := Clock+1$ ;
        $State(e) := State(e) \oplus Intention\_list(A)$ ;
        $(e) \in Intention\_list(A) \implies Last(e) := Clock$
    **fi**
**end**

**Program 14.2** Backward conflict validation algorithm.

Program 14.2 describes a backward validation algorithm. Each object stores $Last(e)$, the most recent logical time of committing a transaction which executed $e$. Each object keeps $First(A,e)$ for each active transaction $A$, the logical time when $A$ first executed $e$. Only if for all $e$ in the intentions list of $A$, and for every $e'$ in conflict with $e$, $First(A,e) > Last(e')$, will transaction $A$ be validated. This condition is given by the following equation

$$\forall e \in Intention\_list(A) \wedge e \succ_o e' \implies First(A,e) > Last(e') \qquad (14.8)$$

The run-time cost of forward and backward validation methods are comparable. An advantage of the forward validation approach is that all transactions, including those which fail, consider serializable states. An advantage of the backward validation approach is that it restarts only in favor of committed transactions. Forward validation, on the other hand, starts in favor of any active transaction. This fact gives an important advantage in overhead cost for backward validation methods in a highly concurrent system.

## 14.4 Broadcasting Solutions

So far we have considered local processes which retain consistency. Another way of achieving consistent state is by atomic broadcasts [44]. An atomic broadcast demonstrates the same atomicity properties as defined for an atomic action in Definition 43.

- It is *indivisible*, such that either the messages are successfully delivered to all the receivers or to none of them.
- It is *serializable*, ensuring that all the messages are delivered to all the receivers in the same order.
- It is *recoverable*, ensuring that either the external effects of all the steps in the broadcast occur (the broadcast completes) within a known time period or none of the steps are completed.

### 14.4.1 Model

Let us consider a system of connected nodes, but not necessarily a fully connected one. The nodes (sources or destinations of messages) and the links (which can contain relaying nodes) may become faulty. Although the system is not necessarily fully connected, a removal of a faulty node does not partition the system. Let us denote the set of nodes (vertices) $\mathcal{V}_p$ and the set of links (edges) $\mathcal{E}_p$.

Each faultless source passes a message to all its neighbors, that is, all the nodes it is connected to. A new message received by a faultless node is relayed to its neighbors. This model of data distribution is often called a *diffusion* wave.

**Fault classification.** Faults can be divided to three classes [44]:

- Exclusion faults which cause nodes not to react to invocation requests.
- Timing faults which cause nodes' reaction to invocation requests to be either too early or too late.
- Byzantine faults which affect nodes in an unpredictable manner.

It is apparent that the last class of faults is the most general one and thereby the most difficult to deal with. We adopt the methodology of [44] to present the algorithms by gradually increasing algorithm complexity.

**Assumptions.**

- Any message sent from node $p$ to node $q$ ($p, q \in \mathcal{V}_p$) is processed by $q$ in at most $\delta$ time units, as measured by clock-$p$ or by clock-$q$.

- Clocks of distinct nodes differ from each other in at most $\varepsilon$ time units.
- The maximum number of faulty nodes is $\pi$, and the maximum number of faulty links is $\lambda$.
- In the presence of $\pi$ faulty nodes and $\lambda$ faulty links, the worst case message delay in the network is $D_{\pi,\lambda}$

Let us consider a specific case, in which a message is sent from a node $p$ at local time $t_p$ through $n$ relaying links and received at a node $q$ at local time $t_q$. Here the latency can be expressed as:

$$-\varepsilon \leq t_q - t_p \leq \varepsilon + n\delta \quad (14.9)$$

**Broadcast layer.** The broadcast layer consists of three tasks:

1. Task *start* initiates the broadcast.
2. Task *relay* forwards the message to an adjacent node.
3. Task *end* delivers the message to the application task.

Let us consider the major variables used in these tasks

- The message to be sent is a value $\sigma$ with type $\Sigma$.
- A local history variable $H$ keeps track of ongoing broadcasts by associating with a time instant $t$, a function $H(t)$ of type $V_p \to \Sigma$ which records the value $\sigma \in \Sigma$ broadcast by $p \in V_p$ at clock time $t$. The history is modified by $\oplus$ and $\ominus$ operations, denoting concatenation and removal respectively.

### 14.4.2  Tolerating exclusion faults

Program 14.3 retains consistency under conditions of exclusion faults which cause nodes not to react to invocation requests. It includes the three broadcast layer tasks, *start*, *relay*, and *end*, with the first two running indefinitely and uses the last being scheduled by them.

The program uses the **send-all** and **send-all-but** broadcast services for information distribution and uses the **receive** service for accepting the broadcast data. It uses the **get** and **deliver** services for communication with the application. We note that the *relay* task forwards recently accepted messages to its neighborhood, as long as they are not outdated.

Program 14.3 uses a local **schedule** service for the timely awakening of the *end* task. The interval timer for this awakening depends on the variable $\Delta$. Let $\Delta$ be defined as

$$\Delta = \pi\delta + D_{\pi,\lambda} + \varepsilon \quad (14.10)$$

*start*::
    **var** $t$ :time; $\sigma : \Sigma$; $p : \mathcal{V}_p$;
    **begin**
        **while** True **do**
            **get**($\sigma$);       from application
            $t$:=clock; $p$:=myId;
            **send-all**($t, p, \sigma$) ;
            $H$:=$H \oplus (t, p, \sigma)$ ;
            **schedule**(*end*,$t + \Delta, t$)
        **od**
    **end**
*end*($t$:time)::
    **var** $p : \mathcal{V}_p$; $v : \mathcal{V}_p \to \Sigma$;
    **begin**
        $v$:={ $h_i$ }, $\forall i : h_i = (t, p, \sigma) \in H$ ;
        **while** $|v| \neq \emptyset$ **do**
            $p$:=min($v$) ;
            **deliver**($\sigma$) **of** $(t, p, \sigma)$;
            $v$:=$v \ominus (t, p, \sigma)$
        **od** ;
        $\forall p, \sigma$: $H$:=$H \ominus (t, p, \sigma)$
    **end**
*relay*::
    **var** $t, \tau$ : time; $\sigma : \Sigma$; $p : \mathcal{V}_p$; $e : \mathcal{E}_p$;
    **begin**
        **while** True **do**
            **receive**($t, p, \sigma, e$) ;       from $e$
            $\tau$:=clock;
            **if** $\tau > t + \Delta$ **then** TOO LATE
                **else if** $\exists (t, p, \sigma) \in H$ **then** OLDY
                **else**
                      **send-all-but**($e, (t, p, \sigma)$ ) ;
                    $H$:=$H \oplus (t, p, \sigma)$;
                    **schedule**(*end*,$t + \Delta, t$)
        **od fi fi**
    **end**

**Program 14.3** Broadcast algorithm for tolerating exclusion faults.

It can be shown for this algorithm [44] that:

1. If a message timestamped $t_p$ is inserted in the history of at least one correct processor, then all correct processors insert this message in their history variables by $t_p + \Delta$ of their clocks.

2. This broadcast algorithm is indivisible, serializable, and recoverable.

3. This broadcast algorithm does not tolerate timing faults.

*start*::
    **var** $t$ : time; $\sigma : \Sigma$; $p : \mathcal{V}_p$; $k : 1 \ldots n$;
    **begin**
        **while** True **do**
            **get**($\sigma$);      from application
            $t$:=clock; $p$:=myId; $k$:=1;
            **send-all**($t, p, k, \sigma$) ;
            $H$:=$H \oplus (t, p, \sigma)$ ;
            **schedule**(*end*,$t + \Delta$,$t$)
        **od**
    **end**
*end*($t$:time)::
    **var** $p : \mathcal{V}_p$; $v : \mathcal{V}_p \to \Sigma$;
    **begin**
        $v$:={ $h_i$ }, $\forall i : h_i = (t, p, \sigma) \in H$ ;
        **while** $|v| \neq \emptyset$ **do**
            $p$:=min($v$) ;
            **deliver**($\sigma$) **of** $(t, p, \sigma)$;
            $v$:=$v \ominus (t, p, \sigma)$
        **od** ;
        $\forall p, \sigma$: $H$:=$H \ominus (t, p, \sigma)$
    **end**
*relay*::
    **var** $t, \tau$ : time; $\sigma : \Sigma$; $p : \mathcal{V}_p$; $e : \mathcal{E}_p$;
    **begin**
        **while** True **do**
            **receive**($t, p, \sigma, e$) ;      from $e$
            $\tau$:=clock;
            **if** $\tau < t - k\varepsilon$ **then** TOO EARLY
            **else if** $\tau > t + k(\delta + \varepsilon)$ **then** TOO LATE
            **else if** $\tau > t + \Delta$ **then** TOO LATE
            **else if** $\exists (t, p, \sigma) \in H$ **then** OLDY
            **else**
                **send-all-but**($e, (t, p, k + 1, \sigma)$ ) ;
                $H$:=$H \oplus (t, p, \sigma)$;
                **schedule**(*end*,$t + \Delta$,$t$)
        **od fi**
    **end**

**Program 14.4**   Broadcast algorithm for tolerating timing faults.

### 14.4.3   Tolerating timing faults

Program 14.4 retains consistency under conditions of timing faults, faults which cause nodes' reaction to invocation requests to be either too early or too late. It uses the same services as Program 14.3, however, with an important modification.

In order to endure timing faults, this program takes into account the time a message spends in the path of the relays which transfer it. Hence, an additional variable is passed in every broadcast, denoted $k$, and is incremented by each relay on its path. In this algorithm, the local clock examined upon the message reception, is bounded as follows

$$t - n\varepsilon \leq \tau \leq t + n(\delta + \varepsilon) \tag{14.11}$$

Moreover, a message cannot spend more than $\pi(\delta + \varepsilon)$ before being accepted by the first correct processor. Therefore, let $\Delta$ be defined as

$$\Delta = \pi(\delta + \varepsilon) + D_{\pi,\lambda} + \varepsilon \tag{14.12}$$

It can be shown for this algorithm [44] that:

1. If a message timestamped $t_p$ is inserted in the history of at least one correct processor, then all correct processors insert this message in their history variables by $t_p + \Delta$ of their clocks.

2. This broadcast algorithm is indivisible, serializable, and recoverable.

3. This broadcast algorithm does not tolerate Byzantine faults.

### 14.4.4 Tolerating Byzantine faults

Program 14.5 describes the broadcast algorithm which tolerates Byzantine faults. It uses signature for authentication and thus extracts forged(), duplicates(), and FAULTY SENDERs. The interval timer for awakening depends on the variable $\Delta$ defined in Equation 14.12, as in Program 14.4.

Both the signature for authentication and the sequence of relays are added here to the communicated messages. Another change with respect to Programs 14.3 and 14.4 is that in this algorithm, $H$ is of type

$$time \rightarrow (\mathcal{V}_p \rightarrow (\Sigma \cup \{\emptyset\})) \tag{14.13}$$

The empty value ($\emptyset$) is permissible for $\sigma$ in addition to values in $\Sigma$, as a result of the authentication.

It can be shown for this algorithm [44] that:

1. If a value $\sigma$ timestamped $t_p$ is inserted in the history of at least one correct processor, and there exists no correct processor which has set its $(t_p, q,)$ to $(t_p, q, \emptyset)$ by $t_p + \Delta$ of its clock, then all correct processors insert $\sigma$ in their history variables.

```
start::
    var t :time; σ : Σ; p : V_p; x :p_message;
    begin
        while True do
            get(σ);         from application
            t:=clock; p:=myId;
            (x,t,p,σ):=sign(t,p,σ) ;
            send-all(x,t,p,σ) ;
            H:=H⊕(t,p,σ) ;
            schedule(end,t + Δ,t)
        od
    end
end(t:time)::
    var p : V_p; v : V_p → (Σ ∪ {∅});
    begin
        v:={ h_i }, ∀i :  h_i = (t,p,σ) ∈ H ;
        while |v| ≠ ∅ do
            p:=min(v) ;
            for (t,p,σ) if σ ≠ ∅ then
                deliver(σ);
            v:=v⊖(t,p,σ)
        od ;
        ∀p,σ:  H:=H⊖(t,p,σ)
    end relay::
    var t,τ : time; σ : Σ; p,q : V_p; e : E_p;
    var P : sequence_of_V_p; x,y : p_message;
    begin
        while True do
            receive(x,t,σ,P,e) ;        from e
            τ:=clock;
            authenticate(x,t,p,σ,P) ;
            if forged(x) then IGNORE
            else if duplicates(P) then DUPLICATES
            else if τ < t − |P|ε then TOO EARLY
            else if τ > t + |P|(δ + ε) then TOO LATE
            else if τ > t + Δ then TOO LATE
            else
                q:=first(P) ;
                if ∃(t,,)∈ H ⋀ ∃(,q,)∈ H then
                    if ∃(,,σ)∈ H then OLDY
                    else if ∃(,q,∅)∈ H then FAULTY SENDER
                    (t,q,):=(t,q,∅) ;
                else
                    H:=H⊕(t,q,σ);
                    schedule(end,t + Δ,t)
                fi  fi
            cosign(x,y) ;
            send-all-but(e, (y,t,p,σ) )
        od
    end
```

**Program 14.5**  Broadcast algorithm tolerating Byzantine faults.

2. This broadcast algorithm is indivisible, serializable, and recoverable.

## 14.5 Concluding Remarks

The issue of consistency has been considered from the aspect of the permissible states of a system as well as from replicated data modification. Assurance of consistency is the maintenance of system states which are correctly reachable, while reflecting a correct history of events at the system interfaces. The importance of this assurance is further emphasized in the presence of multiple computation threads and multiple address spaces, and expanded further in the presence of faults.

Assurance of serializability criteria through locking and other atomicity mechanisms for transactions have been demonstrated, while examining various aspects of indivisibility and recoverability. This chapter has also examined how such assurances are obtained under faulty partitioning, through both optimistic and pessimistic strategies.

The cost of maintaining the system consistently increases as the knowledge of the behavior of faults decreases. This is demonstrated in the last section of this chapter, which compares the atomic broadcast solutions under fail-stop assumptions, latency uncertainty, and Byzantine faults of unpredictable behavior.

# Chapter 15

# Safe Systems

Let us start the discussion on safe systems with a look at a common misconception regarding the relation between reliable systems and safe systems. Many consider a safe system reliable and vice versa. We claim that this point of view is incorrect and misleading. The reliability of a system is characterized by its progress (or liveness) properties while the safety characteristics of a system allow its progress to be totally blocked, as long as its avoidance of failure requirements are satisfied.

There is a significant difference between the reliability goals of a system and its safety goals. We can describe a reliability goal as "the assurance of achieving a property," while the a safety goal is better described as "assurance of not achieving a property." In other words, the reliability properties consist of the system's "dos" and safety properties of its "don'ts."

In many cases the coexistence of safety and liveness properties are contradictory. As an example, consider a system which can achieve its target state only through exposure to an almost-unsafe path. Such a path is one transition away from an unsafe state. Real systems, however, may allow the use of an almost-unsafe path with extra caution. Therefore, caution measures must support monitoring unsafe activities, which may or may not occur. These measures detect undesirable behavior upon or before occurrence.

In the following sections we discuss safety measures and properties and consider the balancing of reliability and safety requirements. Our goal here is to set up an explicit environment for achieving both safety and reliability properties giving up neither operational requirements nor allowing the blocking of the system.

## 15.1 Safety Measures

We have previously discussed our error and failure model for the system (see Section 1.3). Let us recall that a *system* maintains an observable behavior, for which an *error* is a behavior generated by a *fault*. Our system model allows a fault to be permanent or transient. However, a fault does not necessarily generate errors, especially in systems with redundant subsystems in which we exclude results obtained from detected faulty elements. A fault is merely a potential source of an error.

The logical and temporal correctness is essential for the liveness properties (efficient operation) of the system. In addition, the system must essentially satisfy requirements of safety properties. Safety properties are system descriptions that must be satisfied throughout the system's lifetime. These descriptions can be given in the system state space, defining a nonempty set of states denoted "unsafe." They can also be given as assertions which must be maintained false. A safe system, therefore, can be defined as a system which satisfies its safety requirements.

Real-life systems, especially those defined as mission-critical systems, may be of the type for which the risk of being in an unsafe state cannot be permitted. In many cases it is insufficient to guarantee safety properties of a non-faulty system. In these cases, the need to support and guarantee safety properties must also be considered in the presence of failures.

As mentioned in previous chapters, we assume that faults can occur both in the system's hardware and software components, and that we cannot always distinguish hardware and software failures. Since memory failures which occur during instruction-fetch cycles of the processor generate erroneous instruction executions, they are equivalent to software failures. A processor whose program counter is inappropriately altered produces an out-of-order execution of instructions, as does a design fault.

Since faults can generate unsafe states in our system, we may also be required to deal with the probability of handling of unsafe states and not just their avoidance. Therefore, we need to extend the measures of resiliency to faults with-respect-to safety properties.

DEFINITION 45. A safe system is $n$-*resilient* at $m\%$ to *faults* if

1. there exist at least $n + 1$ distinct configurations of its subsystems,
2. any $n$ faults in $n$ of the $n + 1$ configurations maintains a safe system, and
3. the highest probability of any set of $n$ faults in $n$ of the $n + 1$ configurations does not exceed $m\%$. □

The second condition in this definition does not imply that exactly one fault occurs in each configuration. Furthermore, each configuration can have any number of faults, independent of what happens in other configurations.

We now show how this type of resilient configuration can be achieved, either in time (repetitions) or physically (alternatives).

### 15.1.1 The system model

Each computer-based system contains hardware and software subsystems which carry out the computation. The system can include sensors and actuators in addition to its computerized components.

The computerized part can be broken down in a graph model into hardware and software computation nodes. From the failure mode point of view, we partition the nodes into independent groups that ensure damage containment. Let us examine how this separation is defined in the graph model. Let us recall our definition of a computerized system from Section 1.2.1. A set of software elements $V_p = \{p_1, \ldots, p_n\}$ are related to each other through a set of logical links $\mathcal{E}_p$, to form a graph

$$\mathcal{G}_p = (\mathcal{V}_p, \mathcal{E}_p)$$

A set of hardware elements $V_P = \{P_1, \ldots, P_m\}$ are related to each other through a set of physical links $\mathcal{E}_P$, to form a graph

$$\mathcal{G}_P = (\mathcal{V}_P, \mathcal{E}_P)$$

An allocation graph is a bijective mapping of the resources graph ($\mathcal{G}_P$) to the computation graph ($\mathcal{G}_p$). The allocation graph, therefore, describes the allocation of the system resources at a given point of time. The system's computerized part obeys a specific allocation scheme. This scheme is a set of distinct allocation graphs, each of which is a block $\mathcal{S}_{i,I}$. We recall from graph theory (e.g., [61]) that a graph can be disconnected by the removal of a single node (called a cutpoint) or a single edge (called a bridge), and that the graph fragments held together by the cutpoints and bridges are called blocks.

We note here that distinct graphs mean no resource sharing at a given point of time. The scheme extends the graph "snap-shot" description, adding to it another dimension (time) through the ordered sequencing of the allocation graphs.

The union of the allocation graphs is the entire operational part of the system described in $\mathcal{G}_p$ and $\mathcal{G}_P$. Hence, the set of these blocks can be described as

$$\{\mathcal{S}_{i,I}\} = (\mathcal{V}_i, \mathcal{E}_i, \mathcal{V}_I, \mathcal{E}_I) \tag{15.1}$$

where $\mathcal{V}_i \subset \mathcal{V}_p$, $\mathcal{E}_i \subset \mathcal{E}_p$, $\mathcal{V}_I \subset \mathcal{V}_P$, and $\mathcal{E}_I \subset \mathcal{E}_P$. The block $\mathcal{S}_{i,I}$ is the allocation of the hardware resources subgraph

$$\mathcal{G}_I = (\mathcal{V}_I, \mathcal{E}_I)$$

to the software computation subgraph

$$\mathcal{G}_i = (\mathcal{V}_i, \mathcal{E}_i)$$

An important case of these blocks is the one in which for any distinct blocks, say $\mathcal{S}_{i,I}$ and $\mathcal{S}_{j,J}$ with $i \neq j$ and $I \neq J$, the following segregation holds:

$$\mathcal{V}_i \cap \mathcal{V}_j = \emptyset$$
$$\mathcal{E}_i \cap \mathcal{E}_j = \emptyset$$
$$\mathcal{V}_I \cap \mathcal{V}_J = \emptyset$$
$$\mathcal{E}_I \cap \mathcal{E}_J = \emptyset \qquad (15.2)$$

where $\emptyset$ is the empty set. Figure 15.1 gives an example of such blocks. $\mathcal{G}_i$ consists of two processes (proc1 and proc2) which activate three servers (serv1 to serv3). These five objects are allocated along with resources in $\mathcal{G}_I$ which consist of two CPUs (cpu1 and cpu2) and two memory pages (mem1 and mem2), and the processors communicate via a communication line (com1). $\mathcal{G}_j$ consists of proc3, proc4, serv4, and serv5. It is allocated with $\mathcal{G}_J$: cpu3 and mem3. We note that $\mathcal{S}_{i,I}$ and $\mathcal{S}_{j,J}$ are completely isolated from each other.

A subset of this containment scheme can allow physical links to be used in different blocks, and still achieve a high degree of isolation. But Equation 15.2 defines maximum isolation: distinct computations that share neither functions, information, nor hardware resources. Thus, it provides a sufficient condition for a complete damage containment scheme through a complete segregation.

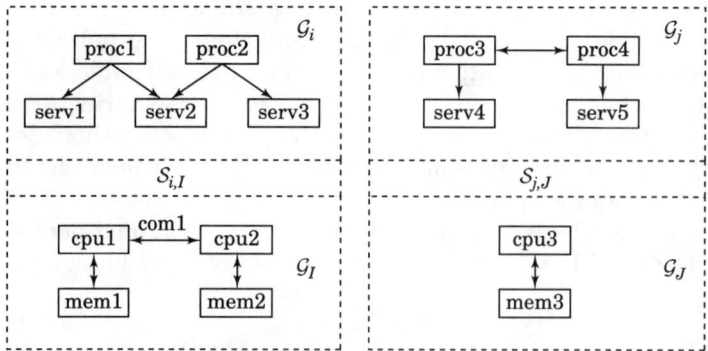

**Figure 15.1** Damage containment in a system.

### 15.1.2 Safety state monitors

In Section 10.3, we showed some classes of system state monitoring. Determining the acceptability of a state is performed there by a comparison test on the value of the system's state. We distinguish between three classes of comparison tests:

- testing a property with respect to an *a priori* known value,
- testing a property with respect to a measured value, and
- testing a property indirectly by a measurement of another property.

These tests require state detectors which act as state comparators. We note that for each distinct block (which is one of our isolated allocation subgraphs), we need an independent monitor. If the monitor of $S_{i,I}$ is included in it, they share a common failure mode. However, if the monitor of $S_{i,I}$ is included in $S_{j,J}$ ($i \neq j$, $I \neq J$) or vice versa, they still establish fault-isolated blocks. If there is a failure in $S_{i,I}$, the monitor which detects it in $S_{j,J}$ is not affected due to this isolation. On the other hand, a fault in $S_{j,J}$ may fail us in determining the state of $S_{i,I}$.

A question that immediately rises is "how can the monitor detect it with no intrusion?" We know that for sensing the system state, the monitor must communicate by some means with the system. It is imperative that we define these monitoring relations as "read-only" relations. Furthermore, these relations must not alter the block's logical and temporal properties. For example, a monitor must never enter a "wait" state for data collection, since it can delay the execution of the members of its block with which it shares resources. Another important property of a monitoring relation is that the deterministic properties of execution are indifferent to the data collected by the link.

A non-invasive relation between the monitor and the monitored block is mandatory for both software ($G_i$) and hardware ($G_I$) components. However, data access cannot be eliminated. Since any access to the monitored process state variables is liable for contention, the monitored process must be responsible for providing its status to external monitoring. One possible solution is to maintain a guard within the monitored process whose task is to update the accessible variable and transmit it properly if so required. Such a guard demonstrates a deterministic behavior and its progress can be sensed through time-stamping of the variable it provides.

The relation between the monitor and the block being monitored can therefore be summarized as follows:

- The monitor is allowed a read-only access from the monitored point of view.

- The monitoring access does not intrude in the execution of the monitored task.
- The monitor includes no parts with undeterministic behavior, in particular those which depend on collected data.

Consequently, we must modify our above definition for fault-isolated blocks of Equation 15.2.

DEFINITION 46. The block $S_{i,I}$ is *fault isolated* if

$$\forall S_{j,J} \ (j \neq i, J \neq I):$$
$$\mathcal{V}_i \cap \mathcal{V}_j = \emptyset$$
$$\mathcal{E}_i \cap \mathcal{E}_j = \psi$$
$$\mathcal{V}_I \cap \mathcal{V}_J = \emptyset$$
$$\mathcal{E}_I \cap \mathcal{E}_J = \Psi \qquad (15.3)$$

where $\psi$ and $\Psi$ may be nonempty and include only monitoring links. □

It is important to note that $\mathcal{M}_i \subseteq \mathcal{V}_j$ and $\mathcal{M}_I \subseteq \mathcal{V}_J$, or in other words, the hardware and software vertices which carry out the monitoring of a specific fault-isolated block are separated from its vertices. Although a monitor is not necessarily a single node, it is always a subgraph with hardware and software components. If such a monitor is constructed of independent objects distributed in various blocks of the system, this fault isolation definition holds for all blocks in the system and monitoring relations that exist between the blocks which host these objects and the monitored block.

Figure 15.2 demonstrates an example of fault-isolated blocks which share monitoring links. If we compare it to the case illustrated by Figure 15.1, we note that $\mathcal{M}_i$ which monitors $\mathcal{G}_i$ is a part of $\mathcal{G}_j$ and uses the same resources $\mathcal{G}_I$ as the rest of the block. $\mathcal{M}_j$ on the other

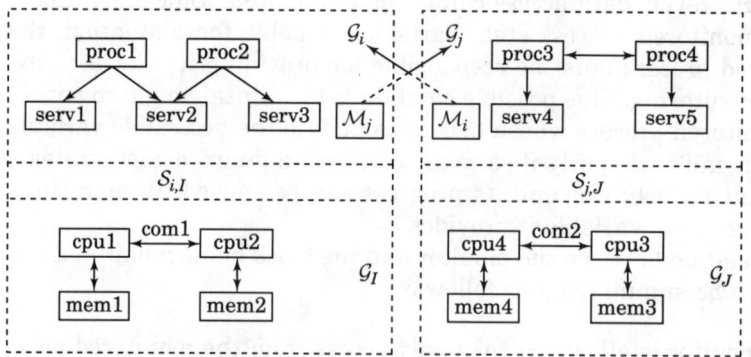

**Figure 15.2** Monitoring fault-isolated blocks.

hand requires additional resources, denoted in Figure 15.2 as cpu4 and mem4, and com2 for communication with the rest of $\mathcal{G}_J$. We note that the blocks $\mathcal{S}_{i,I}$ and $\mathcal{S}_{j,J}$ are fault-isolated as defined by Equation 15.3.

### 15.1.3 Safe states

A safe system with a specific degree of resiliency to faults must provide a distinct part of the configuration with which to maintain the safety requirements imposed. Only a limited number of systems are sufficiently static to allow design-time verification of safety properties in the presence of faults. Today's and certainly future systems will employ a high degree of dynamics in their resource management and computation commitments as reflected in the guarantees they provide upon job acceptance. In these systems, a static analysis of faults can be correct only for a specific allocation graph.

We note that since our system's size is unbounded and can have a dynamic and open nature along with a long life period, the statistics of its building blocks' reliability may actually become reality. Therefore, if the system has safety requirements which are critical, they must be dynamically satisfied as well as guaranteed. This dynamic monitoring requirement for safety-critical states emphasize the acceptability tests we have discussed above. These monitors must assure "detectably-acceptable" safe states, that is, the safe state and the ability to detect its changes must be assured.

A system constructed of configurations according to Definition 45 which are made of fault-isolated blocks (as defined in Definition 46) is a fault-resilient safe system. The vertices and the links of different blocks relate to each other as defined by Equation 15.3. Accordingly, monitors which assure detectably-acceptable safe states of a particular block can only be included in other[1] blocks.

In real-life systems, the observability of their overall behavior is limited. Furthermore, sometimes detecting an unsafe state leaves no way of recovering from it, since such states may be irreversible. In these cases we must detect states which lead to unsafe states, or unsafe "regions" which contain unsafe as well as safe states. It would have been desirable to detect transitions from safe to unsafe states or regions, but state transitions in real systems may have no observability at all. They certainly have no in-transition recovery process. For these reasons, the detection of monitoring regions which contain un-

---

[1] Not necessarily blocks which contain only monitors.

safe states is a most practical one. Adopting this approach, we denote hereafter an unsafe state as a state in an unsafe region.

The definition of a region can be derived from the system state space as the union of all states within a given distance from unsafe states. The distance can be defined, as in graph theory, by the number of transitions from one state to another. Other methods of weighting the distance may be adequate as well. Another example of defining such regions can be taken from the continuous world. We can protect a pressure vessel which bursts at a given pressure $P_0$ by activating a relief valve by sensing any pressure measurement at the region $[P_0 - \Delta, P_0]$. Unsafe regions are generally application dependent, and may be influenced by the system dynamic response in recovery operations. However, they all support the detection of states at a given distance from unsafe states.

## 15.2 Safety Aspects in Resiliency

The resiliency to the faults of a safe system depends on four major requirements:

1. the existence of a sufficient number of distinct configurations of its subsystems,
2. the ability of each of these configurations to achieve the system's objectives in a faultless case,
3. the isolation of these distinct configurations to maintain the system safe in the presence of faults, and
4. the probability of having faults in these configurations does not exceed its specified limits.

However, it must be emphasized that the system resiliency is limited, and locally there may exist no knowledge regarding the number of faults which have already occurred. Therefore, detection of faults that threaten the system safety must activate constraining actions regarding the local fault along with recovery actions if so required. Since monitors which guard the safe state of the system are treated as parts of the system's blocks, their faults must be considered too.

### 15.2.1 System's distinct configurations

Equation 15.1 describes the blocks which decompose the system as the set $\{S_{i,I}\}$ of allocating software modules and logical links ($i$) with hardware resources and physical links ($I$). This allocation is generally valid within a bounded time interval, and consequently a fault (especially a transient fault) may disappear in a repetition of this allocation.

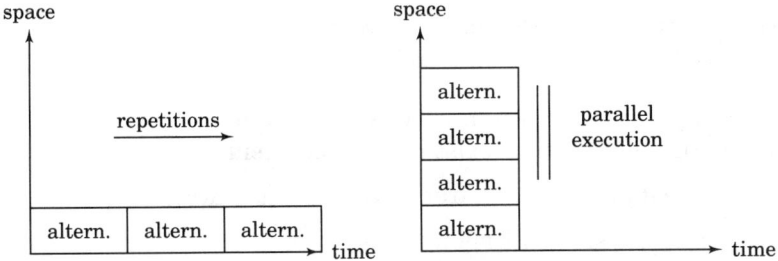

**Figure 15.3** Temporal and physical distinct configurations.

In a wide variety of transient faults in hardware and software, temporal redundancy achieved by repetition of allocation and execution of an object is as good as physical redundancy. The most obvious example of such a case is that of transient faults in communication, which contaminate a bounded size of data. In Section 10.2 the issue of temporal versus physical contamination is discussed, and based on this discussion a transient communication fault can be viewed as a contaminated physical link within a specific allocation instance.

While a temporal redundancy is sufficient for transient faults, this is not the case for permanent faults. A permanent fault in a hardware resource or link cannot be recovered by a repeated allocation and execution of the specific block. Neither can a permanent software fault (e.g., an instruction interpreter distorted) be recovered by repetition.

Figure 15.3 describes the difference between temporal and physical distinct configurations. The temporally differentiated configurations are repetitions of an execution carried out by the same resources, while the physically differentiated configurations are executed in parallel[2] by distinct resources.

The distinctness of the configurations originates in the requirement specified for fault isolation. Physically distinct configurations must endure a permanent fault in hardware and software blocks. In this regard, it is evident that neither a vertex nor an edge in the resources and computation graphs can be shared by distinct configurations. It is imperative to compose the vertices and edges of the graphs with physical entities independent of time.

On the other hand, the distinctness imposed on the temporally differentiated configurations forces discrete *allocation graph instances* in the vertices and edges of the configuration graphs. Here, the same physical resource, software module, or logical link can be used in dis-

---

[2]Parallel executions do not necessarily start and end at the same time, except for the temporally restricted systems.

tinct configurations, but this usage is restricted to distinct allocation sessions. A distinct allocation session requires:

- all the components (vertices and edges) must have been verified and validated in the form they exist in the consistent source,
- loading of modules and data from a consistent source,
- no overlap of the session time interval,
- initialization of resources for usage,
- self-test or BIT (built-in-test) of vertices, and
- handshake or session initialization for the links.

In other words, faults in run-time language library routines, communication, data upset, or memory disruption are sufficiently isolated with this definition of distinctness.

### 15.2.2 Faulty configurations

A major concern in fault tolerance is the question "Who may become faulty?" in the system. On one hand, the implementation is a collection of items assembled together. These items are entitled components, forming together the most detailed specification of the system. On the other hand, mainly for simplification, we describe the system in terms of higher level abstractions or assemblies. Such a simplification is helpful for description as well as for maintenance organization activities, but can become unreasonably expensive if adopted for our fault model. On one extreme, we can choose "the computer" as a single component, leaving no way of using the distributed nature of its components for redundancy and back-up attributes. On the other hand, extending the fault model to the "molecule" level, brings details which do not help fault analysis and may become distracting.

The isolation of distinct configurations to maintain the system safe in the presence of faults is the issue which guides the components derivation.

DEFINITION 47. System *component* satisfies the following properties:

1. It is testable for correctness at component level.
2. It has been verified and validated for correctness at some point of time.
3. It is separately accessible.
4. It has an isolated failure mechanism.
5. It is replaceable by some maintenance activity. □

## Safe Systems

From this standpoint, replicas of a software component which have been verified and validated in the form they exist in the consistent source and which are generated from a consistent source are distinct components. Integrated circuits that share a bus (address or data) with no buffers are not, since some of their failure mechanisms are not isolated.

Therefore, the level of abstraction we adopt for our system's fault model originates in the way in which a faulty component affects others. Hardware elements isolated by buffers can form system's components, since they obey the above characteristics. Memory pages which reside on different board sections and which are properly buffered are also distinct components. On the other hand, objects whose execution depends on data from an initiating broadcast message, have a common mode error source in that broadcast. It takes a very robust semantic link to overcome this type of coupling, requiring some type of data redundancy within the message (e.g., error code).

The probability of having a fault in the hardware components of the system can be estimated and calculated in various ways. Computer based reliability estimators provide tools based on fairly complicated models (e.g., [117, Chap. 9, ]). The probability of having a fault in the software components of the system depends significantly on the engineering practices employed in their development. These practices embrace approaches of design along with validation and verification. However, the probability figure is very sensitive to these practices, as it is in cases of introduction of new technologies in integrated circuits. For safety-critical objects we therefore recommend a combination of practices, some of which are software engineering validation methods and some of which are formal and axiomatic. Using this approach, we can reduce the failure rate expected after validation, by an order of magnitude.

### 15.2.3 Resiliency compensation for safety

According to Definitions 45, 46, and 47, the resiliency of a system depends on its components. From maintenance and recovery points of view, the components must be separately accessible, testable for correctness at the component level, and allow replacements to take over. Definition 47 further requires a component to have an isolated failure mechanism.

However, a real-life system can contain some components in its configurations that do not satisfy the requirement of failure-isolated mechanisms. These components introduce common-mode failures into the system's architecture, and consequently violate the requirements expressed in Definition 46. Figure 15.4 gives an example of common-

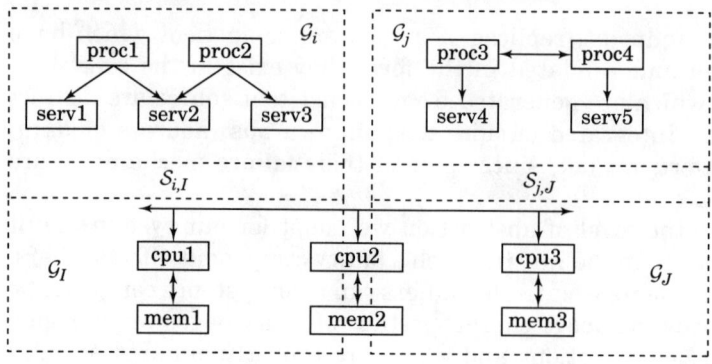

**Figure 15.4** Common-mode faults.

mode faults in failures of resources cpu2, mem2, and the communication bus. In this example we have

$$\mathcal{G}_I \cap \mathcal{G}_J = \{cpu2, mem2, bus\} \neq \emptyset \qquad (15.4)$$

In order to cope with common-mode faults, we need to find a way in which we can monitor these error sources within the degree of resiliency required by our safety requirements. We note here that this type of resiliency is less robust than the one defined in Definition 45. Such a definition favors the safety requirements at the expense of the reliability requirements. Consequently, we can define the minimum relation required for fault monitoring.

DEFINITION 48.   The blocks $\mathcal{S}_{i,I}$ and $\mathcal{S}_{j,J}$ with $i \neq j$ and $I \neq J$ are *fault-monitoring isolated* if

$$\mathcal{V}_i \cap \mathcal{V}_j = \alpha$$
$$\mathcal{E}_i \cap \mathcal{E}_j = \psi$$
$$\mathcal{V}_I \cap \mathcal{V}_J = \beta$$
$$\mathcal{E}_I \cap \mathcal{E}_J = \Psi \qquad (15.5)$$

where $\psi$ and $\Psi$ may be nonempty and include only monitoring links, and $\alpha$ and $\beta$ are either empty or the following holds for them

$$\forall x \in \{\alpha \cup \beta\} : \mathcal{M}_x \cap \mathcal{V}_i = \mathcal{M}_x \cap \mathcal{V}_j = \emptyset \qquad (15.6)$$

where $x$ is a common-mode fault component in $\{\mathcal{V}_i \cup \mathcal{V}_j \cup \mathcal{V}_I \cup \mathcal{V}_J\}$ and $\mathcal{M}_x$ is its monitor. □

Henceforth, the definition of being resilient to faults only with respect to safety (and not necessarily reliability) follows.

DEFINITION 49. A system is *safety-resilient* to $n$ *faults* if

1. there exist at least $n + 1$ fault-monitoring isolated configurations of its subsystems, and
2. any $n$ faults in the $n + 1$ configurations maintain a safe system. □

In order to better understand the above principles, a detailed design example is appropriate.

## 15.3 Design Examples

Let us look into an example of designing a process controlled plant. The plant processes materials which are poured through controlled inlets. The energy supply to the process (heat and electricity) are also controlled. Figure 15.5 describes a sensing and actuating system for this controlled plant. The control system contains three major elements:

1. the material and energy inlets,
2. a sensor which provides indication on materials' levels within the plant, and
3. a control mechanism, implemented by $\{G_i, G_I\}$.

We note that the control mechanism implements two control loops: an external loop of the production and an internal loop of sensor tuning. We have specified no resiliency factors for this implementation, because so far we did not introduce any safety requirement to this design.

### 15.3.1 Monitoring the safety

Let us now introduce an additional fact regarding the processing plant. Due to its high energy processes, there exist situations in which the plant could explode if control is lost. These situations do not occur as long as the system functions properly, since the control loops guard the

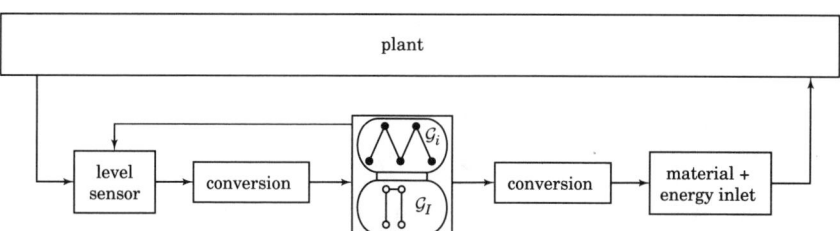

**Figure 15.5** Sensing and actuating system.

defined production path. However, if an explosion occurs, the results may be disastrous.

Therefore, we impose an additional requirement on the system design:

- given a failure in the controlled plant, at least two additional independent failures must occur before the plant becomes unsafe.

The rationale behind this requirement is to overcome a possible fault in the recognition mechanisms which identify the unsafe state and initiate recovery action.

This requirement follows the extension of the system to include at least two additional monitoring loops. The fact that the system has to recover from two failures, one of which can be in a monitor, requires this extension. These two monitors detect out-of-bound behavior of the plant, which may lead to an unsafe state. They do not participate in the production control mechanism, and they are only required to deactivate energy when a production control failure results in an unsafe state.

In our example, these monitors measure pressure and temperature in the plant's pressure tanks. Each measurement is independent and its results are supplied to an isolated computerized subsystem. The temperature sensor is processed by $\{G_j, G_J\}$, and the pressure is treated by $\{G_k, G_K\}$. In addition, they monitor $\{G_i, G_I\}$ which can provide indications of other faults. Some of these indications may be relevant much before pressure or temperature exceed their limits. Consequently, these monitors allow us to react to unsafe states with commands that disable the heaters or release the relief valves to suppress the blow-up risk. Figure 15.6 describes the two additional monitoring channels.

Let us find what relations are needed for the trio: $\{G_i, G_I\}$ $\{G_j, G_J\}$ and $\{G_k, G_K\}$. According to Equation 15.3 these blocks will be fault isolated if

1. their vertices are disjoint

$$\mathcal{V}_i \cap \mathcal{V}_j = \emptyset$$
$$\mathcal{V}_i \cap \mathcal{V}_k = \emptyset$$
$$\mathcal{V}_j \cap \mathcal{V}_k = \emptyset$$
$$\mathcal{V}_I \cap \mathcal{V}_J = \emptyset$$
$$\mathcal{V}_I \cap \mathcal{V}_K = \emptyset$$
$$\mathcal{V}_J \cap \mathcal{V}_K = \emptyset \qquad (15.7)$$

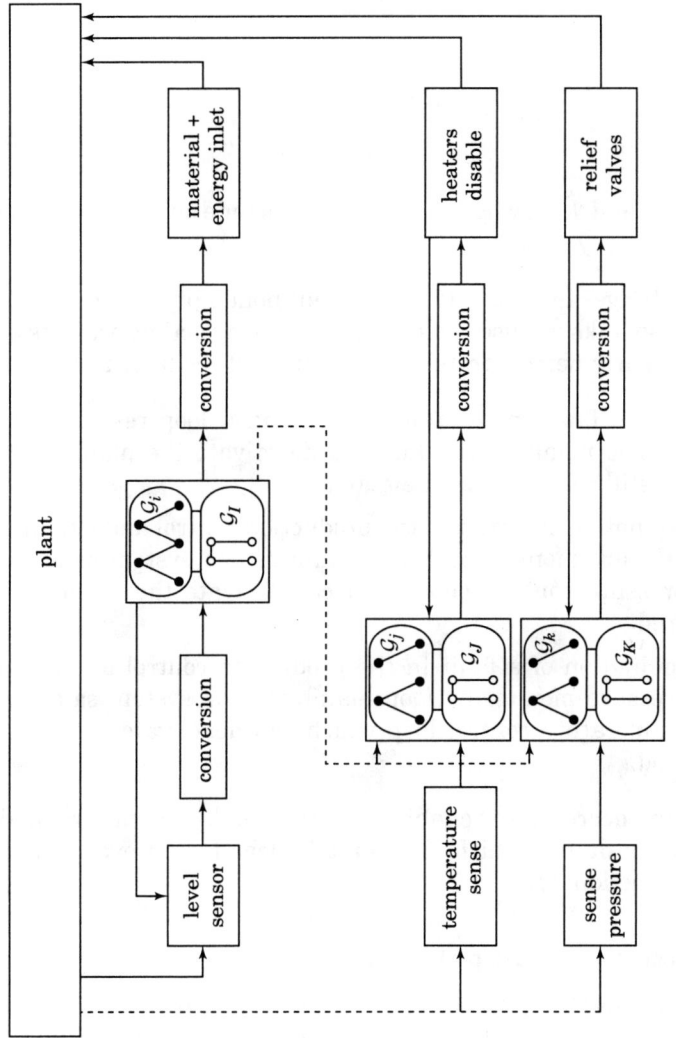

**Figure 15.6** The example system with monitoring blocks.

2. their edges

$$\mathcal{E}_i \cap \mathcal{E}_j = \psi_{i,j}$$
$$\mathcal{E}_i \cap \mathcal{E}_k = \psi_{i,k}$$
$$\mathcal{E}_j \cap \mathcal{E}_k = \psi_{j,k} = \emptyset$$
$$\mathcal{E}_I \cap \mathcal{E}_J = \Psi_{I,J}$$
$$\mathcal{E}_I \cap \mathcal{E}_K = \Psi_{I,K}$$
$$\mathcal{E}_J \cap \mathcal{E}_K = \Psi_{J,K} = \emptyset \quad (15.8)$$

where $\psi$ and $\Psi$ may be nonempty and include only monitoring links or else empty sets.

The dotted edges in Figure 15.6 represent nonempty $\psi_{i,j}$, $\psi_{i,k}$, $\Psi_{I,J}$ and $\Psi_{I,K}$. These links are used solely for monitoring activities. This design example clearly demonstrates the system is 2-resilient to faults:

1. Having no fault in the production control loop results in a safe plant. The monitors can fail and deactivate the plant, but their failure still results in a safe state.

2. A combination of a fault in the production control and a fault in the temperature monitoring loop maintains the system safe through the pressure control loop which can activate the relief valves if required.

3. A combination of a fault in the production control and a fault in the pressure monitoring loop maintains the system safe through the temperature control loop which can deactivate the heaters if required.

We note the decrease in reliability introduced by the failure modes of the monitors due to which the plant is deactivated even when it is properly functioning.

### 15.3.2  Decreasing failure probability

The safety requirement that any two independent failures cannot result in an unsafe plant can be expressed in terms of Definition 45 as a requirement to be 2-resilient at $m\%$ to faults. In order to decrease $m$ (the highest probability of any set of such two faults) we must analyze the actuation path of the control loop. The description of the plant in Figures 15.5 and 15.6 depicts the single-point pass of commands through a single serial block. This block which actually commands the material and energy inlets establishes a single-point failure in the system, with faults whose consequences are the undesirable unsafe states. We note this block because it is the one which generates the actual command to the system: both in energizing it and in controlling the material flow into it. Hence, this block is also the actual generator of the unsafe states, although it is not necessarily the cause for these states. However, a single fault in this block can be sufficient for an unsafe state, and consequently the probability of such a fault may be high.

A significant reduction of the probability of such faults can be achieved in multiplying the material and energy inlet control, as proposed in Figure 15.7. An independent dual-redundant implementation of this block reduces the probability of failure from $m\%$ to $m^2 \cdot 10^{-2}\%$. The implementation of the two must maintain serialization of the independent commands in such a way that any halt request of the two will be dominant in case of a conflict between them. A proper place to implement the dual-redundant inlet controllers is near the heaters and relief valves controllers. Consequently, the dotted line of monitoring $\{G_i, G_I\}$ is not a solely monitoring line any more, but rather an operational line for transferring the inlet commands.

One can also obtain an intuitive understanding of the above implementation reasoning through a commonsense approach. If there is a trouble, we need to close the inlets through which material and energy are input to the plant. It reduces risk because we have two separate halt commands to the process. It is redundant to the energy disablement and the pressure relief since it provides a transition to a safe state. It is economical, since it saves materials in case of a failure, whereas the monitors just remove danger by relieving pressure and heat. It simplifies recovery procedures, since one does not have to deal with spent material and the system state is well defined. Figure 15.7 describes the modification of the actuating channel to handle this approach. We note that we separate the IO processors $\{G_m, G_M\}$ and $\{G_n, G_N\}$ of $\{G_j, G_J\}$ and $\{G_k, G_K\}$. This is done at that point due to the resource load imposed on $\{G_J\}$ and $\{G_K\}$ when they have to deal with the sensors, the disablement or the relief, and the inlets. In addition, $\{G_m, G_M\}$ and $\{G_n, G_N\}$ indirectly take over the independent monitoring of $\{G_i, G_I\}$, through $\{G_j, G_J\}$ and $\{G_k, G_K\}$ respectively.

Let us find what relations are needed for the quintuple: $\{G_i, G_I\}$, $\{G_j, G_J\}$, $\{G_k, G_K\}$, $\{G_m, G_M\}$ and $\{G_n, G_N\}$. According to Equation 15.3, fault isolation of the system's three blocks ($\{G_j, G_m, G_J, G_M\}$, $\{G_i, G_I\}$, and $\{G_k, G_n, G_K, G_N\}$) is achieved if

1. their vertices are disjoint

$$\mathcal{V}_i \cap (\mathcal{V}_j \cup \mathcal{V}_m) = \emptyset$$

$$\mathcal{V}_i \cap (\mathcal{V}_k \cup \mathcal{V}_n) = \emptyset$$

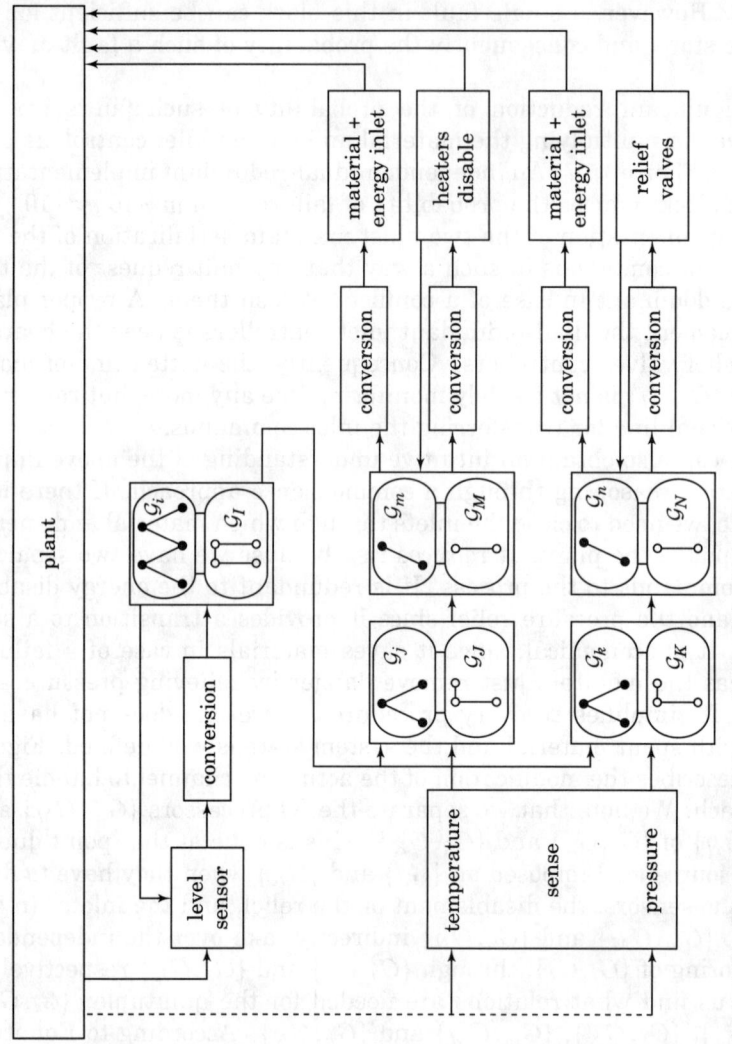

**Figure 15.7** The example system with modified actuation.

$$(\mathcal{V}_j \cup \mathcal{V}_m) \cap (\mathcal{V}_k \cup \mathcal{V}_n) = \emptyset$$
$$\mathcal{V}_I \cap (\mathcal{V}_J \cup \mathcal{V}_M) = \emptyset$$
$$\mathcal{V}_I \cap (\mathcal{V}_K \cup \mathcal{V}_N) = \emptyset$$
$$(\mathcal{V}_J \cup \mathcal{V}_M) \cap (\mathcal{V}_K \cup \mathcal{V}_N) = \emptyset \tag{15.9}$$

**2.** their edges satisfy

$$\mathcal{E}_i \cap \mathcal{E}_m = \psi_{i,m}$$
$$\mathcal{E}_i \cap \mathcal{E}_n = \psi_{i,n}$$
$$(\mathcal{E}_j \cup \mathcal{E}_m) \cap (\mathcal{E}_k \cup \mathcal{E}_n) = \psi_{j,m,k,n}$$
$$\mathcal{E}_I \cap \mathcal{E}_M = \Psi_{I,M}$$
$$\mathcal{E}_I \cap \mathcal{E}_N = \Psi_{I,N}$$
$$(\mathcal{E}_J \cup \mathcal{E}_M) \cap (\mathcal{E}_K \cup \mathcal{E}_N) = \Psi_{J,M,K,N} \tag{15.10}$$

where $\psi$ and $\Psi$ can be nonempty and include only monitoring links or else empty sets.

### 15.3.3 Coping with common mode faults

Let us now suppose there exists a common mode fault $\{G_l, G_L\}$ in the pair $\{G_j, G_J\}$ and $\{G_k, G_K\}$. Let us analyze our system to see how it can cope with such a fault without redesigning the trio $\{G_l, G_L\}$, $\{G_j, G_J\}$ and $\{G_k, G_K\}$. Figure 15.8 describes a solution for modifications due to the common mode failure of the computing part. In this solution no additional blocks are needed, and we monitor $\{G_l, G_L\}$ by $\{G_m, G_M\}$ and $\{G_n, G_N\}$. In addition, the relationship defined in Equations 15.9 and 15.10 is maintained.

Giving up the resiliency with respect to reliability, we focus our attention to the safety requirements given by Definition 49. We need our configurations to be fault monitored according to Equation 15.5 in Definition 48. This relation is straightforward and we leave its verification as an exercise to the reader.

The additional relations needed for the trio $\{G_l, G_L\}$ by $\{G_m, G_M\}$ and $\{G_n, G_N\}$ are defined based on the monitoring Equation 15.3. In other words, we obtain two fault isolated blocks if the vertices satisfy

$$\mathcal{V}_l \cap \mathcal{V}_m = \emptyset$$
$$\mathcal{V}_l \cap \mathcal{V}_n = \emptyset$$
$$\mathcal{V}_m \cap \mathcal{V}_n = \emptyset$$
$$\mathcal{V}_L \cap \mathcal{V}_M = \emptyset$$

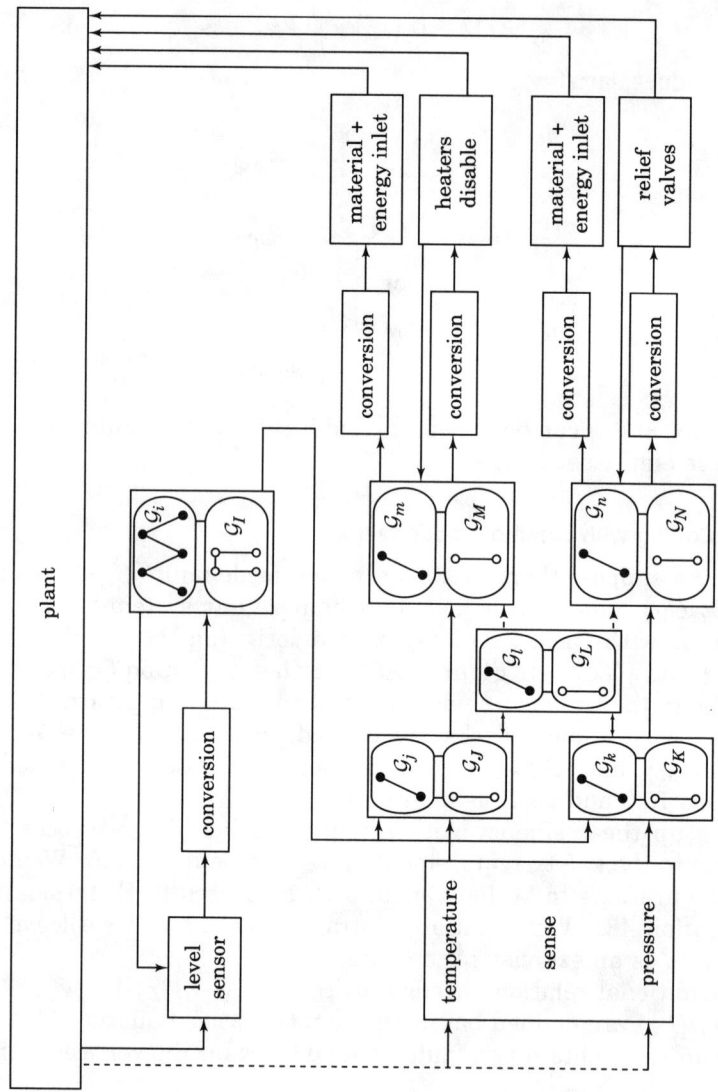

**Figure 15.8** The example solution to common mode fault.

$$\mathcal{V}_L \cap \mathcal{V}_N = \emptyset$$
$$\mathcal{V}_M \cap \mathcal{V}_N = \emptyset \tag{15.11}$$

and the edges satisfy

$$\mathcal{E}_l \cap \mathcal{E}_m = \psi_{l,m}$$
$$\mathcal{E}_l \cap \mathcal{E}_n = \psi_{l,n}$$
$$\mathcal{E}_m \cap \mathcal{E}_n = \emptyset$$
$$\mathcal{E}_L \cap \mathcal{E}_M = \Psi_{L,M}$$
$$\mathcal{E}_L \cap \mathcal{E}_N = \Psi_{L,N}$$
$$\mathcal{E}_M \cap \mathcal{E}_N = \emptyset \tag{15.12}$$

where $\psi$ and $\Psi$ can be nonempty and include only monitoring links or else empty sets.

Equations 15.9, 15.10, 15.11 and 15.12 express the relations required for the system described in Figure 15.8 which satisfies the safety requirements along with the reliability requirements.

## 15.4 Engineering Practices

Various engineering practices serve the system development in the assurance of its reliability and safety requirements. Hardware designs usually practice methods of qualification that assure withstanding loads and environmental conditions. The software application design is not yet that mature. Nevertheless, there are several engineering practices which are very productive and we briefly describe them.

### 15.4.1 Qualification

The principles which guide hardware qualification schemes are mainly testing the system in extreme states of load and environment conditions. Nominal load and environmental requirements are generally tested within the standard deviation ($1\sigma$) of their range. During qualification tests, these requirements are set at higher degrees, stressing the design to its limits. It is important to note that performance figures are generally specified to hold within the $1\sigma$ range. In qualification, we test load and environment parameters at their $3\sigma$ ranges for endurance rather than performance. However, safety requirements must be satisfied in all ranges of load and environmental conditions.

The qualification of a system with respect to its resiliency to faults is done with failure mode and effect analysis (**FMEA**). This method is extremely difficult in complex systems, and for it, automated testing seems more appropriate. There exist some experimental ways of verifying the resiliency of a system to faults via fault injection (e.g., [27, 29]). Faults can be inserted or forced exhaustively or statistically, to test the resiliency with respect to various types of faults. The results of such tests can be used to evaluate reliability figures of the system, as well as to validate the satisfaction of safety requirements. While for achieving reliability evaluation one can use the target system, this is not possible for validation of safety requirements.

### 15.4.2 Software engineering

There are three major regions of error sources in software components in the system:

1. operating system and environment,
2. compilers,
3. application programs.

In order to properly validate the satisfaction of the requirements, one must consider all of them.

Operating system errors can arise in a number of aspects. Its support functions of process activations, scheduling, hardware interfaces and physical support, interprocess communication, and over all its resource management make it extremely vulnerable to faults. It is a very sensitive part in any system, mainly due to the common-mode faults it can cause. Validation of safety properties of the operating system employed in the system's environment is complicated, especially due to the absence of detailed information (other than user manuals) available for the system designer's staff. In complex systems, it is therefore essential to have a guaranteed knowledge of the important properties of the operating system. Some operating systems (e.g., [94]) provide a formal proof of some of their important properties. When such guarantees do not exist, one can only rely on the experience gained with the specific operating system in other systems.

Compiler errors can arise from two sources:

1. Errors in the compiler design (permanent)
2. Errors generated during compilation (transient).

The validation of a compiler having no permanent faults is as complicated as for the operating system. Here again, the absence of detailed information available to the system designer is counterproductive. Again, in the case of having no guarantees for having no specific permanent faults, one can only rely on the experience gained in other systems. On the other hand, run-time errors which are generated during compilation can be overcome by multiple compilation sessions.

The application design must use two major software engineering practices when one deals with safety-critical systems. First, the design must take into account the contamination effects and derive bounds for them. Object-oriented approach principles (see Section 1.2.3) serve this practice quite well. Secondly, independent verification and validation practices must be employed. These practices achieve independence in two forms.

- An in-house independent group that serves as a validation group exercises the design of another group.
- An external independent group traces the design in various abstraction levels.

These methods are not perfect. In the first, the two groups may share a "successful delivery" goal, and end up with an incomplete test coverage. In the second, a group which was hired to criticize may try to find unrealistic faults just in order to prove that they did criticize. However, a combination of the two within their limits is beneficial.

Formal validation is necessary at various abstraction levels. High-level model allow validation of concepts and interfaces. In this category we find algorithm validation and various performance measures. Code-level validation is required in most designs, especially in various nonstandard implementations. Bit-level validation is carried out with respect to low-level language implementation, assuring that the proper target data is in place. This type of validation is cumbersome and unnecessary in most cases. Nevertheless, it is strongly required in very sensitive front ends. For example, in our design example, one can require bit-level verification of the activation commands. Here, one can assure that the case of "always activate the process" and "never paralyze the plant" does not exist.

## 15.5 Concluding Remarks

Safe systems must obey their safety requirements. Their resiliency can be expressed either with respect to the system reliability goals, as given by Definition 45, or with respect to the system safety requirements, as defined in Definition 49. The latter allows coping with common-mode faults, but reduces the system's functional reliability while increasing its safety. We have discussed the sometimes contradicting requirements of safety and reliability throughout this chapter.

We have built our fault model based on our graph-based model of the system. We have defined both the components and their interrelation characteristics which constitute safe behavior. We have introduced the monitoring of relations between isolated components for run-time safety verification as well as for resiliency compensation. A detailed design example demonstrated how a safe system is constructed, and how common-mode faults are dealt with.

The main ideas we have emphasized in this chapter are as follows.

- Isolation of components is not the major issue in safety-resiliency. Isolation of faults is, and its conditions have been discussed in depth.
- Monitors must be employed in order to provide run-time verification of the system's safe state.
- Independent monitors can compensate safety-resiliency reduction.
- Engineering practices assure that the implementation follows the desirable design concepts.

The chapter has concluded with a brief review of major engineering practices which must be exercised in the design of safe systems.

# Chapter 16

# Fault Tolerant Allocation

In order to execute any program it is essential that all necessary resources be made available to it in a timely manner. For a system these resources include hardware resources, such as processors, memory, communication devices and IO devices, and software resources such as shared databases and other server components. When we consider fault-tolerant operations, the allocation of resources must be consistent with the requirements of such operations. In this chapter, we present the basic concepts of fault tolerant allocation of resources.

## 16.1 Problem Definition

There have been numerous studies about many dimensions of the resource allocation problem. In a number of these studies, the goal of the allocation has been an optimization of some metric of the execution performance. The model which has been most often used in these cases reflects an allocation of processes to processors. In this model, both the set of processes and the set of processors have been subjected to some inter-set relations and intra-set optimality constraints. In addition, the nature of each of the processors has been homogeneous, indivisible, and self-contained.

Let us recall our system model defined in Section 1.2.1 for sets of processes and processors. A set of processes $\mathcal{V}_p = \{p_1, \ldots, p_n\}$ are related to each other through a set of logical links $\mathcal{E}_p$, to form a graph

$$\mathcal{G}_p = (\mathcal{V}_p, \mathcal{E}_p)$$

A set of processors $\mathcal{V}_P = \{P_1, \ldots, P_m\}$ are related to each other through a set of physical links $\mathcal{E}_P$, to form a graph

$$\mathcal{G}_P = (\mathcal{V}_P, \mathcal{E}_P)$$

Allocating processes to processors is a function whose domain is the set $\mathcal{V}_p$ and whose range is $\mathcal{V}_P$. We can consider the allocation optimization problem as the selection of the function which minimizes a particular metric. As we show in the following sections, most of the metrics relate in some way to $\mathcal{G}_p$ and $\mathcal{G}_P$ through cost functions that are affected by $\mathcal{E}_p$ and $\mathcal{E}_P$.

## 16.2 Definitions and Formulation

Let us start the description of our proposed solution with a description of our model of a distributed computation and the problem of allocating resources to such a distributed computation.

### 16.2.1 Model description

Let us consider a model in which computation is a system constructed from objects and resources. Let these objects and their relations be those defined in Section 1.2.3. The objects that participate in a computation are related to each other via semantic links pointed to by the object joints. In that respect, resources can also be viewed as objects. However, we distinguish between the two for differences in fault tolerance properties that relate to the monotonicity of faults. The distinction also relates to properties that determine damage containment in case of a fault. For these reasons we define a resource as an element which requires no other services, and whose failure mode is monotonic. Objects can require services from resources or other objects, and their failures can be either monotonic or transient.

The properties of the resources can allow us to model the system elements in terms of resource segments. For example, we can model one specific memory page as a resource, if we can detect a failure at this level of resolution. We may even be able to trigger an off-line recovery at the same level. On the other hand, if we cannot detect the failure at the page level, we can model the entire memory at a given locality as a resource, or even the whole locality (i.e. the processors, the memory, the devices, etc.) as a single resource. Therefore, we note the flexibility of that model, supporting a complete resolution spectrum of the system elements.

We allocate required resources to executing and "to-be-executed" objects, each having its own joint. These resources are physically linked according to their geographic and hardware constraints. However, in addition to resources, objects may need services that are provided by other objects, which in turn may need other services and resources, and so on. We represent this relationship by a graph, where the objects and resources are nodes, and the relations between them are directed arcs. Note that the resources are always the leaf-nodes, since a resource is not expected to need services from other resources.

The distinction between transient and monotonic faults, as expressed in our object/resource model, allows the use of two recovery mechanisms. We denote the most common one as *temporal redundancy*, in which we execute a "retry" effort upon a fault detection. This mechanism is perfectly suited for faults whose existence may be a transient phenomenon. It also permits roll-back recovery. Real-time constraints can conflict with temporal redundancy, however, because the time needed for recovery may not exist. Furthermore, in the case of a monotonic failure, retrying is ineffective. In such cases only *physical redundancy* can increase the system resiliency. Roll-forward recovery and the N-version programming are examples of such redundancy.

In Figure 16.1 we give an example of the two mechanisms. Objects $a$ and $b$ are allocated with temporal redundancy, while object $c$ has a physical redundancy in object $d$. In the model that follows, resources are to be subject only to physical redundancy, while the redundancy of objects is defined by the computation designer. We note again that the reason for this originates in the monotonic failures within the resources.

One major obstacle that the allocation and relocation mechanisms must overcome is shown in Figure 16.2. Although objects $B_1$ and $B_2$ are physically redundant, as are objects $C_3$ and $C_4$, the allocation in the figure results in a 0-resilient computation. Any failure of one of the four resources results in a computation fault, since both redundant threads depend on all four resources. If $B_1$ is allocated with $C_3$, and $B_2$ with $C_4$, the outcome is a 1-resilient computation.

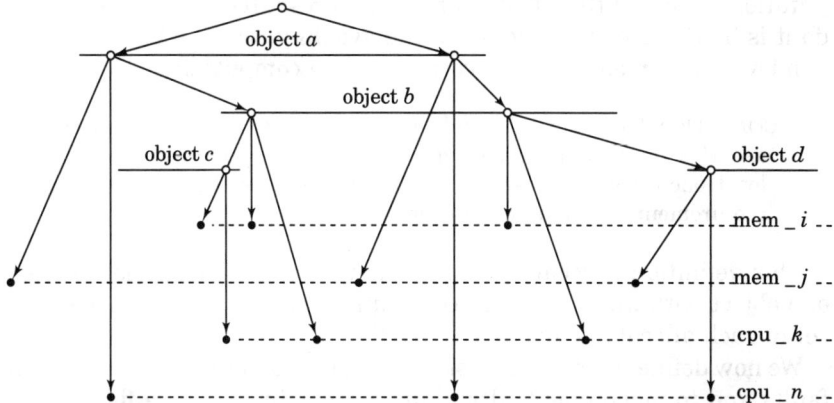

**Figure 16.1** Temporal $(a,b)$ and physical $(c/d)$ redundancy.

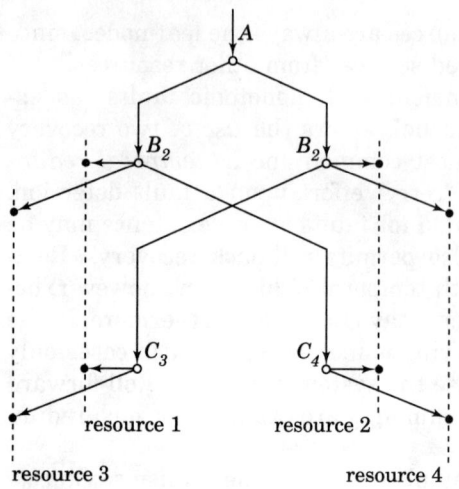

**Figure 16.2** Wrong use of resorces: 0-resiliency.

### 16.2.2 Conditions and formulation

Let each executable object instance $p$ have a set of resource requirements $\{R_i^{(p)} : 1 \leq i \leq k\}$ and a set of service requirements $\{S_i^{(p)} : 1 \leq i \leq n\}$, forming its *dependency set*, which we denote as $DS_p$. We note that we restrict $p$ with a time constraint $TC_p$, which implies a projected time constraint on each member of its dependency set. Each projection is a result of the temporal relation between $p$'s execution and its requirements. A service requirement can be executed by another executable object instance, which can be chosen out of a set of alternatives. Let us refer to our definitions in Section 12.4 regarding the dependency set (Definition 37), the dependency graph (Definition 38), and the reachability set (Definition 39).

There are ways to guarantee the absence of a conflict among computation elements that share servers and resources [91]. One way to do it is by the use of calendars, while avoiding conflicts between each window of occurrence and its corresponding computation interval.

> CONDITION 24. An object $p$ is *allocatable* if its resource requirements are *allocatable* and if for each of its service requirements, there is at least one *allocatable* service alternative (in case the set of its service requirements is not an empty set). □

This definition is recursive, implying that there must exist at least one object with an empty service requirement set in the reachability set of each allocatable object.

We now define the resiliency of an allocated computation to transient faults and to monotonic faults. But in order to do so, we first define two special subgraphs of an allocated dependency graph.

DEFINITION 50. An *allocation graph* of an object is a subgraph of the object's dependency graph in which only allocatable objects and the schedulable resources are represented. □

Note that when the allocation graph of object $p$ includes the object $p$ itself, it also contains all the resource requirements and all the service requirements of $p$.

DEFINITION 51. An *allocation alternative* of an object is a subgraph of the object's allocation graph in which for every service requirement only one service alternative is represented. □

Due to the definition of the allocation graph, the service alternatives represented in the allocation alternative are clearly allocatable.

We now have the tools that allow us to phrase the conditions for an allocation resiliency. The allocation resiliency we are interested in is of the type that can guarantee a certain degree of connectivity of our computation graph. Thus, we want to define resiliency to monotonic faults as well as to transient faults. In addition, we recall the example in Figure 16.2 and we show how the following conditions avoid the obstacle presented there.

CONDITION 25. An allocation for the execution of object $p$ is $n$-*resilient* to *monotonic faults* if $p$ is allocatable, and there exist at least $n + 1$ distinct allocation alternatives whose intersection with each other contains at most the node $p$. □

The close relation of that condition to Condition 20 is self evident.

Let us emphasize that this condition excludes any resource requirement or service requirement of $p$ from taking part in the intersection. This allows us to base the execution of each distinct allocation alternative on different resources and services. Furthermore, each allocation alternative demonstrates a complete damage containment from the other alternatives.

Let us now examine the condition for an allocation to be resilient to transient faults. We recall that we allow multiple executions of instances of the same object to achieve temporal redundancy. Thus, we need to define first the allocation of each of these instances.

DEFINITION 52. An *allocatable instance* of an allocation alternative $\mathcal{A}_p$ of an object $p$ is the tuple $(\mathcal{A}_p, TC_p)$, where $TC_p$ is the particular time constraint reserved for this allocation alternative. □

Using the above definition and recalling that a physical redundancy is also a temporal redundancy, we have the following condition.

CONDITION 26. An allocation is $n$-*resilient* to *transient faults* if the computation is $k$-resilient to monotonic faults, and each of its $k + 1$ allocation alternatives, $0 \leq i \leq k$, have $\tau_i$ distinct allocatable instances, such that

$$\sum_{i=0}^{k} \tau_i \geq n. \tag{16.1}$$

□

This condition is clearly related to Condition 21, as in the physical redundancy case.

The practical implication of the above condition can be stated informally in terms of the following allocation philosophy. Let us assume we invoke an allocator to achieve an objective of a given resiliency to transient faults. We may find out that the number of distinct allocatable instances at a given allocation alternative cannot support it. If so, we can use another allocation alternative to allocate additional allocatable instances. Thus, we use physical redundancy to support requirements for temporal redundancy.

## 16.3 Allocation Algorithm

In this section we introduce our allocation algorithm, which is based on the definitions and conditions we have introduced above. But instead of providing the detailed algorithm, we use a condensed version of the algorithm to introduce the principles on which it works.

Considering the *dependency graph* defined in Section 16.2.2, we define a leaf node as a node whose dependency set has an empty service requirement set. Recall that in the dependency graph, the nodes are the executable objects we want to allocate. As we will see later, a leaf node plays a special role in this allocation algorithm, and is responsible for generating the *"yes"* answers to the messages.

The *state* of an executable object during allocation can be *allocatable* or *non-allocatable*. An executable object is *allocatable* when it satisfies the allocatability condition as specified in Condition 24. Even when an executable object is non-allocatable, it is assumed to be capable of responding to the algorithm performed by the allocator.

We assume that the allocation algorithm is executed by allocators, each of which is invoked to test the satisfaction of Condition 24 by a particular object.[1] Therefore, we start by defining the invocation messages used in the algorithm, and go on to describe the principles of the algorithm.

---

[1] The assumption does not restrict the generality of the algorithm, but instead, enriches its possible implementations. Allocators can be different instances of the same allocator (e.g., a recursive call), or different allocators executing concurrently.

### 16.3.1 Message types used

- *ALLOCATE (from, whom, physical redundancy, temporal redundancy, to)* is the initiator message.
  - *from* - initiating object ID
  - *whom* - **set** of alternative object_ SAPs to be allocated
  - *physical redundancy* - degree of physical redundancy
  - *temporal redundancy* - degree of temporal redundancy
  - *to* - receiving allocator ID

- *ALLOC_ REQ (of, tag, from, whom, level, to)* is the query message.
  - *of* - initiator ID
  - *tag* - tag number of this *of*'s computation session
  - *from* - object_ SAP that requests the service.
  - *whom* - object_ SAP whose service is requested.
  - *level* - degree of temporal redundancy requested.
  - *to* - receiving allocator ID.

- *ALLOC_ REP (color, of, tag, from, whom, $\Delta$level, to)* is the feedback message.
  - *color* - yes / no
  - *of* - initiator ID
  - *tag* - tag number of this *of*'s computation session
  - *from* - object_ SAP that replies
  - *whom* - object_ SAP which requested the service from *from*
  - $\Delta$*level* - degree of temporal redundancy in debt
  - *to* - receiver ID

### 16.3.2 Principles of allocation initiation

The following algorithm rule is implemented as an interface between the user who wants to initiate an allocation session and the allocator. This initiation rule can be implemented as either a special service access-point of the allocator or a dedicated server of another type. When we initiate an allocation session, we must specify its fault tolerance objectives, which are its set of alternatives in which the computation can be carried out. The initiation rule tries to reach the fault tolerance objectives by requesting allocation of constrained computation alternatives (from the set defined above). These computation alternatives must adhere to the physical and temporal redundancy defined by the user.

We tag alternative subgraphs to allow concurrent allocation of dependency subgraphs while maintaining a null intersection among these subgraphs. To support the distributed allocation scheme, the computation *ID* and the graph *tag* are spread with the requests throughout the graph. We use the computation ID and tag as the means to control the connectivity of the graph.

The initiator (*me*) sends enough *ALLOC_ REQ (...)* messages to allocate members of the alternative set defined by "*whom*", and *me* now has to wait for eventual answers. In order to have a higher degree of concurrency, an artificial object joint is created. We note that *me* can be an allocator of the operating system, which we can use to serve many applications. Thus, this artificial joint serves the need of keeping *me* active while waiting to collect the answers when they arrive. It also allows choosing another alternative when the answer is negative.

Any decrease in physical redundancy is implicitly prevented by the algorithm. The physical redundancy is controlled through the insertion function, that does not reserve in a particular object-joint, two requests with the same *ID* and different *tag*s. This property adheres to the null intersection requirement in Condition 25. Let us summarize these principles in an algorithmic form, to phrase the initiation rule.

- Upon receiving *ALLOCATE (from, whom, physical redundancy, temporal redundancy, me)*::
  1. Create an artificial object (ROOT) whose dependency set consists of an empty set of resource requirements, and a set of service requirements whose cardinality equals the physical redundancy level required plus one. Distribute the alternatives of *whom* among these service requirements.
  2. For every service requirement in ROOT do:
     - Select the first service alternative in the service requirement.
     - Send ALLOC_ REQ for allocating the selected service alternative, distinguishing each service requirement with a different tag. The ALLOC_ REQ asks for the temporal redundancy required, imposes the requested time constraint, and designates ROOT as the initiator of the allocation request.

### 16.3.3 Principles of an algorithm for allocator

The following algorithm rules are implemented in all allocator instances in the system. They consist of actions responding to an *ALLOC_REQ (...,me)* message (allocation request), and actions responding to an *ALLOC_ REP (...,me)* message (allocation reply).

An allocator receives a *ALLOC_ REQ (...,whom,...,me)* message the allocation of an executable object (*whom*). This object, *whom*, must have the allocatability property holding for its resources, for each of its "to-be-executed" instances. If they are allocatable, it forwards *ALLOC_ REQ (...)* messages to allocate its service requirements in its

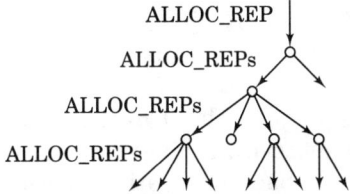

**Figure 16.3** Forward wave of ALLOC_REQ messages.

dependency set. This *forward wave* of *ALLOC_REQ ( ... )* messages proceeds, propagating these messages until the propagation cannot continue. The forward wave stops if a requesting message reaches either an executable object which is *non-allocatable* or a leaf executable object which has no service requirements. Figure 16.3 describes the propagation of *ALLOC_REQ ( ... )* messages.

Now let us examine how the *ALLOC_REP ( ... )* messages are generated. Consider the case where we request the allocation of an executable object *whom*, and the allocator verifies it (or its resources) to be *non-schedulable*. Then, there is no point in verifying the *allocatability* of the resource requirements of *whom*. Therefore, the allocator generates an *ALLOC_REP (no, ... )* message to the object which requested its service. On the other hand, let us consider the case where we request to allocate a leaf object *whom*, and the allocator verifies it and its resources to be *schedulable*. Having no resource requirements in *whom*'s dependency set, the allocator can generate an *ALLOC_REP (yes, ... )* message to the object which requested its service.

Figure 16.4 describes the way in which the forward wave generates the backward wave. We attached indices to the messages in order to help the ordering of their occurrences.

The *backward wave* of *ALLOC_REP ( ... )* messages propagates in the following way. Let us start with the positive reply that requires some conditions to hold. First, the object's resource requirements must have been *allocatable*. Second, this object must have received *all* the answers expected with a positive "color." Only then, can the allocator

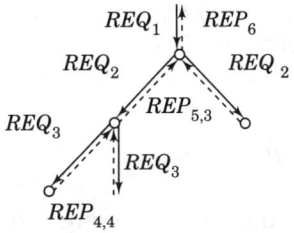

**Figure 16.4** Backward wave of ALLOC_REP messages.

send back a positive answer message *ALLOC_ REP (yes, ... ,prev, ... )* to the object that had requested the services. Thus, each node performs a boolean AND of all the positive answers. On the other hand, let us consider the case where a requesting object exhausts all the service alternatives for any particular service request. Then, it cannot meet its requirements, and it sends back a negative answer message *ALLOC_ REP (no, ... ,prev, ... )*. We note that in the latter case some services may have already been reserved (in particular the object itself and the resources). We must remove these reservations to release them for other possible requests.

Between the two cases of generating the "yes" and "no" replies, we need to respond to other *ALLOC_ REP ( ... )* messages. We have to note positive replies until all expected answers are verified positive. On the other hand, we must invoke other alternatives as long as the object has not exhausted its service alternatives and the objectives are not satisfied.

Let us introduce the above principles in two reaction rules. The first presents the required reaction to an allocation request (the forward wave). The second presents the reaction to a reply (the backward wave).

- *Upon receiving ALLOC_ REQ (of, tag, from, whom, temporal redundancy level, me)*::
  1. Iterate $my\_level$ successful iterations, trying to reserve an execution interval for *whom* in the calendar, and for its resource requirements in their calendars. The number of iterations is bounded by the required temporal redundancy level.
  2. If no iteration was successful, send ALLOC_ REP answering *no*.
  3. Otherwise, if *whom* is a leaf-object (having no service requirements), send ALLOC_ REP answering *yes*, indicating how many missing temporal redundancy instances there are according to $my\_level$.
  4. Otherwise (not being a leaf-object) do the following for every service requirement in *whom* dependency set.
     - Select the first service alternative in the service requirement.
     - Send ALLOC_ REQ for allocating the selected service alternative, asking for the temporal redundancy $my\_level$.
  5. Update *whom* joint to include the proper information to deal with replies.

- *Upon receiving ALLOC_ REP (color, of, tag, from, whom, $\Delta level$, me)*::
  1. If the color is *yes*, and all the required temporal redundancy instances have been allocated, then mark this service requirement as *done*.

2. Otherwise, not having enough temporal redundancy instances, if there is another possible service alternative in the service requirement, do the following.
   - Select the next service alternative in the service requirement.
   - Send ALLOC_ REQ for allocating the selected service alternative, requiring the unsatisfied temporal redundancy level (up to $my\_level$).
3. However, if there are no more service alternatives at that service requirement, the following two cases are distinguished.
   - If no alternatives at all at that requirement have been allocated, then send ALLOC_ REP answering *no* to the object that required the service of *whom*. In that case release *whom*, its resources, and the rest of the requirements.
   - If some alternatives at that requirement have been allocated, then decrease the level of temporal redundancy viewed by *whom*, to the lowest between its current view and the view seen by *from*. Then, mark this requirement as *done*.
4. If all service requirements are *done*, send an ALLOC_ REP with positive answer to the object that required the service of *whom*, indicating the level of temporal redundancy as limited by *whom*'s view or its requirements.

Finally, we note the way in which the degree of temporal redundancy is maintained, in order to satisfy Condition 26. The temporal redundancy achieved by the object itself and its resources are bounded by the one requested from the service requirements. If a service alternative cannot satisfy the degree required by a requesting object, an additional alternative is invoked to satisfy the debt (the remaining redundancy), and so on as long as there are alternatives. The requesting object is informed about the final debt in case one could not reserve the required redundancy. The sum of the redundancy achieved by the alternatives of a service requirement establishes the degree of that service requirement. The lowest degree achieved by a member of the service requirements is the one reserved. That way the informed requesting object can try and increase the degree by requesting another alternative. The principle here is to use *physical* redundancy when no more *temporal* redundancy can be achieved.

## 16.4 Concluding Remarks

The allocation scheme presented in this chapter satisfies specified fault tolerance objectives through the use of a set of alternatives in which the service can be carried out. These alternatives can provide physical or temporal redundancy, as defined by the application.

In [94], we have suggested that an algorithm of this type allows recovery of the resiliency after a fault occurs. If there are unused resources in the system that can support the continuation of the execution of objects, why not use them. A reallocation of an unused alternative as a substitute to the faulty one can be easily implemented with the tools described above for the allocation. The unit which detects the faulty server, invokes the allocator as if the faulty object replied to a negative ALLOC_REP, and thus triggers the search of another alternative. If such an alternative is found, the computation initiator is only informed about the recovery via a positive ALLOC_REP message. Otherwise, it is informed with a $\Delta$ level that results from the fault.

Part

4

# Fault Tolerance in Real-Time Systems

The last part of this book examines fault tolerance issues in real-time systems. Real-time systems set an additional challenge to fault tolerance, as recovery must satisfy the time constraints of the failing process. This issue is therefore the first to be examined in this part, emphasizing the resource allocation point of view. Two additional issues which are specific for real-time systems are then discussed, since they are crucial for real-time applications: fault-tolerant communication under real-time constraints and time services which tolerate faults. The special requirements of real-time systems call for a special treatment of fault tolerance techniques for such systems.

# Chapter 17

# Allocation in Real-Time Systems

Allocation of resources in "next-generation" real-time operating systems requires several features in addition to those demonstrated by current systems. These additional requirements increase the complexity of the resource allocation mechanisms. Allocation is closely related to scheduling, and the two are based on *time* considerations, rather then on a static priority scheme.

As described in Chapter 16, the allocation is fault-tolerance motivated, in order to support the application's reliability goals. Furthermore, distributed system issues and adaptive behavior requirements further increase the complexity of the mechanisms. The temporal restrictions complicate the mechanisms even more. Each computation we want to schedule within the specified time constraints, needs resources at time intervals derived from these constraints. These resources can be of the hardware type (e.g., processors, memory segments, communication, IO devices, etc.) or the software type (e.g., shared data base).

We have proposed [94] an architecture in which we handle object manipulation such that access-time behavior supports the objectives of real-time constraints. We propose an allocation scheme that accomplishes the hard real-time requirement guaranteeing meeting the deadline of each accepted job. In addition, this allocation scheme enhances fault tolerance, while supporting both damage containment and resiliency. It does this in cooperation with the schedulability verification mechanism introduced in [91], and with an object architecture, in which for each object there exists a *calendar* management that relates time to its execution. Another feature of this scheme is that it can be used for reallocation that increases the resiliency back to its previous value after a failure occurred.

```
type time_constraint = construct
    {   Id: computation identifier ;
        tag: thread indicator of computation Id ;
        level: redundancy index ;
        tc: convex_time_interval ;
        back_slack, for_slack : real ;
        P : non_convex_time_interval ;
        freq : real ;
        state : integer } ;
type resource_requirement = construct
    {   $\mathcal{R}_R^{in}$, $\mathcal{R}_R^{out}$: temporal relations ;
        R : convex_time_interval } ;
        ↑resource ;
type service_alternative = construct
    {   $\mathcal{R}_s^{in}$, $\mathcal{R}_s^{out}$: temporal relations ;
        s : convex_time_interval } ;
        ↑object_SAP ;
type service_requirement = set of service_alternatives ;
type schedule_type = (preemptive, non-preemptive) ;
type Answer_Wait_Indicator = (off, on, done) ;
variables
    calendar : ordered set of time_constraints ;
    Sched_Type : schedule_type ;
    dependency_set : set of k resource_requirements
                     and n service_requirements ;
auxiliary variables
    wait_set: set of construct
    {   prev : ↑object_SAP ;
        Id : computation identifier ;
        tag : thread indicator of computation Id ;
        $TC_{me}$ : time_constraint ;
        my∆level : redundancy index ;
        s[n] : set of n service_alternatives ;
        Ans[n] : set of n Answer_Wait_Indicators } ;
```
**Program 17.1** Typical joint variables in a calendar.

The object model is described in Section 1.2.3. A system which includes several objects maintains the relations between them. The temporal properties of each relation are expressed as either convex or nonconvex time intervals in a calendar within the relevant joint. Relations between these intervals can express absolute precedence or overlapping, or relations between the start and end points of the intervals [94]. Program 17.1 depicts typical joint variables in a calendar.

When the allocation is carried out, reservations are made in the proper calendars to support future schedulability. As in the nonreal-time case, we allocate required resources to executing and "to-be-executed" objects, each having its own joint and calendar.

## 17.1 Additional Conditions and Formulation

In order to verify allocatability, we need first to assure schedulability. The schedulability conditions establish the means for defining object allocatability that does not violate a feasible schedule. Therefore, we need to extend Condition 24 in Section 16.2.2 as follows.

> CONDITION 27. An object $p$ is *allocatable*, if it is *schedulable*, its resource requirements are *schedulable*, and for each of its service requirements there is at least one *allocatable* service alternative (in case the set of its service requirements is not an empty set). □

Condition 27 is recursive. It indicates that the resources required by an object must be schedulable and at least one service alternative must be allocatable for this object to be allocatable. When schedulability of a resource is confirmed, reservation is made in favor of the requiring object, thereby supporting its allocatability.

## 17.2 Allocation Algorithm under Real-Time Constraints

We can now implement this conditional verification for real-time objects. Clearly, we can do it by using the allocation principles defined in Section 16.2.2 and modifying the algorithm presented in Section 16.3.

When we initiate an allocation session, we must specify its fault-tolerance objectives, its set of alternatives in which the computation can be carried out, and the timing constraints for this computation. The insertion of these specifications into an object's joint and the proper calendar is performed by a procedure entitled INSERT_TC. Each request points to the identification of the object which submitted it (*ID*). Furthermore, since there might exist multiple requests of the same object, each one is labled by a *tag*. INSERT_TC does not reserve in a calendar, two requests with the same *ID* and different *tag*s.

An allocator receives a *ALLOC_ REQ* ( ... ,*whom*, ... ,*me*) message to allocate an executable object (*whom*). This object, *whom*, must have the *schedulability* property which holds for itself and for its resources, for each of its "to-be-executed" instances. If it is schedulable, it forwards *ALLOC_ REQ* ( ... ) messages to allocate its service requirements in its dependency set.

This forward wave of allocatability and schedulability verification works in the same way as in Section 16.3. However, schedulability is verified for the object itself and its resources, and allocatability is verified for its service alternatives.

Real-time properties propagate through the time constraints sent in the *ALLOC_ REQ ( ... )* messages to the service requirements and through the timing requests imposed on the resource requirements. These constraints are projections of the incoming timing constraint. The projections are made according to the required temporal relations between the invoker's constraint and those imposed on the requirements. We assume that these relations are known in advance, and that they are *convergent*, as defined in Definition 53.

DEFINITION 53. A *convergent* temporal relation sequence, is a sequence of temporal relations $(\mathcal{R}_{xy}, \boxed{\ldots}, \mathcal{R}_{yx})$ that satisfies

$$x\mathcal{R}_{xy}y \ldots \mathcal{R}_{zx}x' \ll x \vee x\mathcal{R}_{xy}y \ldots \mathcal{R}_{zx}x' = x \vee$$

$$\vee\, x\mathcal{R}_{xy}y \ldots \mathcal{R}_{zx}x' \uparrow x \vee x\mathcal{R}_{xy}y \ldots \mathcal{R}_{zx}x' \downarrow x$$

for time intervals $x, y$. □

A convergent sequence of temporal relations assures bounded behavior of the sequence. If a constraint with a finite occurrence interval is related through such a sequence to itself, the result is contained by that interval. In other words, reentrant invocation originated through external dependencies results in a projected constraint contained by the original constraint. Let us describe it with an example. Let $x$ be the original constraint imposed on an object. This object requires a service from another object. The temporal relation between $x$ and the constraint of the service, $y$, is $\mathcal{R}_{xy}$. Let us assume that the service itself requires another service, and let it be given by the original object constrained by $x'$. Let the relation between $y$ and $x'$ be $\mathcal{R}_{yx}$. This situation can be described as $x\,\mathcal{R}_{xy}\mathcal{R}_{yx}\,x'$. For a convergent $\mathcal{R}_{xy}\mathcal{R}_{yx}$ we are guaranteed that the occurrence interval of $x'$ is contained within the occurrence interval of $x$.

Thus, we must have a different approach than in the nonreal-time case:

- *Upon receiving ALLOC_ REQ (of, tag, from, whom, temporal redundancy level, me)*::
  - ...[1]
  - Send ALLOC_ REQ for allocating the selected service alternative, asking for the temporal redundancy *my_ level*, projecting the proper time constraint according to the temporal relation between *whom* and the service.

---

[1] See page 362 for similar steps.

As in Section 16.3, *backward wave* of *ALLOC_ REP ( ... )* messages propagates from the leaves to the root. But since local schedulabilty is essential, first the object and its resource requirements must have been found *schedulable* and only then are alternatives tested to be *allocatable*.

Again, we have a slightly different approach as compared to the nonreal-time case:

- *Upon receiving ALLOC_ REP (color, of, tag, from, whom, $\Delta$level, me)*::
  - ...[2]
  - Send ALLOC_ REQ for allocating the selected service alternative, requiring the unsatisfied temporal redundancy level (up to *my_ level*), projecting the proper time constraint according to the temporal relation between *whom* and the service.

### 17.2.1  Message types used

- *ALLOCATE (from, whom, TC, physical redundancy, temporal redundancy, to)* is the initiator message:
  - *from* - initiating object ID.
  - *whom* - **set** of alternative object_ SAPs to be allocated.
  - *TC* - time constraint.
  - *physical redundancy* - degree of physical redundancy.
  - *temporal redundancy* - degree of temporal redundancy.
  - *to* - receiving allocator ID.
- *ALLOC_ REQ (of, tag, from, whom, level, TC, to)* is the query message:
  - *of* - initiator ID.
  - *tag* - tag number of this *of*'s computation session.
  - *from* - object_ SAP that requests the service.
  - *whom* - object_ SAP whose service is requested.
  - *level* - degree of temporal redundancy requested.
  - *TC* - time constraint.
  - *to* - receiving allocator ID.
- *ALLOC_ REP (color, of, tag, from, whom, $\Delta$level, TC, to)* is the feedback message:
  - *color* - yes / no.
  - *of* - initiator ID.

---

[2]See page 362 for similar steps.

- *tag* - tag number of this *of*'s computation session.
- *from* - object_ SAP that replies.
- *whom* - object_ SAP which requested the service from *from*.
- $\Delta level$ - degree of temporal redundancy in debt.
- *TC* - time constraint.
- *to* - receiver ID.

### 17.2.2  Local and external variables

In the algorithm presented here we use some of the variables defined for the joint of an object (see Program 17.1 and the following local variables:

- my_ level: the degree of temporal redundancy of this object
- my$\Delta$level: the debt in temporal redundancy of this object
- $\Delta$level: the debt in temporal redundancy of the service requirement
- $TC_{me}$: time constraint of this object
- $R_i$: a resource requirement
- $\mathcal{R}_{R_i}$: the temporal relation between this object and resource requirement $R_i$
- $TC_i$: the time constraint of the requirement as projected from $TC_{me}$ using the temporal relation $\mathcal{R}$
- $S_i$: a service requirement
- $\mathcal{R}_{S_i}$: the temporal relation between this object and service requirement $S_i$
- $s_j^{(i)}$: a service alternative of service requirement $S_i$
- ($k, n$ are the number of requirements for resources and services, respectively)

### 17.2.3  The allocation algorithm

*Upon receiving ALLOCATE(from,whom,TC,phys_deg,temp_deg,me)::*

**begin**
/* Creating an artificial ROOT and setting its parameters */
Let *whom* be associated with $\{p_1, \ldots, p_k\}$ with $k > phys\_deg$ ;
Construct a nonvolatile auxiliary object *ROOT* with the following:
1. $DS_{ROOT,TC_{ROOT}} = \{ \{R_i^{(ROOT)} : 1 \leq i \leq k\}$
$\{S_i^{(ROOT)} : 1 \leq i \leq n\} \}$
where $\{R_i^{(ROOT)} : 1 \leq i \leq k\} = \phi$
and $S_i^{(ROOT)} = \{< s_j^{(ROOT)(i)}, TC >: s_j^{(ROOT)(i)} \in \{p_1, \ldots, p_k\} \}$.

2. $ID \leftarrow ROOT$.
3. $prev \leftarrow from$.
4. $TC_{me} \leftarrow TC$.
5. my$\Delta$level$\leftarrow 0$.
6. $s[i] \leftarrow s_1^i$, for $1 \leq i \leq phys\_deg$.
7. $Ans[i] \leftarrow off$, for $1 \leq i \leq phys\_deg$.
      /* Distribute the request for the nonintersecting
         subgraphs, each of which denoted with a different $tag$.
   **for** $tag \leftarrow 1$ **to** $phys\_deg$ **step** 1 **do** send
       ALLOC_REQ(ROOT,tag,ROOT,$s_1^{(ROOT)(tag)}$,temp_deg,TC,
           allocator) ;
   **od**
**end** □

*Upon receiving ALLOC_REQ(ID,tag,from,whom,temp_deg,TC,me)::*

**begin**
    my_level$\leftarrow 0$; I_M_O_K$\leftarrow$true;
    $TC_{me} \leftarrow$ construct($whom, TC, ID, tag$);
    $TC_{me}.level \leftarrow$ my_level; $TC_{me}.state \leftarrow$ idle;
    **while** (my_level<temp_deg)$\wedge$(I_M_O_K) **do**
             /* Temporal redundancy reservations */
        **if** INSERT_TC($whom, TC_{me}$) **then**
                /* Reserve necessary resources for $whom$ */
            $i \leftarrow 1$; R_is_O_K$\leftarrow$true ;
            **while** $(i \leq k)\wedge$(R_is_O_K) **do**
                $TC_i \leftarrow$ project ($\mathcal{R}_{R_i}, TC_{me}$) ;
                $TC_i.level \leftarrow$ my_level ;
                **if** INSERT_TC($R_i, TC_i$) **then**
                    $i \leftarrow i + 1$ ;
                **else** /* cannot get them all: release guaranteed subset */
                    R_is_O_K$\leftarrow$false ;
                    **for** $q \leftarrow 1$ **to** $i$ **step** 1 **do**
                        REMOVE_TC ($R_q, TC_q$) ;
                    **od**
                    I_M_O_K$\leftarrow$false ;
                    REMOVE_TC ($whom, TC_{me}$) ;
                **fi**
            **od** /* resource reservation terminated */
        **else**
            I_M_O_K$\leftarrow$false ;
        **fi**
        **if** (I_M_O_K) **then**
            my_level$\leftarrow$my_level+1 ;
            $TC_{me}.level \leftarrow$ my_level ;
        **fi**
    **od**

```
my△level ← my_level − temp_deg ;
if (my_level=0) /* failed to reserve whom */ then send
   ALLOC_REP(no, ID, tag, whom, from, my△level, TC_me, allocator) ;
else /* my_level> 0, something was reserved */
   if (∀S_i ∈ DS(whom) : S_i = φ) /* leaf-object */ then send
      ALLOC_REP(yes, ID, tag, whom, from, my△level, TC_me,
                allocator) ;
   else /* non-leaf-object: invoke allocation of serv
              requirements */
      prev ← from ;
      for i ← 1 to n step 1 do
         TC_i ← project (R_{S_i}, TC_me) ;
         send ALLOC_REQ(ID, tag, whom, s_1^{(whom)(i)},
                         my_level, TC_i, allocator) ;
         s[i] ← s_1^{(whom)(i)} ;
         Ans[i] ← off ;
      od
         /* In case allocator is reentrant: store in whom joint */
         store ID, prev, TC_me, my△level, s[i = 1, . . . , n],
               Ans[i = 1, . . . , n] ;
   fi  fi
end □
```
```
Upon receiving ALLOC_REP(color, ID, tag, from, s_j^{(whom)(i)}, △level, TC, me)::
```
**begin**
```
Restore auxiliary variables according to ID, tag, s_j^{(i)}
         /* prev, TC_me, my△level, s[i = 1, . . . , n], Ans[i = 1, . . . , n] */
if (color = yes) then
 if (△level= 0) then
    Ans[i] ← done ;
 else /* △level< 0 : more alternatives needed,
                     some already reserved */
    Ans[i] ← on ;
 fi
fi
if (△level< 0)∧(j < M^{(whom)(i)}) then
   TC_i ← project (R_{S_i}, TC_me) ;
   send ALLOC_REQ(ID, tag, whom, s_{j+1}^{(whom)(i)}, −△level, TC_i,
                   allocator) ;
   store s[i] ← s_{j+1}^{(whom)(i)} ;
elseif (△level< 0)∧(j = M^{(whom)(i)})∧(Ans[i] = on) then
   my△level ← min(my△level, △level) ;
   Ans[i] ← done ;
elseif (△level< 0)∧(j = M^{(whom)(i)})∧(Ans[i] = off) then
      /* no alternative reserved release other requirements */
   send UNLOAD(whom, TC_me) ;
         /* release for others [91] */
```

```
        if (ID ≠ whom) then /* climb up to try again */ send
            ALLOC_REP(no, ID, tag, whom, prev, myΔlevel, TC_me,
                     allocator) ;
        else /* ROOT failed to be allocated */ send
            ALLOC_REP(no, ROOT, tag, s[i], ROOT, myΔlevel, TC_me,
                     prev) ;
      fi
    else /* error in algorithm */
    fi
    if(∀i : Ans[i] = done) then
      if (whom ≠ ID) then send
          ALLOC_REP(yes, ID, tag, whom, from, myΔlevel, TC_me,
                   allocator) ;
      else /* ROOT is properly allocated */ send
          ALLOC_REP(yes, ROOT, tag, ROOT, ROOT, myΔlevel, TC_me,
                   prev) ;
              /* temporal redundancy debt in myΔlevel */
          delete ROOT;
      fi fi
end □
```

## 17.3 Concluding Remarks

The allocation scheme presented in this chapter satisfies the specified time constraints in addition to fault-tolerance objectives which are achieved through the use of a set of alternatives in which the service can be carried out. To achieve that, each object maintains a calendar, in which reservations are made for the services it guarantees. This scheme supports reallocation for resiliency recovery after failure, in a similar manner as the nonreal-time algorithm presented in Chapter 16.

Chapter

# 18

# Protocols for Real-Time Communication

In real-time communication we require a timely delivery of messages as well as correct data transfer between sender and receiver. In many cases a message which arrives too late is useless even though the data transfer is correct. We can, therefore, associate deadlines with real-time messages, as we do with any other real-time computation activity. Association of deadlines with messages has various interpretations in the literature. Since communicating processes may get their timing data from different clocks, the reference of the deadlines must be properly defined. Furthermore, if both send and receive actions have deadlines, how one relates to the other must be defined.

This chapter starts with a description of two approaches that deal with real-time communication at the data-link layer of the layered communication model. Each of the two approaches utilizes a different level of the data link layer. We start with a protocol that governs access to the communication medium. The example we introduce deals with a medium that allows contention between different senders. We then describe activities at the logical-link layer, emphasizing the real-time issues of synchronous communication protocols.

## 18.1 Protocols with Contention

Carrier sense multiple access (CSMA) protocols are widely used as medium access control (MAC) protocols. These protocols allow contention between senders that use the communication medium. The set of computation elements (processes) that share the use of the communication medium can have conflicts on gaining control over it. When two processes transmit simultaneously over this medium, the information is jammed, causing a *collision*. CSMA-CD protocols assume such collisions are detectable (CD stands for collision detection).

A special class of CSMA protocols, which demonstrates advantages in its relatively low collision ratio, includes the *virtual-time* CSMA protocols (VTCSMA [103]). In this class, each of the communicating nodes maintains two clocks: a real-time clock $T_P(t)$ and a virtual-time clock $T_V(t)$. The real-time clock runs continuously based on a monotonic algorithm. When the virtual clock runs, its speed $a_V(t)$ is higher than that of the real-time clock $a_P(t) \approx 1$. However, the virtual clock runs only when the communication medium is idle. Once the communication medium becomes busy, the virtual clock stops and is synchronized with the real-time clock when the communication medium becomes idle again. The use of these two clocks at each communicating node supports handling "send time" of messages according to a time based criterion.

Let us examine a protocol, called VTCSMA-L [136], which associates a deadline with each message. Let us denote the time in which a message $M$ must reach a destination $j$ by $d_M$ (the deadline of $M$). In addition, let us denote the message length (in time units) by $L_M$ and the estimated communication delay between sender $i$ and destination $j$ by $\widehat{\mu}_i^j$. Therefore, the latest time to start sending this message is

$$begin_{max}(M) = d_M - \widehat{\mu}_i^j - L_M \qquad (18.1)$$

The laxity of sending $M$ at time $t$ is then

$$x_+^M = d_M - \widehat{\mu}_i^j - L_M - t \qquad (18.2)$$

The VTCSMA-L protocol achieves a network-wide transmission policy of minimal-laxity-first, using the following principle. When the sender's virtual clock equals $begin_{max}(M)$, it sends the message $M$. Since $a_V(t) > a_P(t)$, the virtual clock shows "time to send has arrived" before the real-time clock, as long as the medium is idle. The ratio $a_V(t)/a_P(t)$ is a protocol parameter, whose value can be determined from the system peak load.

Let us now review the protocol's rules. Let "wait_queue" be a queue of messages waiting for transmission. We assume that the messages in the queue are ordered according to their current laxities (the lowest laxity at the front). This assumption is handled by the *insert* command in the algorithm. Let the medium detector show the value busy or the value idle, according to the medium status. We also assume that a collision detector works continuously. The following four rules implement this protocol.

1. *Upon request to transmit $(M, d_M)$* ::

   **begin**
   $\quad begin_{max}(M) \leftarrow d_M - \widehat{\mu}_i^j - L_M$ ;
   $\quad$ **if** (medium=idle) $\land$ $(T_V(now) \geq begin_{max}(M))$
   $\quad\quad$ **then** /* Immediate transmission */
   $\quad\quad\quad$ send$(M)$ ;
   $\quad\quad$ **else** /* Insert to the queue according to laxity */
   $\quad\quad\quad x_+^M \leftarrow begin_{max}(M) - T_P(now)$ ;
   $\quad\quad\quad$ insert $(M, x_+^M)$ ;
   $\quad$ **fi**
   **end**

2. *Upon detection of (medium=idle)* ::

   **begin** $\quad$ /* Reset virtual clock and reset indicator */
   $\quad T_V(now) \leftarrow \rho \leftarrow T_P(now)$ ;
   $\quad\quad$ /* Remove late messages which are lost */
   $\quad$ **for** $\forall i \in$ wait_queue **do**
   $\quad\quad$ **if** $begin_{max}(i) > T_P(now)$
   $\quad\quad\quad$ **then**
   $\quad\quad\quad\quad$ remove $i$ from wait_queue ;
   $\quad$ **od fi**
   **end**

3. *Upon collision detection while transmitting $M$* ::

   $\quad\quad$ /* Employ a random recovery to retransmit
   $\quad\quad\quad$ as a contention resolution rule,
   $\quad\quad\quad$ assuming there is still time to do it */
   **begin** $\quad$ /* – pick a new $begin_{max}(M)$ from interval, */
   $\quad begin_{max}(M) \leftarrow random(T_P(now), d_M - \widehat{\mu}_i^j - L_M)$ ;
   $\quad\quad$ /* – then, update laxity, */
   $\quad x_+^M \leftarrow begin_{max}(M) - T_P(now)$ ;
   $\quad\quad$ /* – then, insert to the queue properly. */
   $\quad$ insert $(M, x_+^M)$ ;
   **end**

4. *Virtual-clock ticks* ::

   **begin**
   $\quad$ **while** medium$\neq$busy **do**
   $\quad\quad T_V(now) \leftarrow a_V(now) \times (T_P(now) - \rho)$ ;
   $\quad\quad$ candidate $\leftarrow$ front(wait_queue) ;
   $\quad\quad$ **if** $begin_{max}(candidate) = T_V(now)$
   $\quad\quad\quad$ **then** /* Send by minimal-laxity-first */

                    remove *candidate* from wait_queue ;
                    send *candidate* ;
        **od fi**
    **end** □

The algorithm does not guarantee that messages will meet their deadlines because of inappropriate contention resolution. However, its solution employs a time-based reasoning which is equivalent to scheduling shared resources in minimal-laxity discipline. The proper tuning of the ratio $a_V(t)/a_P(t)$ can give "near optimal" results in systems with small load variance.

## 18.2  Synchronous Protocols

Some languages require synchronous communication protocols, enforcing a *rendezvous* of the sender and the receiver of each message [23, 66]. The receiver must wait for the message in order to receive it. The sender must wait for ensuring the correct arrival of the message at the receiver.

In the Ada[1] language [23], parallel processes are called *tasks*. Each task may have some *entries*, which are called from other tasks. Two tasks interact by first synchronizing, then by exchanging information and finally by continuing their individual activities. This synchronized meeting to exchange information is called *rendezvous*. Let us examine the influence of the way we implement this interaction on a system's real-time properties.

The functionality required from the interaction mechanism must be served both locally and remotely. Therefore, assuming there are kernel primitives that support this mechanism, we can require the following behavior from the primitives to support the rendezvous concept [53].

1. When a task invokes the kernel to interact with another task, the invoked primitive executes in the processor on which the calling task resides.
2. If the called task resides on the same processor as the calling one, all the required information concerning the called task is available.
3. If the called task resides on a different processor, the primitive which has been invoked sends a request to that processor. In this

---

[1]Ada is a registered trademark of the U. S. Department of Defense (Ada Joint Project Office).

request it specifies the calling processor, the called task, and the requested operation. The request service procedure at the remote site will then invoke the kernel primitive which corresponds to the requested operation, completing the interaction.

There are some implementations that support the above requirements. The comparison criteria can be minimizing system overhead and the task-blocking time.

"*Server*" rendezvous is the first implementation we consider. The calling task remains suspended until the called task executes the *accept* body. For this implementation, a single copy of the accept body is sufficient. This copy should be stored in the private memory of the accepting processor. In order to complete the rendezvous, the mechanism invokes the scheduler *twice* if the entry calling execution precedes the accept execution, and *three times* otherwise. We note that each scheduler invocation can result in a context switch. A remote invocation requires *one* or *two* inter-processor request signals, requiring *two* or *four* scheduling operations, respectively. Server rendezvous can carry out parameter passing through shared memory.

The second implementation we consider is the "*procedural call*" rendezvous. Here, the calling task always executes the accept body. There are two ways in which one can obtain accessibility of the accept body. The first is by keeping an exact copy of the accept body on each private memory of a processor that runs the calling task. The second is by storing the accept body in the shared memory. The shared memory solution resembles the "server" solution in communications cost. The replication solution can be ineffective or impossible if a resource needed for the accept body is only available to a particular processor. In this implementation, no special mechanism of parameter passing is needed, since the caller executes the accept body.

"*Order of arrival*" rendezvous [53], is an implementation that reduces the scheduling points required. Here, the accept body is a part of the execution of the last task which joins the rendezvous. In single processor systems, only *two* scheduling points are needed. In the case of tightly-coupled multiple processor system, *one* inter-processor request signal and *two* scheduling operations are needed to complete a rendezvous. However, the difficulties in resource allocation we have pointed out in the "procedural call" approach, exist here too.

The differences between the three rendezvous implementation approaches emphasize the significance of the compiler-level implementation. It has a significant effect on the temporal behavior of programs as well as on the communications load of a system.

However, in all these approaches the basic problems of guaranteeing timely message delivery are not resolved. Let us, therefore, add to the above mechanisms criteria for communication correctness. Due to the nature of synchronous communication, these criteria differ slightly from those of asynchronous communication. We note that both sender and receiver must agree on correctness before each continues its computation. We can state the correctness of synchronous communication in two conditions:

1. the sender and the receiver must agree on the correctness of the data transfer, and
2. the agreement must be reached within the respective deadlines.

Let us examine a system where a process $i$ initiates a rendezvous with process $j$, in order to transmit a message $M$ from $i$ to $j$. Each process has access to a local clock. Recalling the time provided from these clocks, as defined earlier

$$T_p(t) = a_p(t)C_p(t) + b_p(t), \; p = i, j \qquad (18.3)$$

we can define safeness of on-time delivery. Let us assume that the sender $i$ has a deadline $d_i^{(M)}$ to conclude on either success or failure of sending $M$ ($d_i^{(M)}$ is in real-time terms, not in local time). Respectively, $j$ has a deadline $d_j^{(M)}$ to conclude on the success of receiving $M$. Let us consider three protocols that take these deadlines into account[2].

The *clock drift* protocol compensates for the the clock drift and the estimated delays by constraining the send execution

$$begin_{max}(send_M) = (1 - \delta a_i)(d_j^{(M)} - \max(\mu_i^j) - L_M) - \delta b_i \qquad (18.4)$$

Note that $j$ must have advertised its deadline, or else $i$ cannot know $begin_{max}(send_M)$ that satisfies $j$. Any later start of sending $M$ cannot assure that $j$ will receive it before $d_j^{(M)}$.

The *clock rate* protocol uses an advertisement of a relative period that $j$ is willing to wait, say $\omega_j$. Let $T_i(0)$ be the time in which $i$ received $\omega_j$. Compensating for the clock rate differences and the delay constraints the send execution

$$begin_{max}(send_M) = \qquad (18.5)$$

$$T_i(0) - (1 + \delta a_i)(\max(\mu_i^j) + \max(\mu_j^i) + L_M) - \delta b_i + \max(\frac{a_i(t)}{a_j(t)})\omega_j$$

---

[2]These three protocols are examined in local time interpretations in [86].

Here one compensates for the delay of the advertisement, as well as for the clock differences.

The *last call* protocol carries out no advertisement. The receiver $j$ just sends a "last call" message to the sender $i$ notifying the sender the time after which satisfying $d_j^{(M)}$ cannot be assured. The receiver is therefore constrained by

$$begin_{max}(last\_call) = (1 - \delta a_j)(d_j^{(M)} - \max(\mu_i^j) - \max(\mu_j^i) - L_M) - \delta b_j \tag{18.6}$$

In the above three protocols, both the sender and the receiver can make provisions for exception handling after a deadline miss. This is also the case with these protocols when using local deadlines instead of the real-time ones [86]. On a deadline miss detection, a process invokes the exception handling, being unable to satisfy the two conditions of communication correctness. Let us note that the failure to meet the deadlines, although using these protocols, can still originate at a lower level protocol. For example, the use of the contention protocol at the the MAC level, can still create collisions that can fail the logical-link protocol time provisions.

## 18.3 Bounded Semantic Links

Another issue which relates to the data-link layer and its interfaces is clearly presented in a sender-receiver relation. The total order of "send precedes receive" is evident. However, consider the case in which the sender drops a message at time $t_1$, and the receiver only needs it at $t_2 > t_1$. If no other object is involved in this relation, the two objects must have valid thread executions simultaneously. On the other hand, if there exists a link object which overlaps both, there is no need to satisfy the simultaneity requirement.

This type of a link must be protected against failures in content and in size, namely a *semantic* protection. The semantic correctness must therefore be verified at drop time as well as at delivery time with a default value delivered in case of a failure in the semantic test.

The approach of using bounded semantic links can be found in various real-life systems. There are systems (e.g., [33]) in which messages are dropped at links without even specifying the receiver, and the object for which the message is intended and timely picks it up and reacts to it.

In an object-based architecture, objects relate to each other through time constrained invocations, whose temporal behavior we have examined. However, timing analysis is only one aspect. Let us now add to the invocation its logical requirements. We first examine the case of an invocation of a local object, and propose a passive link to support these logical requirements. We then examine the case of a remote invocation, and extend the link to an active one. We do it through the use of an *agent* object for the remote service. In local and remote invocations, the invoking object transfers a parameter block to the invoked object. We consider the parameter block as a typed message, defined by the proper service access point of the invoked object. In that sense, the use of one-way typed messages for invocations seems to suit this need. One way to implement such invocations at the invoking object program is [107]:

send < *typed_message* > to < *object* > at < *time_constraint* >

We note that the mechanism which supports these invocations needs to support additional tasks as well, as will be discussed.

### 18.3.1  Passive links

Semantic links establish the relations between objects. Each invocation relation has both temporal and logical properties. The temporal properties determine the duration in which the link is active. The logical properties determine the invocation type, based on the parameters transferred from the invoking object to the invoked one. Through these properties, the semantic link becomes the justification in our object-based architecture. It is located between the justifier and the justificand, pointed at by both and pointing to both. After the binding process initializes the context, these pointers allow a one-step addressing mode. The invoking object updates the link directly. The invoked object gets the parameters directly from the link.

Now, let us consider a case of an invoking object whose time constraint does not intersect the time constraint of the invoked one. One of the tasks of the semantic link is, therefore, to buffer the parameter block of this invocation. Figure 18.1 shows an example of the disjoint execution intervals of the invoking object ($TC_x$) and the invoked on ($TC_y$). The temporal relation between these computations is

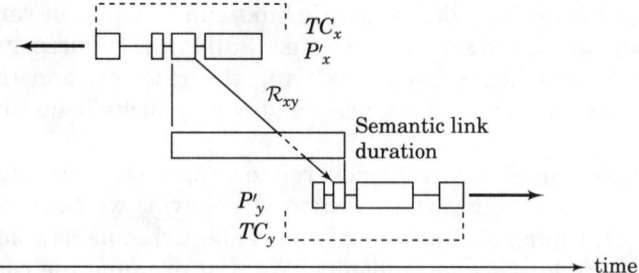

**Figure 18.1**  Duration of semantic link.

$\mathcal{R}_{xy}$. This relation involves the third subinterval of $P_x$ and the second subinterval of $P_y$. Therefore, the semantic link must cover the earliest possible $P_x$ (denoted $P'_x$) and the latest possible $P_y$ (denoted $P'_y$). Note that the temporal relation $\mathcal{R}_{xy}$ is a non-convex one, and specifies the required subintervals which the link serves[3].

The strong type checking in a semantic link is helpful in reducing exceptions on the invoked object side. Let us assume that upon link creation, the link contains a default value for each of its internal elements. The invoking object updates elements in the link only after they pass the type checks. Therefore, there is no need for end-to-end acknowledgements in this mechanism, an acknowledgement which requires the execution of the invoking and invoked objects to overlap.

The simplest way to implement such a local link is to use a passive buffer. The invoking object and the invoked one share the buffer. The invoking object updates the buffer (writer access) and the invoked object uses the buffer as its input parameter block (reader access).

### 18.3.2 Agents

Distributed computations require a stronger support in their semantic links. The changes are due to two major characteristics of distributed systems: communication requirements for remote relations, and system heterogeneity. However, all of the previous arguments still hold. Therefore, we consider the distributed support as a logical extension of the local semantic link, rather than a different type of link. The advantages of such an approach is in the coherent view the user can obtain for both local and remote invocations.

Although the user does not distinguish between local and remote invocations, the operating system does. In Chapter 1 we examined some remote service considerations, and presented the *agent* object. Since the agent resides on the invoking object's site, the invocation of the agent is treated locally. Thus, we achieve the required transparency of executing remote invocations, in the maintenance of this server representative. However, the need to deal with the remote execution and the communication issues is now the agent's responsibility. The agent must agree with the "parent" object on a feasible schedule for the invocation, and reserve the time in the remote object calendar. In addition, the agent must provide the required interface for communication and remote resource reservation. The communication

---

[3] In case we define relations between the convex windows of occurrence then the link duration is $TC_x \uplus TC_y$.

servers involved in this invocation must verify their feasibility accordingly. Each agent, therefore, has both the parent remote service and the communication servers as members of its dependency set.

We want the agent to play its role in interfacing remote servers while supporting the system heterogeneity, along with achieving the above transparency. Since different localities can have different interpretations to the same data, the agent must provide "on line" translation to the strongly typed communication. The agent, as a representative of the remote server in the local site, is certainly the proper choice for that translation.

In order to support the remote relations, the semantic link is therefore an active link and the agent object is the source of this active behavior. The invoking object invokes the agent with a local link. During the allocation binding phases, the agent verifies that both the communication servers and the remote service have the capability of performing the proper execution. We note here that since the semantic link is size-bounded and strongly typed, the parameter block transfer is bounded too.

## 18.4 Fault-Tolerant Real-Time Communication

In fault-tolerant real-time communication, the reliability of the data transfer is accomplished at the link level as well as the transport level. At the link layer, the protocol supports recovery from failures in transferring a single block over a single link. At the transport layer, the protocol supports end-to-end recovery from failures.

The synchronous protocols impose difficulties in recovery strategies, since they require the sender and receiver threads to execute simultaneously for both the regular communication session and the recovery from a failure. Implementation of such a scheme at the data-link level imposes six or more scheduling constraints: on sender, link and receiver, two or more sessions (regular and recoveries) for each. Asynchronous protocols are easier since they are released from the requirement of simultaneity.

The scheme of agents acting on behalf of the application allows application-transparent enhancement of the communication session. Under this scheme, the sender agent can fork the transmission and carry it out obtaining either temporal (retries on the same link) or physical (transmit via different links) redundancy. Clearly, the receiving agent is coordinated to join the incoming transmissions. Furthermore, semantic integrity tests can be performed by the agent, since it is an independent process. The receiving agent can carry out an agreement protocol of some type to overcome the required type of behavior of faults. Such a protocol may include the insertion of default values for cases a quorum is not obtained within the imposed time constraint.

## 18.5 Concluding Remarks

The importance and criticality of the communication subsystem in the implementation of fault tolerance for real-time systems, has been discussed in this chapter. However, the common communication protocols used today do not demonstrate the requirements which are needed by real-time applications, even without the fault tolerance objectives. Therefore, our presentation of the issue has been directed to the system architects, since they do not have off-the-shelf solutions available today.

It is essential that the appropriate communication mechanisms be provided in any distributed system. When such system is to be used for real-time applications, severe constraints must be placed on the design and the implementation of the communication system. We have proposed the mechanisms of the semantic link and the agent to provide the communication support in a distributed heterogeneous environment.

# Chapter 19

# Fault-Tolerant Time Services

In a comprehensive approach to the construction of a fault-tolerant time-server, one has to start with providing the means for a local resynchronization of each time server in the system. This resynchronization updates the server's parameters, and thus the interpretation of the local clock. However, one must introduce additional facilities, in the form of clock broadcasts and participant-forum establishment. This additional support satisfied the requirements for fault tolerance. Combining these facilities results in a comprehensive solution of a system-wide distributed time service. Let us consider an example of such facilities: the local resynchronization, the distributed clock-exchanges, and the construction of a complete service [83].

## 19.1 Algorithm I: Local Resynchronization

Consider using the following algorithm to resynchronize the time-server clock $T_p(t)$, constructed from the clock $C_p(t)$ by the linear combination $a_p(t)C_p(t) + b_p(t)$. The resynchronization algorithm updates the coefficients of the linear combination while maintaining a monotonic behavior of $T_p(t)$. In this mechanism, we can regard $a_p(t)$ and $b_p(t)$ as the interpretation of $p$-clock $C_p(t)$ by a time-server clock $T_p(t)$. The update is done according to an "ideal" clock $C_p^{(i)}(t)$, an improved version of $C_p(t)$, generated and maintained according to the protocols described later in this chapter. Updating $a_p(t)$ is really changing the pace of $T_p(t)$. Let us define formally this interpretation.

- Let $T^{(0)}, T^{(1)}, \cdots$ be an unbounded increasing sequence of times such that
$$\forall j : T^{(j+1)} - T^{(j)} \leq J \quad (19.1)$$

In other words, we resynchronize at least every $J$ time units.

- Let $C_p^{(0)}, C_p^{(1)}, \ldots$ be a sequence of $p$-clocks.
- For $i > 0$ let $t_p^{(i)}$ be the Universal Time, such that $C_p^{(i-1)}(t_p^{(i)}) = T^{(i)}$.
- Let $t_p^{(0)}$ be the Universal Time, such that $C_p^{(0)}(t_p^{(0)}) = T^{(0)}$.

The service clock $T_p$, for $t \geq t_p^{(0)}$, is calculated by

$$T_p(t) = a_p(t)C_p(t) + b_p(t) \qquad (19.2)$$

where $a_p(t)$ and $b_p(t)$ are coefficients defined as:

- $t_p^{(0)} \leq t \leq t_p^{(1)}$ (initial coefficients):

$$a_p(t) = 1, b_p(t) = T^{(0)} - C_p(t_p^{(0)})$$

- $t_p^{(i)} < t \leq t_p^{(i+1)}, i > 0$ (coefficients update):

$$a_p(t) = 1 + \frac{C_p^{(i)}(t_p^{(i)}) - T_p(t_p^{(i)})}{J} \qquad (19.3)$$

$$b_p(t) = b_p(t_p^{(i-1)}) + (a_p(t_p^{(i-1)}) - a_p(t_p^{(i)}))C_p^{(i)}(t_p^{(i)}) \qquad (19.4)$$

This algorithm shows a bounded error of the time service clock $T_p(t)$, if the "ideal clock" it is using ($C_p^{(i)}(t)$) has a bounded correction property [83]. In other words, if $C_p^{(i)}(t)$ satisfies
$\forall j, k$ with $j < k$ and $t_p^{(k)} - t_p^{(j)} < J$:

$$\sum_{i=j}^{k-1} |C_p^{(i+1)} - C_p^{(i)}| < \sigma_p, \qquad (19.5)$$

then for $t_p^{(i)} \leq C_p^{(i)}(t) \leq t_p^{(i+1)}$

$$|C_p^{(i)}(t) - T_p(t)| < \frac{e}{e-1}\sigma_p \qquad (19.6)$$

## 19.2 Example

Let us consider the case of an inaccurate $p$-clock $C_p(t)$ whose rate at $t = t_p^{(0)}$ is $c$. $C_p(t)$ drifts from this rate linearly. Thus, it generates the clock rate

$$\frac{d}{dt}C_p(t) = c + \rho(t - t_p^{(0)}) \qquad (19.7)$$

where $c$ and $\rho$ are constants. Using this clock-drift model, let us examine the behavior of this resynchronization algorithm. We assume in our example that the "ideal" clock $C_p^{(i)}(t)$ is the universal time. In other words, we assume a perfect correction of an imperfect clock, and examine the resynchronization behavior.

The algorithm starts with

$$lla(t) = 1$$
$$b(t) = T_p^{(0)} - C_p(t_p^{(0)}) \tag{19.8}$$

and holds these values for $t_p^{(0)} < t \leq t_p^{(1)}$. With these assumptions we have continuity with respect to the initial conditions $T_p(t_p^{(0)}) = T_p^{(0)}$.

Let us now consider the behavior of the clock error between resynchronizations. During the period that precedes the first resynchronization at $t_p^{(1)}$, the local clock drifts from correctness. Just prior to the resynchronization, it shows

$$C_p(t_p^{(1)}) = c(t_p^{(1)} - t_p^{(0)}) + \frac{1}{2}\rho(t_p^{(1)} - t_p^{(0)})^2 + C_p(t_p^{(0)}) \tag{19.9}$$

The server at that time provides

$$llT_p^{(1)} = 1 \cdot (c(t_p^{(1)} - t_p^{(0)}) + \frac{1}{2}\rho(t_p^{(1)} - t_p^{(0)})^2 + C_p(t_p^{(0)}))$$
$$+ (t_p^{(0)} - C_p(t_p^{(0)}))$$
$$= t_p^{(1)} + E(t_p^{(1)}) \tag{19.10}$$

The error term $E(t_p^{(1)})$ can be expressed as

$$E(t_p^{(1)}) = (c-1) \cdot (t_p^{(1)} - t_p^{(0)}) + \frac{1}{2}\rho(t_p^{(1)} - t_p^{(0)})^2 \tag{19.11}$$

This error includes both the linear and the quadratic terms that result from our drift model.

The first update of coefficients occurs at $t_p^{(1)}$. Having assumed a perfect clock correction, we have $C_p^{(i)}(t_p^{(1)}) = t_p^{(1)}$. Then,

$$lla(t) = 1 + \frac{t_p^{(1)} - T_p^{(1)}}{J} = 1 + \frac{E(t_p^{(1)})}{J} \tag{19.12}$$

$$b(t) = T_p^{(0)} - C_p(t_p^{(0)}) - \frac{E(t_p^{(1)})}{J} t_p^{(1)} \tag{19.13}$$

Notice that the correction of $a(t)$ compensates the clock-rate inaccuracy in the following way: if no update occurs in the following $J$ time units, then

$$T_p(t_p^{(1)} + J) = C_p^{(i)}(t_p^{(1)}) + J \tag{19.14}$$

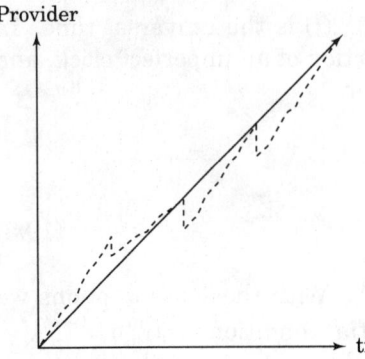

**Figure 19.1** Example results of resynchronization algorithm.

Substituting the error term in the new values of the coefficients, yields

$$lla(t) = 1 + \frac{(c-1)\cdot(t_p^{(1)} - t_p^{(0)}) + \frac{1}{2}\rho(t_p^{(1)} - t_p^{(0)})^2}{J} \tag{19.15}$$

$$b(t) = T_p^{(0)} - C_p(t_p^{(0)}) - \frac{(c-1)\cdot(t_p^{(1)} - t_p^{(0)}) + \frac{1}{2}\rho(t_p^{(1)} - t_p^{(0)})^2}{J} t_p^{(1)} \tag{19.16}$$

for $t_p^{(1)} < t \leq t_p^{(2)}$. We note that with $t_p^{(1)} = t_p^{(0)} + J$ and $\rho = 0$, we have $a(t) = c$.

Figure 19.1 describes a particular[1] implementation of the above resynchronization example. We note the property of resynchronizing is such that the equality to the ideal clock (the diagonal in this figure) is obtained $J$ time units after the resynchronization.

## 19.3 Algorithm II: Byzantine Clock Broadcast

Let us now consider the construction of a local "ideal clock" reference for a synchronizing process $p$, $C_p^{(i)}$. Our computations are modeled as a set of processes $\mathcal{V}_p = \{p_1, \ldots, p_n\}$ that are related to each other through a set of logical links $\mathcal{E}_p$, to form a graph

$$\mathcal{G}_p = (\mathcal{V}_p, \mathcal{E}_p)$$

A set of processors $\mathcal{V}_P = \{P_1, \ldots, P_m\}$ are related to each other through a set of physical links $\mathcal{E}_P$, to form a graph

$$\mathcal{G}_P = (\mathcal{V}_P, \mathcal{E}_P)$$

---

[1] In this example resynchronization takes place at the same time we update the local clock, exactly every $J$ time units. The ideal clock is the real-time, denoted as the diagonal in the figure. Further resynchronization cycles are identical to the last two presented in the figure.

The concept described in this section, presents the idea of *time providers* as a preferable alternative to time servers. Members of the intersection of the sets $\mathcal{E}_p$ and $\mathcal{E}_P$ are the paths through which we can synchronize $\mathcal{V}_p$. Time providers send information to synchronizing processes through these paths. The providers employ a broadcast discipline to distribute this information. Receiving processes derive from this broadcast, if it is adequate, the necessary parameters to construct locally their "ideal clock."

Let $\Gamma_p$ represent the set of $p$-clocks received by a synchronizing process $p$. Each $\Gamma_p$ is of course initially empty, having received no messages. Time providers broadcast the information for these $p$-clocks to their neighbors. Each broadcast contains an interval out of which the ideal $p$-clock are derived. We denote the left end point of a time interval $U$ by $\triangleleft(U)$, and the right end point by $\triangleright(U)$. In addition to the interval, each broadcast contains information about the path throughout which the message traveled so far.

A time provider $j$ broadcasts to every $p$-clock, using the following restricted diffusion for the broadcast:

1. $j$ sends an interval $U^{(j)}$ to all its neighbors. Let $U_p^{(j)}$ denote $U^{(j)}$ as obtained by process $p$. We show later (Algorithm III) how $j$ constructs[2] the set of $U_p^{(j)}$ it broadcasts.

2. Let $p$ receive a broadcast with an interval $R$ along a path $\pi$ at time $t$. After the reception, $p$ sets the $p$-clock $I_p^\pi$ to a value $R^\pi$ at time $t$. Since $I_p^\pi$ is a $p$-clock, its pace equals that of the local clock. Then, $p$ adds $I_p^\pi$ to $\Gamma_p$. After updating its own $\Gamma_p$, $p$ relays $R$ to each of its neighbors (say $q$), unless one of the following restrictions holds.

    - $q$ is on the path $\pi$, or
    - $\exists U \in \Gamma_p : \triangleleft(U) \geq \triangleleft(R_p^\pi) \bigwedge \exists V \in \Gamma_p : \triangleright(V) \geq \triangleright(I_p^\pi)$, or
    - the length of $\pi$ exceeds a given $k$.

    The first restriction prevents unnecessary duplication. The second prevents sending intervals which will not add to what is already known. The third restriction eliminates outdated information.

3. The above steps are repeated until a time after which no more useful messages can arrive. Then, $p$ sets $UT_p^{(j)}$ to be the $p$-clock whose $\triangleleft$ and $\triangleright$ are the left-most and the right-most maxima of all the $p$-clocks $I_p^\pi$.

---

[2]If both $j$ and $p$ are nonfaulty, $U_p^{(j)}(t_p^{(i)})$ contains the "universal time" of $t_p^{(i)}$.

## 19.4 Algorithm III: Complete Time-Service

Now we can construct the complete time server using the previous two algorithms for resynchronization and broadcast. The following time-service algorithm starts with an agreement between the participants on the set of providers they are going to choose. Then, once this set is established, a sequence that constructs the time service follows. The providers broadcast and the receivers accept, forward, filter faults, and resynchronize.

1. The members $p \in \Pi_i$ use an agreement protocol (e.g., [78, 81]) to choose a set of time providers $\{j_i\}$, numbered from 1 to $m$. Let the following assumptions describe the clock synchronization environment.
   - Let $\tau$ be the delay of the broadcast algorithm.
   - Let $UT_p^{(j_i)}$ denote an interval for a $p$-clock broadcast by $j_i$.
   - Let $\chi$ be a bound on the provided interval, such that $\chi \geq \frac{1}{2} \|UT_p^{(j_i)}\|$.
   - Let $\rho$ denote an upper bound on the provider's clock drift-rate, $\rho \geq |1 - \frac{d}{dt}C_{j_i}(t)|$.
   - Let $H$ be the minimal time between consecutive broadcasts of $UT^{(j_i)}$ by all $j_i$'s.
   - Let $k$ be the maximal path length allowed for the broadcast algorithm (Algorithm II).

2. For a sequence of *predetermined* times $T^{(i)}$ and time providers $j_i$:
   a) Provider $j_i$ executes Algorithm II
      - when $\triangleright(UT^{(j)}) = T^{(i)} - \tau - k(\chi + \rho H)$.
      
      The algorithm uses a $U^{(j_i)}$ that is equal to the current value of $UT^{(j_i)}$ in order to broadcast an interval for a $p$-clock $UT_p^{(j_i)}$ to every $p \in \Pi_i$.
   b) Each $p \in \Pi_i$ screens the faulty intervals out of the $m$ it should have received with a defined function $A^f$. It uses the non-faulty intervals to set an "assumed to be correct" $p$-clock
   
   $$C_p^{(i)} \leftarrow A^f(UT_p^{(1)}, .., UT_p^{(m)})$$
   
   For example [83], $A^f(UT_p^{(1)}, .., UT_p^{(m)})$ can be the average of the multiset of $m - 2f$ numbers obtained by taking the *midpoints* of all the $UT_p^{(i)}$ ($i = 1, 2, \ldots, m$), and omitting the $f$ lowest and $f$ highest of them.
   c) Each $p$ uses Algorithm I (resynchronization) to compute $T_p$.

## 19.5 Achievement of Fault Tolerance

There are two major contributors to the achievement of fault tolerance objectives in this algorithm. The first is the agreement on the forum set $\{j_i\}$, a subset of $\Pi_i$, out of which we synchronize the local time provider. The second is the selection of the sufficient quorum, through $A^f(UT_p^{(1)},..,UT_p^{(m)})$, which we use to synchronize the $p$-clock. However, these two contributors are a large factor in the cost involved in executing the algorithm.

The larger the forum $\{j_i\}$ is, the smaller an effect of a single faulty provider becomes. We note that in these cases we can use a larger quorum by increasing $m$. Thus, the effect of a faulty provider is reduced by the averaging process. However, a large forum increases the communication traffic significantly, especially in a broadcast discipline.

The quorum selection function, $A^f(UT_p^{(1)},..,UT_p^{(m)})$, influences the fault tolerant properties in some ways. The criterion of selection characterizes some of these properties. Let us consider the selection chosen in the previous section in omitting the $f$ lowest and $f$ highest of the $m$ $UT_p^{(i)}$'s. This selection[3] assumes a symmetric distribution of the erroneous providers. In that case, we can ignore the lowest and highest $UT_p^{(i)}$'s, since they are expected to be symmetrically distributed. In case of a strongly biased population, this selection process performs poorly. For example, when all the faulty providers (say $2f$) are low, half of them are considered and $f$ correct providers are ignored.

The parameters $m$ and $f$ also significantly influence the characteristics of the quorum selection function. We note that when we increase $m$, we can increase $f$ and still stay with an averaging population of the same size. Hence, an additional robustness can be derived from increasing these parameters, but this increase causes a higher communication cost. Furthermore, a larger quorum may require a longer time to collect enough broadcasts from the forum members.

---

[3]The selection in the algorithm [83] has been chosen mainly due to its simplicity.

# Part 5

# Epilog

Part 5

Epilog

# Chapter 20

# Conclusion

Design of fault-tolerant systems poses many new and interesting challenges for the system designers. It is essential that the designer maintain a comprehensive view of multiple aspects and disciplines. Such designs require that not only the proper functioning of the system form the basis of the design but also that the malfunctioning of the system be addressed. We believe that it is only through such considerations that the reliability and safety of systems of tomorrow can reach the required levels of expectation.

The topics covered in this book are based on our experience in the industrial and academic worlds in study and design of such systems. The topics we, and our colleagues, required in our efforts are addressed in this book. There is no detailed discussion of theoretical concepts which a reader may wish to explore by going to the vast literature that exists and some of which is referenced here. Many of the techniques and algorithms in the literature are refinements of those presented here. Note that a majority of papers published in this area are excellent presentations of a well studied local problem. While a system designer must be familiar with the local problems in many areas, the comprehensiveness of his view is critical for his success.

We believe that the complexity of the systems will continue to increase as more and more systems begin using the capabilities of computers in their operations. As our reliance on such systems also increases, their reliability and safety become crucial requirements. New techniques for understanding the fault phenomenon and its relationships to failures are necessary. A direction for some of these have been presented in this book.

## Chapter 20

# Conclusion

# Bibliography

[1] Arora, R., S. Rana, and M. Gupta, "Distributed Termination Detection Algorithm for Distributed Computations," *Inf. Proc. Letters*, vol. 22, no. 6, pp. 311–314 (May 1986).

[2] Boksenbaum, C., M. Cart, J. Ferrie, and J. F. Pons, "Concurrent Certification in Distributed Database Systems," *RR No 13, CRIM*, Université de Montpellier (September 1984).

[3] Carvalho, O. and G. Roucairol, "On Mutual Exclusion in Computer Networks," *Comm of ACM*, vol. 26, no. 2, pp. 146–147 (February 1983).

[4] Chang, E. J. and R. Roberts, "An Improved Algorithm for Decentralized Extrema-Finding in Circular Configurations of Processors," *Comm of ACM*, vol. 22, no. 5, pp. 281–283 (May 1979).

[5] Chandy, K. M., J. Misra, and L. Haas, "Distributed Deadlock Detection," *ACM TOCS*, vol. 1, no. 2, pp. 144–156 (May 1983).

[6] Chandy, K. M. and J. Misra, "The Drinking Philosophers Problem," *ACM TOPLAS*, vol. 6, no. 4, pp. 632–646 (October 1984).

[7] Dijkstra, E. and C. Scholten, "Termination Detection for Diffusing Computation," *Inf. Proc. Letters*, vol. 11, no. 1, pp. 1–4 (August 1980).

[8] Dijkstra, E., W. Feijen, and A. Van Gasteren, "Derivation of a Termination Detection Algorithm for Distributed Computation," *Inf. Proc. Letters*, vol. 16, pp. 217–219 (June 1983).

[9] Dolev, D., M. Klawe, and M. Rodeh, "An O(n log n) Undirectional Distributed Algorithm for Extrema Finding in a Circle," *Journal of Algorithms*, vol. 3, pp. 245–260 (1982).

[10] Hirschberg, D. S. and J. B. Sinclair, "Decentralized Extrema Finding in Circular Configuration of Processors," *Comm of ACM*, vol. 23, no. 11, pp. 627–628 (November 1980).

[11] Lomet, D. B., "A Practical Deadlock Avoidance Algorithm for Database Systems," *Proc. ACM SIGMOD Conference on Management of Data*, pp. 122–127 (1977).

[12] Lomet, D. B., "Coping with Deadlock in Distributed Systems," *Database Architecture*, pp. 95–105, North Holland (1979).

[13] Madduri, H. and R. Finkel, "Extension of the Banker's Algorithm for Resource Allocation in a Distributed Operating System," *Inf. Proc. Letters*, vol. 19, no. 1, pp. 1–8 (July 1984).

[14] Maekawa, M., "A $\sqrt{N}$ Algorithm for Mutual Exclusion in Decentralized Systems," *ACM TOCS*, vol. 3, no. 2, pp. 145–159 (May 1985).

[15] Misra, J. and K. M. Chandy, "Termination Detection of Diffusing Computation in CSP," *ACM TOPLAS*, vol. 4, no. 1, pp. 37–43 (January 1982).

[16] Misra, J., "Detecting Termination of Distributed Computations Using Markers," *Proc. of 2nd Conference on Principles of Distributed Computing*, pp. 290–294, Montreal (August 1983).

[17] Murphy, S. L. and A. U. Shankar, "A Note on the Drinking Philosophers Problem," *ACM TOPLAS*, vol. 10, no. 1, pp. 178–188 (January 1988).

[18] Rana, S. P., "A Distributed Solution of the Distributed Termination Problem," *Inf. Proc. Letters*, vol. 17, pp. 43–46, North Holland (July 1983).

[19] Rosenkrantz, D. J., R. E. Stearns, and P. M. Lewis, "System Level Concurrency Control in Distributed Database," *ACM TODS*, vol. 3, no. 2, pp. 178–198 (June 1978).

[20] Shavit, N. and N. Francez, "A New Approach to Detection of Locally Indicative Stability," *Proceedings of ICALP*, ed. L. Kott, pp. 344–358, Rennes, France (July 1986).

[21] Suzuki, I. and T. Kasami, "An Optimality Theory for Mutual Exclusion Algorithms in Computer Networks," *Proc. of 3rd Int. Conference on Distributed Computing*, pp. 365–370, Miami (October 1982).

[22] Topor, R., "Termination Detection for Distributed Computation," *Inf. Proc. Letters*, vol. 18, pp. 33–36 (January 1984).

[23] "Reference Manual for The Ada Programming Language," U.S. DOD (ANSI) MIL-STD 1815a-1983 (February 1983).

[24] Agrawala, A. K. and S.-T. Levi, "Objects Architecture for Real-Time, Distributed, Fault Tolerant Operating Systems," *IEEE Workshop on Real-Time Operating Systems*, pp. 142–148, Cambridge, MA (July 1987).

[25] Allchin, J. E. and M. S. McKendry, "Synchronization and Recovery of Actions," *2nd ACM SIGACT-SIGOPS Symposium on Principles of Distributed Computing*, Montreal (August 17–19, 1983).

[26] Apt, K., N. Francez, and W. De Roever, "A Proof System for Communicating Sequential Processes," *ACM TOPLAS*, vol. 2, no. 3, pp. 359–385 (July 1980).

[27] Arlat, J., M. Agura, L. Amat, Y. Crouzet, J. Fabre, J. Laprie, E. Martins, and D. Powel, "Fault Injection for Dependability Validation," *IEEE TOSE*, vol. 16, no. 2, pp. 166–182 (February 1990).

[28] Avizienis, A., "The N-Version Approach to Fault Tolerant Software," *IEEE TOSE*, vol. 20, no. 1, pp. 1491–1501 (December 1985).

[29] Barton, J., E. Czeck, Z. Segall, and D. Sweviorek, "Fault Injection Experiments Using FIAT," *IEEE TOC*, vol. 39, no. 4, pp. 575–582 (April 1990).

[30] Ben-Ari, M., *Principles of Concurrent Programming*, Prentice-Hall International, Englewood Cliffs, NJ (1982).

[31] Birman, K. P., T. A. Joseph, T. Raeuchle, and A. El-Abbadi, "Implementing Fault Tolerant Distributed Objects," *IEEE TOSE*, vol. SE-11, no. 6, pp. 502–508 (June 1985).

[32] Birman, K. P. and T. A. Joseph, "Reliable Communication in an Unreliable Environment," TR85-694, Department of Computer Science, Cornell University, Ithaca, NY (July 1985).

[33] Boasson, Maarten, "Architecture, Software and Complexity," *IEEE International Conference on Computer Systems and Software Engineering, CompEuro 90*, pp. 92–96, Tel Aviv (May 8–10, 1990).

[34] Bochmann, G., *Distributed Systems Design*, Springer-Verlag, Berlin, Germany (1983).

[35] Booch, G., Object-Oriented Development," *IEEE TOSE*, vol. SE-12, no. 2, pp. 211–221 (February 1986).

[36] Brinch, Hansen P., *Operating Systems Principles*, Prentice Hall, Englewood Cliffs, NJ (1973).

[37] Brinch, Hansen P., *The Architecture of Concurrent Programs*, Prentice Hall, Englewood Cliffs, NJ (1973).

[38] Chandy, K. M., J. C. Browne, C. W. Dissly, and W. R. Uhrig, "Analytic Models for Rollback and Recovery Strategies in Data Base Systems," *IEEE TOSE*, vol. SE-1, no. 1, pp. 100–110 (March 1975).

[39] Charniak, E. and D. McDermott, *Introduction to Artificial Intelligence*, Addison-Wesley Publishing Company, Reading, MA (1985).

[40] Chu, W. W. and L. M-T. Lan, "Task Allocation and Precedence Relations for Distributed Real-Time Systems," *IEEE TOC*, vol. C-36, no. 6, pp. 667–679 (August 1987).

[41] Cooper, E. C., "Replicated Procedure Call," *ACM Operating Systems Review*, vol. 20, no. 1, pp. 44–55 (January 1986).

[42] Cristian, F., "Exception Handling and Software Fault Tolerance," *IEEE TOC*, vol. C-31, no. 6, pp. 531–540 (June 1982).

[43] Cristian, F., "Correct and Robust Programs," *IEEE TOSE*, vol. SE-10, no. 2, pp. 163–174 (March 1984).

[44] Cristian, F., H. Aghili, R. Strong, and D. Dolev, "Atomic Broadcast: from Simple Message Diffusion to Byzantine Agreement," *IEEE Fifteenth Fault Tolerance Computing Symposium (FTCS)*, pp. 200–206 (1985).

[45] Davidson, S. B., H. Gracia-Molina, and D. Skeen, "Consistency in Partitioned Networks," *Computing Surveys*, vol. 17, no. 3, pp. 341–370 (September 1985).

[46] De Marco, T., *Structured Analysis and System Specification*, Yourdon Press, New York, NY (1978).

[47] Dijkstra, E. W., *Cooperating Sequential Processes*, ed. F. Genuys, *Programming Languages*, pp. 43–112, Academic Press, New York, NY (1968).

[48] Dijkstra, E. W., *A Discipline of Programming*, Prentice Hall, Englewood Cliffs, NJ (1976).

[49] Dolev, D., N. A. Lynch, S. S. Pinter, E. W. Stark, and W. E. Weihl, "Reaching Approximate Agreement in the Presence of Faults," *Journal of ACM*, vol. 33, no. 3, pp. 499–516 (July 1986).

[50] El Abbadi, A., D. Skeen, and F. Cristian, "An Efficient Fault Tolerant Algorithm for Replicated Data Management," *ACM Symposium on Principles of Database Systems*, pp. 215–229, Portland, OR (March 1985).

[51] Fekete, A. D., "Asynchronous Approximate Agreement," *The Sixth Annual ACM Symposium on Principles of Distributed Computing*, pp. 64–76, Vancouver, Canada (August 1987).

[52] Ferguson, D., G. Kar, G. Leitner, and C. Nikolaou, "Relocating Processes in Distributed Computer Systems," *IEEE Proceedings of the Fifth Symposium on Reliability in Distributed Software and Database Systems*, pp. 171–177, Los Angeles, CA (January 1986)

[53] Garetti, P., P. Laface, and S. Rivoira, "Multiprocessor Implementation of Tasking Facilities in Ada," *Proceedings of the 12th IFAC/IFIP Workshop of Real Time Programming*, pp. 97–102, Hatfield, UK (March 29–31, 1983).

[54] Gelenbe, E. and D. Derochette, "Performance of Rollback Recovery Systems under Intermittant Failures," *Communication of the ACM*, vol. 21, no. 6, pp. 493–499 (June 1978).

[55] Gelenbe, E., "On The Optimum Checkpoint Interval," *ACM Journal*, vol. 26, no. 2, pp. 259–270 (April 1979).

[56] Gelenbe, E., D. Finkel, and S. K. Tripathi, "Availability of a Distributed Computer System with Failures," *Acta Informatica*, vol. 23, pp. 643–655 (1986).

[57] Gifford, D. K., "Weighted Voting for Replicated Data," *ACM Operating Systems Review*, vol. 13, no. 5, pp. 150–162 (December 1979).

[58] Ginat, D., D. D. Sleator, and R. E. Tarjan, "A Tight Amortized Bound for Path Reversal," *Inf. Proc. Letters*, vol. 31, no. 1, pp. 3–5 (April 1989).

[59] Goldman, K. J. and N. A. Lynch, "Quorum Consensus in Nested Transaction Systems," *The Sixth Annual ACM Symposium on Principles of Distributed Computing*, pp. 27–41, Vancouver, Canada (August 1987).

[60] Haberman, A. N., "Synchronization of Communicating Processes," *Communication of the ACM*, vol. 15, no. 3, pp. 171–176 (March 1972).

[61] Harary, F., *Graph Theory*, Addison-Wesley Publishing Company, Reading, MA (1969).

[62] Harel, D. and A. Pnueli, "On The Development of Reactive Sysytems," ed. K. Apt, Logics and Models of Concurrent Systems, pp. 477–498, Springer-Verlag, Heidelberg, Germany (1985), Weitzman Institute of Science, Rehovot, Israel (January 1985).

[63] Herlihy, M. P., "A Quorum Consensus Replication Method for Abstract Data Types," *ACM TOCS*, vol. 4, no. 1 (February 1986).

[64] Herlihy, M. P., "Optimistic Concurrency Control for Abstract Data Types," *ACM SIGOPS*, vol. 21, no. 2, pp. 33–44 (April 1987).

[65] Hoare, C. A. R., "Monitors: An Operating System Structuring Concept," *Communications of the ACM*, vol. 17, no. 10, pp. 549–557 (October 1974).

[66] Hoare, C. A. R., "Communicating Sequential Processes," *Communications of the ACM*, vol. 21, no. 8, pp. 666–677 (August 1978).

[67] Howard, J. H., "Proving Monitors," *Communication of the ACM*, vol. 19, no. 5, pp. 273–285 (March 1976).

[68] *80960 Designer's Reference Manual*, Intel Corporation, Santa Clara, CA (1988).

[69] ISO 9506, Manufacturing Messaging Specification, Semaphore Management Services, Clause 13, pp. 229–234 (December 1987).

[70] Jain, B.N. and A. K. Agrawala, *Open Systems Interconnection: Its Architecture and Protocols*, rev. ed. McGraw Hill, New York (1993).

[71] Johnson, B. W., *Design and Analysis of Fault Tolerant Digital Systems*, Addison-Wesley Publishing Company, Reading, MA (1989).

[72] Joseph, T. A. and K. P. Birman, "Low Cost Management of Replicated Data in Fault Tolerant Distributed Systems," *ACM TOCS*, vol. 4, no. 1, pp. 54–70 (February 1986).

[73] Kar, G., C. Nikolaou, and J. Reif, "Assigning Processes to Processors: A Fault Tolerant Approach," *Proceedings of 14th International Conference on Fault Tolerant Computing Systems (FTCS)*, pp. 306–309, Kissimmee, FL (June 1984).

[74] Karp, R. M., "Reducibility Among Combinatorial Problems," eds. R. E. Miller and J. W. Thatcher, in Complexity of Computer Computations, pp. 85–104, Plenum Press, New York (1972).

[75] Katzman, J. A., *A Fault Tolerant Computing System*, Tandem Computers Inc. (1977).

[76] Kleinrock, L., *Queueing Systems*, John Wiley and Sons, New York (1975).

[77] Knuth, D. E., *The Art of Computer Programming*, Addison-Wesley Publishing Company, Reading, MA (1973).

[78] Lakshman, T. V. and A. K. Agrawala, "Efficient Decentralized Consensus Protocols," *IEEE TOSE*, vol. SE-12, no. 5, pp. 600–607 (May 1986).

[79] Lamport, L., "Time, Clocks and Ordering of Events in a Distributed System," *Communications of the ACM*, vol. 21, no. 7, pp. 558–565 (July 1978).

[80] Lamport, L., R. Shostak, and M. Pease, "The Byzantine Generals Problem," *ACM TOPLAS*, vol. 4, no. 3, pp. 382–401 (July 1982).

[81] Lamport, L., "Using Time instead of Timeout for Fault-Tolerant Distributed System," *ACM TOPLAS*, vol. 6, no. 2, pp. 254–280 (April 1984).

[82] Lamport, L., "On Interprocess Communication," parts I and II, *Distributed Computing*, vol. 1, no. 2, pp. 77–101, Springer-Verlag (1986).

[83] Lamport, L., "Synchronizing Time Servers," SRC report no. 18, December SRC, Palo Alto, CA (June 1987).

[84] Lampson, B. and H. Sturgis, "Crash Recovery in a Distributed Data Storage System," Technical Report, Computer Science Lab, Xerox PARC, Palo Alto, CA (1976).

[85] Laprie, J. C., J. Arlat, C. Béounes, and K. Kanoun, "Definition and Analysis of Hardware and Software Fault Tolerant Architectures," *IEEE Computer*, vol. 23, no. 7, pp. 39–51 (July 1990).

[86] Lee, I. and S. B. Davidson, "Adding Time to Synchronous Process Communications," *IEEE TOC*, vol. C-36, no. 8, pp. 941–948 (August 1987).

[87] LeLann, G., "Issues in Fault-Tolerant Real-Time Local Area Network," *IEEE Proceedings of the Fifth Symposium on Reliability in Distributed Software and Database Systems*, pp. 28–32, Los Angeles, CA (January 1986).

[88] Levenson, N., "Software Safety: What, Why, and How," *ACM Computing Surveys*, vol. 18, no. 2, pp. 125–164 (June 1986).

[89] Levi, S.-T. and B. D. Plateau, "A Distributed Algorithm for Deadlock and Termination Detection of Distributed Computations," *The Second International Symposium on Computer and Information Science*, ed. Erol Gelenbe, pp. 111–133, Istanbul (October 19–21, 1987).

[90] Levi, S.-T. and A. K. Agrawala, "Objects Architecture: A Comprehensive Design Approach for Real-Time, Distributed, Fault-Tolerant, Reactive Operating Systems," CS-TR-1915, Technical Report, Department of Computer Science, University of Maryland, College Park, MD (September 1987).

[91] Levi, S.-T., "A Methodology for Designing Distributed, Fault-Tolerant, and Reactive Real-Time Operating Systems," Ph.D. Dissertation, Department of Computer Science, University of Maryland, College Park, MD (April 1988).

[92] Levi, S.-T., D. Mossé, and A. K. Agrawala, "Allocation of Real-Time Computations under Fault-Tolerance Constraints," *Proceedings of the Nineth Real-Time Symposium*, pp. 161–170, Huntsville, AL (December 1988).

[93] Levi, S.-T., S. K. Tripathi, S. D. Carson, and A. K. Agrawala, "The MARUTI Hard Real-Time Operating System," *The Fourth Israel Conference on Computer Systems and Software Engineering* (June 5–6, 1989), and *ACM Operating Systems Review*, vol. 23, no. 3, pp. 90–105 (July 1989).

[94] Levi, S.-T. and A. K. Agrawala, *Real-Time System Design*, McGraw Hill, New York, NY (1990).

[95] Liskov, B. H. and A. Snyder, "Exception Handling in CLU," *IEEE TOSE*, vol. SE-5, no. 6, pp. 546–558 (November 1979).

[96] Liskov, B. and R. Scheifler, "Guardians and Actions: Linguistic Support for Robust Distributed Programs," *ACM TOPLAS*, vol. 5, no. 3, pp. 381–404 (July 1983).

[97] Ma, R. P., E. Lee, and M. Tsuchiya, "Design of Task Allocation Scheme for Time Critical Applications," *IEEE Proceedings—Real Time Systems Symposium*, Miami Beach, FL (December 1981).

[98] Ma, R. P., E. Lee, and M. Tsuchiya, "A Task Allocation Model for Distributed Computing Systems," *IEEE TOC*, vol. C-31, no. 1 (January 1982).

[99] Mancini, L., "Modular Redundancy in a Message Passing System," *IEEE TOSE*, vol. SE-12, no. 1, pp. 79–86 (January 1986).

[100] Merlin, P. M. and B. Randell, "State Restoration in Distributed Systems," *Digest of the 8th International Symposium on Fault Tolerant Computing*, FTCS-8, pp. 129–134, Toulouse, France (1978).

[101] Meyer, J. F., "On Evaluating the Performability of Degradable Computing Systems," *IEEE TOC*, vol. C-29, no. 8, pp. 720–731 (August 1980).

[102] Mills, D. L., "DCNET Internet Clock Service," RFC-778, Defense Advanced Research Projects Agency, Information Processing Techniques Office (April 1981).

[103] Molle, M. L., and L. Kleinrock, "Virtual Time CSMA: Why Two Clocks are Better than One," *IEEE Transactions on Communications*, vol. COM-33, no. 9 (September 1985).

[104] Mossé, D. and A. K. Agrawala, "On Fault-Tolerance in Real-Time Environments," CS-TR-????, Technical Report, Department of Computer Science, University of Maryland, College Park, MD (1990).

[105] *MC68020 32-Bit Microprocessor User's Manual* by Motorola Inc., Prentice-Hall, Englewood-Cliffs, NJ (1985).

[106] Trehel, M. and M. Naimi, "A Distributed Algorithm for Mutual Exclusion based on Data Structures and Fault Tolerance," *6th International Phoenix Conference on Computers and Communication*, Scottsdale, AZ, pp. 35–39 (February 1987).

[107] Nehmer, J., "An Object Architecture for Hard Real-Time Systems," CS-TR-2003, Technical Report, Department of Computer Science, University of Maryland, College Park, MD (March 1988).

[108] Nicola, V. F. and J. M. Spanje, "Comparative Analysis of Different Models of Checkpointing and Recovery," *IEEE TOSE*, vol. 16, no. 8, pp. 807–821 (August 1990).

[109] Owicki, S. and D. Gries, "Verifying Properties of Parallel Programs: An Axiomatic Approach," *Communications of the ACM*, vol. 19, no. 5, pp. 279–285 (May 1976).

[110] Parnas, D. L., "On the Criteria to be Used in Decomposing Systems into Modules," *Communications of the ACM*, vol. 15, no. 12, pp. 1053–1058 (December 1972).

[111] Parnas, D. L., "Designing Software for Ease of Extension and Contraction," *IEEE TOSE*, vol. SE-5, no. 2, pp. 128–137 (March 1979).

[112] Peterson, J. L., and A. Silberschatz, *Operating System Concepts*, Addison-Wesley, Reading, MA (July 1984).

[113] Peterson, W. W., *Error Correcting Codes*, MIT Press, Cambridge MA (1961).

[114] Postel, J., *Internet Protocol*, RFC-791, Defense Advanced Research Projects Agency, Information Processing Techniques Office (September 1981).

[115] Postel, J., *Internet Control Message Protocol*, RFC-792, Defense Advanced Research Projects Agency, Information Processing Techniques Office (September 1981).

[116] Postel, J., *Transmission Control Protocol*, RFC-793, Defense Advanced Research Projects Agency, Information Processing Techniques Office (September 1981).

[117] Pradhan, D. K. ed., *Fault Tolerant Computing: Theory and Techniques*, Prentice-Hall, Englewood-Cliffs, NJ (1986).

[118] Pressman, R. S., *Software Engineering: a Pragmatic Approach*, 2nd ed., McGraw Hill, New York, NY (1988).

[119] Randel, B., "System Structure for Software Fault Tolerance," *IEEE TOSE*, vol. SE-1, no. 2, pp. 220–232 (June 1975).

[120] Reed, D. P., "Implementing Atomic Actions on Decentralized Data," *ACM TOCS*, vol. 1, no. 1, pp. 3–23 (February 1983).

[121] Ricart, G. and A. K Agrawala, "An Optimal Algorithm for Mutual Exclusion in Computer Network," *Communication of the ACM*, vol. 23, no. 1, pp. 9–17 (January 1981).

[122] Schlichting, R. D. and F. B. Scneider, "Fail-Stop Processors: An Approach to Designing Fault Tolerant Computing Systems," *ACM TOCS*, vol. 1, no. 3, pp. 222–238 (August 1983).

[123] Schneider, F. B., D. Grie, and R. D. Schlichting, "Fault Tolerant Broadcasts," Science of Computer Programming, vol. 4, pp. 1–15, North Holland (1984).

[124] Serlin, O., "Fault Tolerant Systems in Commercial Applications," *IEEE Computer*, vol. 17, no. 8, pp. 19–30 (August 1984).

[125] Skeen, D., "Nonblocking Commit Protocols," Memorandum no. UCB/ERL M81/11, Electronic Research Lab, College of Engineering, University of California, Berkeley, CA (March 1981).

[126] Skeen, D. and M. Stonebraker, "A Formal Model of Crash Recovery in a Distributed System," *IEEE TOSE*, vol. SE-9, no. 3, pp. 219–228 (May 1983).

[127] Son, S. H. and A. K. Agrawala, "Distributed Checkpointing for Globally Consistent States of Databases," *IEEE TOSE*, vol. SE-15, no. 10, pp. 1157–1167 (October 1989).

[128] Stallings, W., "The DOD Communication Protocol Standards," *SIGNAL magazine*, vol. 40, no. 8, pp. 29–34 (April 1986).

[129] Stallings, W., *Local Networks*, 2nd ed., Macmillan, New York, NY (1987).

[130] Tanenbaum, A. and R. von Renesse, "Distributed Operating Systems," *ACM Computing Surveys*, vol. 17, no. 4, pp. 419–470 (December 1985).

[131] Taylor, D. J. and C. J. Seger, "Robust Storage Structures for Crash Recovery," *IEEE TOC*, vol. C-35, no. 4, pp. 288–295 (April 1986).

[132] Thomas, R. H., "The Majority Consensus Approach to Concurrency Control," *ACM TODBS*, vol. 4, no. 2, pp. 180–209 (June 1979).

[133] Toy, W. N., "Fault Tolerant Design of Local ESS Processor," *Proceedings of the IEEE*, vol. 66, no. 10, pp. 1126–1145 (October 1978).

[134] Watson, R. W., "Distributed System Architecture Model," eds. B. W. Lampson, M. Paul, and H. J. Siegert, *Distributed Systems: Architecture and Implementation–an advanced course*, pp. 10–43, Springer Verlag, Berlin (1981).

[135] Wensley, J. H., L. Lamport, J. Goldberg, M. W. Green, K. N. Levitt, P. M. Melliar-Smith, R. E. Shostak, and C. B. Weinstock, "SIFT: Design and Analysis of a Fault Tolerant Computer for Aircraft Control," *Proceedings of IEEE*, vol. 60, pp. 1240–1255 (October 1978).

[136] Zhao, W. and K. Ramamrithan, "A Virtual Time CSMA Protocol for Hard Real-Time Communication," *Proceedings of Real-Time System Symposium (IEEE)*, pp. 120–127, New Orleans, LA (December 2–4, 1986).

# Index

abort, 237
Abstract Syntax Notation, 83
acceptance test, 246, 253
acknowledge, 288
action, 305
activity, 84
Ada, 380
address space, 106
adjacent subsystem, 71
agent, 385
agreement, 201
    approximate, 221
    commit, 202
    consensus, 212
    voting, 209
allocatability, 356, 358, 369
allocatable instance, 357
allocation, 353, 367
    algorithm, 372
    alternative, 357
    deadlock, 159
    graph, 357
    initiation, 359
allocator, 360
alternative, 278
alternative tag, 359
application layer, 81
arc, 107
ASN, 83
ASR, 12, 234
assertion, 110
atomic action, 97, 236, 309
    data, 98
    nested, 237
atomic broadcast, 322
audit trail, 261
availability, 247, 263, 265

axiomatic approach, 110, 112, 122, 163, 173

backward wave, 361, 371
BIT, 245
bound data, 98
bounded buffer problem, 53, 65
bounded semantic link, 383
bridge, 96
broadcast, 322
Byzantine generals problem, 226

cache coherence, 33
calendar, 356, 367
CCR, 97
    APDU, 100
    protocol, 100
    services, 99
certification, 310
checkpoint, 261
checksum code, 243
CISC, 17
clock drift protocol, 382
clock rate protocol, 382
commit, 97, 202
    three phase, 207
    two phase, 204
communication, 69
    deadlock, 162, 170, 174
    real-time, 377
communication network, 93
    interface, 93
component, 338
computation node, 106
concurrency, 99, 305
    commitment and recovery, 97
connection-based data transfer, 73

# 410    INDEX

connectivity, 280, 357, 360
consensus, 212
consistency, 305
    interactive consistency conditions, 226
consistent state, 237
contamination, 240
    physical effects, 240
    temporal effects, 240
contention RT protocol, 377
conversation, 256, 258
cookie shop problem, 66
crash recovery, 202
critical section, 112
CSP, 58
cyclic code, 244

daisy chain, 288
damage containment, 332, 354
data manager, 216
datagram, 87
DCE, 45
deadlock, 121
    allocation, 159
    communication, 162, 170, 174
    computation, 169
    detection, 169, 192
    detection algorithm, 196
    prevention, 164
    set, 170–172
dependency graph, 280, 358
dependency set, 279, 356
dialogue units, 84
dining philosophers problem, 122
distributed computer, 2, 107
distributed program, 2, 49, 106
distributed system, 2, 93, 108
distribution principles, 4

effectiveness, 247
election, 145
    broadcast, 155
    ring, 146
electronic switch system (ESS), 27
elemental unit, 249, 282, 302
entity, 71
error, 12, 234
    detection, 74
event, 306
    ordering, 115

fail stop, 12
fail-stop process, 239
failure, 12, 235
fault, 12, 235
    injection, 350
    isolation, 334, 340
    permanent, 12
    prevention, 233
    transient, 12
fault-tolerant synchronization, 389
finite projective plane, 212
flow control, 74
forker, 281
forward recovery, 238, 271
forward wave, 361, 369

gateway, 96

HDLC, 88
heterogeneity, 95
hierarchical composition, 2
high availability systems, 24
history, 214, 306

idempotency, 116, 203
IEEE, 90
    layer model, 90
    LLC layer, 93
    MAC layer, 92
    physical layer, 91
indication primitive, 76
information hiding, 4
intention list, 202, 320
internetwork communication, 95
interrupt, 286
    handler, 292
    nesting, 288
    vector, 288
IO automata, 216
IP, 87
IPC, 45, 286

joiner, 281
joint, 9, 210

last call protocol, 383
laxity, 378
layer, 5
    application, 81
    functions, 77
    link, 88
    LLC, 93

INDEX  411

logical link, 93
MAC, 92
network, 86
physical, 89, 91
presentation, 83
session, 84
transport, 84
layered architecture, 70
life cycle, 11
link layer, 88
LLC layer, 93
locking, 307, 320
logical link layer, 93

MAC layer, 92
major synchronization, 100
message passing coprocessor, 22
monitor, 51
    relation, 333
monotonic fault, 354
MPC, 22
MTBF, 12, 248
Multibus-II, 20
multiplexing, 73
mutual exclusion, 121
    time ordering algorithms, 130
    token ring algorithms, 125

network layer, 86
node, 106
NSCP, 246
NVP, 247, 271

object, 7, 306
    allocatable, 356, 358, 369
    invocation, 8, 110
    state, 291
occurrence graph, 259
one copy serializability, 307
ordered partitioning problem, 67
OSI, 81
    application layer, 81
    layer model, 81
    link layer, 88
    network layer, 86
    physical layer, 89
    presentation layer, 83
    reference model, 80
    session layer, 84
    transport layer, 84

pack, 114, 277

packet, 22
packet switching, 87
parity, 242
partitioning, 315
peer protocol, 6
performability, 248
physical layer, 89, 91
physical redundancy, 240, 355, 360
    degree, 359, 371
presentation layer, 83
priority, 287
process, 6, 106
processing node, 106
projective plane, 212
protocol, 69, 72
    specifications, 78

QMR, 39
quad modular redundancy, 39
quiescence, 187
quorum, 214, 216

rank of subsystem, 71
reachability set, 172, 280
read quorum, 216
readers writers problem, 55
reconfiguration, 41, 238
recoverable subgraph, 260
recovery, 74
    block, 246, 297
    point, 237
    procedure, 237
redundancy, 5, 271
    modular, 39, 271
    physical, 271
    temporal, 271
reliability, 247
rendezvous, 380
replica, 209, 277
replicated procedure call, 278
request primitive, 76
residue code, 243
resiliency to faults, 247, 330, 336, 341
    to monotonic faults, 280, 357, 364
    to transient faults, 280, 357
resource, 111
    allocation, 353
restart procedure, 99
RISC, 32
roll back recovery, 237, 253
rollback, 97

safety resiliency, 330, 336, 341
SAP, 9
schedule, 216
selective extinction, 146
semaphore, 50
serial scheduler, 217
serial system, 216
serializability, 278, 306
service-elements, 76
session layer, 84
SIFT, 274
signal, 295
splitting, 73
Stratus, 26
subsystem, 71
Synapse N+1, 30
synchronization, 49, 112
    points, 84
synchronous RT protocol, 380
system, 6, 216

Tandem 16, 24
TCP, 85
temporal redundancy, 240, 355, 363
    degree, 359, 363, 371
temporal relation, 368
    convergent sequence, 370
terminated set, 172

termination, 171
    detection, 169, 171, 192
    detection algorithm, 196
    diffusion algorithm, 174
    Eulerian circuit algorithm, 182
    ring algorithms, 177
    spanning tree algorithm, 179
    time stamp, 185
test and set, 50
thread, 2, 105
TLB, 33
trace, 214, 306
transaction, 237, 309
    manager, 216
transient, 354
transparency, 7
transport layer, 84
trap, 286
troupe, 114, 277
typed message, 95

validation, 310
voting, 209
VTCSMA, 378

write quorum, 216

X.25, 87